Introductory Solid Mechanics

Introductory Solid Mechanics

Saroj Kumar Sarangi

Department of Mechanical Engineering
National Institute of Technology Patna, India

CWP
Central West Publishing

NATIONAL
LIBRARY
OF AUSTRALIA

A catalogue record for this book is available from the National Library of Australia

ISBN (print): 978-1-925823-96-7

Preface

I am pleased to present the text entitled "Introductory Solid Mechanics" to the engineering student community. While preparing this book, special care has been taken to present the subject matter of solid mechanics in an easily understandable style. Large number of worked out problems are given to enable the readers to grasp the subject effectively from learning point of view. At the end of each Chapter, highlights containing important definitions, concept and formula are given followed by short type questions and exercise problems.

I am thankful to my colleagues, friends and students who encouraged me to write this book. Thanks are also due to the engineers, authors and publishers, whose works and text have been a source of inspiration and guidance to me while preparing this book.

I express my gratefulness to "Central West Publishing, Australia" for making every effort to publish the book in a short span of time.

Though every care has been taken in checking the manuscript and proof reading, yet claiming perfection is very difficult. Readers are requested to intimate any errors and other useful suggestions to improve this work for the next Edition.

Constructive criticism shall be highly appreciated.

Dr. Saroj Sarangi
Author

Contents

Chapter 1

Simple Stresses and Strains

Learning Objectives

After going through this chapter, the reader will be able to
- classify stresses into various categories.
- state and apply Hooke's Law.
- compute deformations of loaded bars of uniform as well as tapered cross sections.
- calculate stress intensities caused by applied loads in simple and compound sections.
- solve statically determinate as well as indeterminate problems.
- derive and state the relationship between elastic constants.
- compute the stresses and strains due to temperature changes.

1.1 INTRODUCTION

The methods and techniques generally used in the analysis and design of structures are based upon concepts and principles which the reader would have learnt in a first course in engineering mechanics. While dealing with such concepts and principles in engineering mechanics, the bodies are considered to be rigid. However, materials deform under the action of forces. This chapter presents the basic nature of deformation of materials under load. The material properties describing the deformation under the action of forces are also discussed.

1.2 STRESS

When a material is subjected to external forces, it tends to undergo some deformation. Against this deformation, some internal resistance is offered by the molecules of the material of the body. This internal resistance per unit area offered by the material of the body against external loading is called the intensity of stress (called simply as stress).

Here the load (external force) is considered uniformly distributed over the area. If the internal force is not distributed uniformly over the area, the intensity of stress should be calculated for a small area over which the internal force can be considered uniformly distributed.

Then,

$$\text{stress} = \lim_{\delta A \to 0} \frac{\delta P}{\delta A}$$ where δP is the force resisted by a small area δA.

If the load P is considered uniformly distributed over area A,

$$\text{stress} = \frac{P}{A}$$

Units of stress

In MKS units: kgf/cm^2 where load is in kgf and area in cm^2
In SI units: N/m^2 or Pascal (Pa) where load is in N and area in m^2
$1\ N/m^2 = 1$ Pascal $= 1$ Pa, $10^6\ N/m^2 = 1$ MPa (mega Pascal),
$10^9\ N/m^2 = 1$ GPa (giga Pascal). Also, 1 MPa$= 1\ N/mm^2$

1.1.1 Types of stresses

There are basically two types of stresses: normal stress and shear stress.

Normal stress

It is the stress acting normal (perpendicular) to the cross section of a member. It is denoted by σ (sigma). It is of two types: (a) tensile stress and (b) compressive stress.

Tensile stress: When a member is subjected to two equal and opposite pulls so that it tends to elongate (increase in length), then stress induced at any cross section of the member is called tensile stress. An axially loaded bar in tension is shown in Fig. 1.1. Here, P is the axial tensile load applied, and A is the area of cross section. Thus,

$$\text{Tensile stress} = \frac{\text{Resisting force}}{\text{Area of cross section}}$$

$\sigma = \dfrac{P}{A}$ (Resisting force is equal to the applied force P)

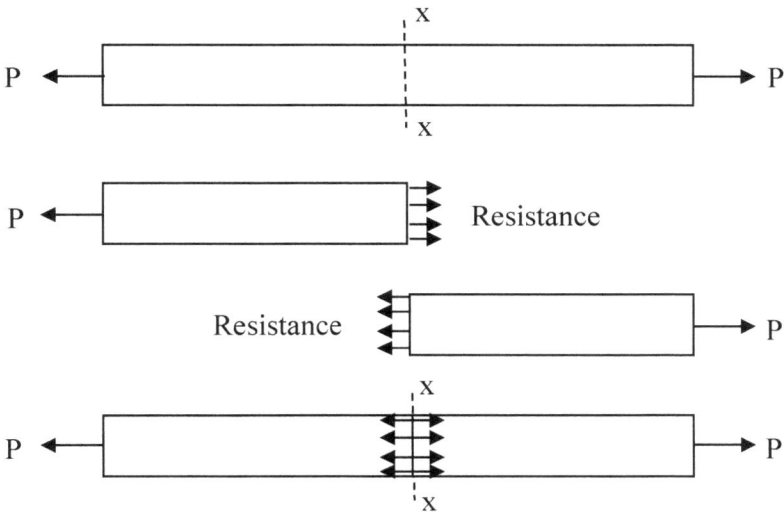

Fig. 1.1

Compressive stress: When a member is subjected to two equal and opposite pushes so that it tends to contract (decrease in length), then the stress induced at any cross section of the member is called

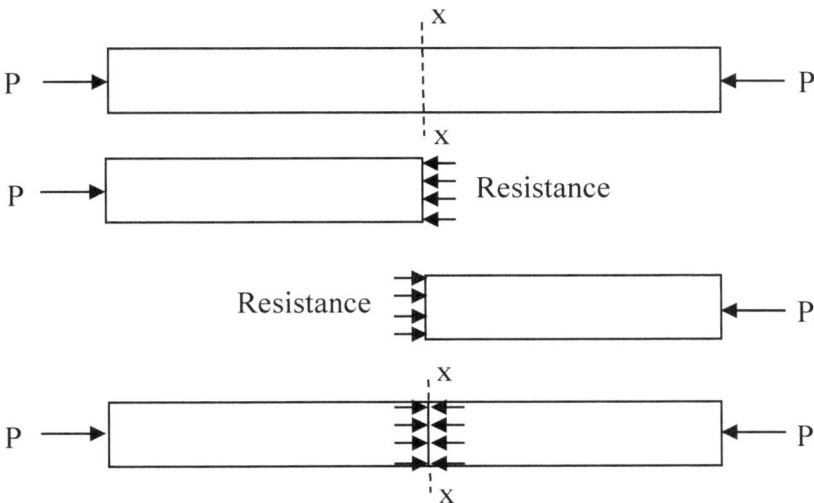

Fig. 1.2

compressive stress. An axially loaded bar in compression is shown in Fig. 1.2. Here, P is the axial compressive load applied, and A is the area of cross section. Thus,

$$\text{Compressive stress} = \frac{\text{Resisting force}}{\text{Area of cross section}}$$

$$\sigma = \frac{P}{A} \text{ (Resisting force is equal to the applied force P)}$$

The tensile and compressive stresses are denoted as σ_t and σ_c respectively.

Shear stress

When a member is subjected to two equal and opposite forces acting tangentially to the section, then the stress induced is called shear stress. It is denoted by τ (Tau).

A rectangular block held at bottom is subjected to a horizontal force at the top edge parallel to side AB, as shown in Fig. 1.3.

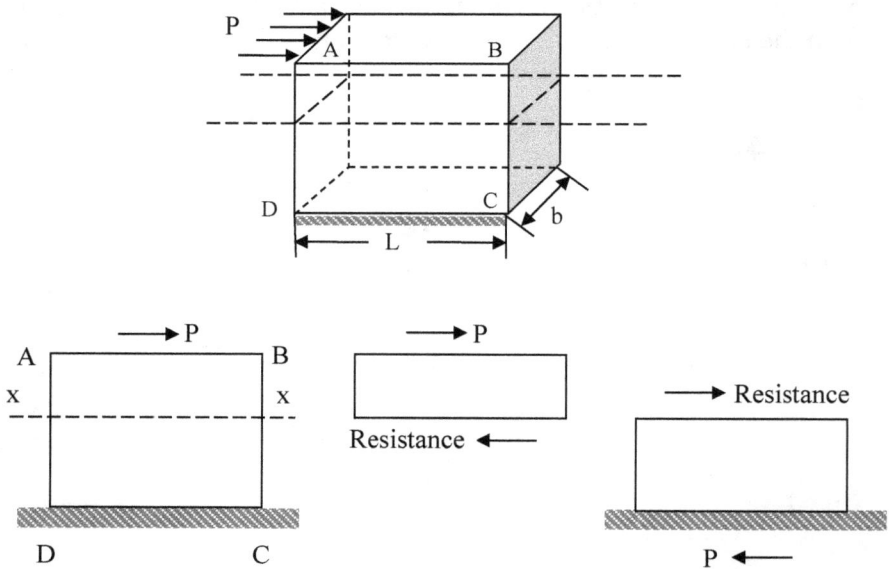

Fig. 1.3

The resistance along the section x-x is called the shear resistance (equal to P in magnitude).

$$\text{Shear stress } (\tau) = \frac{\text{Shear resistance}}{\text{Sheared area}} = \frac{P}{L\,b}$$

Note:
(1) Load is applied on the body whereas stress is induced in the material of the body.
(2) Stresses are of two types (normal stress and shear stress) whatever may be the type of loading on the body (stress perpendicular to cross section → σ and stress tangential to cross section → τ).

1.3 STRAIN

Whenever some external force acts on a body, the body gets deformed. Strain is the measure of this deformation. Like stresses, strains may also be classified into two types: normal strain (denoted by ε) and shear strain (denoted by φ)

Normal strain (ε)

It is the ratio of change in dimension to original dimension.

$$\varepsilon = \frac{\text{change in dimension}}{\text{original dimension}} \quad \text{(No unit)}$$

This normal strain is also called linear strain.

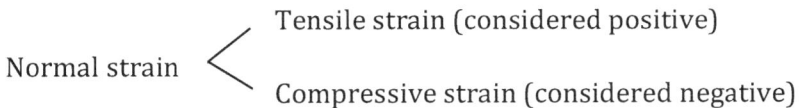

Normal strain $\Big\langle$ Tensile strain (considered positive)

   Compressive strain (considered negative)

If the deformation is measured along the length of the body, then the corresponding strain is called longitudinal strain.

$$\text{Longitudinal stain}, \varepsilon_l = \frac{\text{change in length}}{\text{original length}}$$

An axially loaded bar in tension is shown in Fig. 1.4.

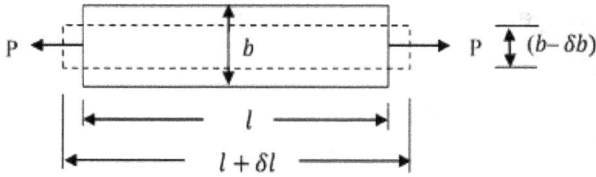

Fig. 1.4

Here, l = original length, $l + \delta l$ = final length

$$\therefore \; \varepsilon_l = \frac{\delta l}{l} \; \text{(dimensionless)}$$

Due to the application of axial tensile load, the body elongates in the direction of stress but in the direction perpendicular to that of stress, there is contraction of the body. This contraction produces lateral strain in the body which is equal to the ratio of decrease in lateral dimension (δb) to the original lateral dimension (b) of the body (Fig. 1.4). Similarly,

$$\text{Volumetric strain}, \varepsilon_v = \frac{\text{change in volume}}{\text{original volume}}$$

$$\text{Superficial strain}, \varepsilon_s = \frac{\text{change in area}}{\text{original area}}$$

Shear strain (ɸ)

Consider a rectangular block held at bottom is subjected to tangential force P as shown in Fig. 1.5. Now the face ABCD will be distorted

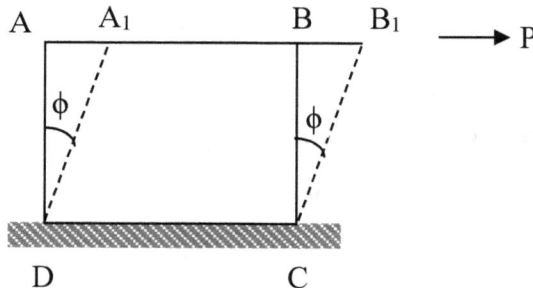

Fig. 1.5

to A_1B_1CD through an angle ϕ. This angular deformation is called shear strain.

Shear strain, $\phi = \dfrac{AA_1 \text{ (or } BB_1)}{AD}$

(As ϕ is a small quantity, $\tan \phi \simeq \phi$ and $\tan \phi = \dfrac{AA_1}{AD}$)

1.3.1 Volumetric strain

Volumetric Strain of a body is the ratio of change in volume to its original volume.

Volumetric strain, $\varepsilon_v = \dfrac{\delta V}{V}$

where δV is the change in volume and V is the original volume of the body.

Also, volumetric strain is the sum of strains in three mutually perpendicular directions.
$\varepsilon_v = \varepsilon_x + \varepsilon_y + \varepsilon_z$, where $\varepsilon_x, \varepsilon_y$ and ε_z are the strains in three mutually perpendicular directions x, y and z, respectively.

To show this, consider a bar of length l, width b and height h as shown in Fig. 1.6.

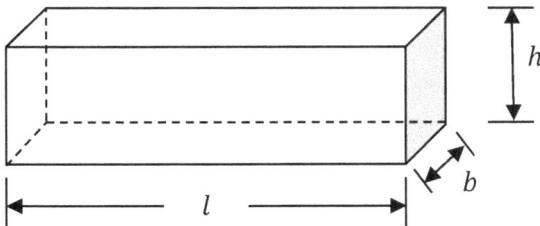

Fig. 1.6

Since volume of the bar is a function of l, b and h

$\delta V = bh\delta l + lh\,\delta b + lb\delta h$
($\delta l, \delta b$ and δh indicate the change in the respective dimension)

$$\frac{\delta V}{V} = \frac{bh\delta l + lh\,\delta b + lb\delta h}{lbh}$$

$$\therefore \ \varepsilon_v = \frac{\delta l}{l} + \frac{\delta b}{b} + \frac{\delta h}{h} = \varepsilon_x + \varepsilon_y + \varepsilon_z$$

1.4 ELASTICITY

Whenever a body is acted upon by external load, it undergoes some deformation. The property by virtue of which the body regains its original shape and size after removal of the external load is called elasticity.

Elastic material: If the deformation disappears completely i.e. the body regains its original shape and size after unloading then the material is said to be elastic.

Inelastic material: If the body does not regain its original shape and size after unloading, then the material is said to be inelastic.

Plastic material: If the body undergoes permanent deformation then the material is said to be plastic.

1.5 HOOKE'S LAW

The law has been established by Robert Hooke. He first established it experimentally. This law is described as follows:

For normal stress - Within elastic limit, normal stress is directly proportional to normal strain.

Mathematically, $\sigma \propto \varepsilon$ and $\sigma = E\,\varepsilon$

where, E is the constant of proportionality known as modulus of elasticity or Young's modulus.

For shear stress - Within elastic limit, shear stress is directly proportional to shear strain.

Mathematically, $\tau \propto \phi$ and $\tau = G\,\phi$

where, G is the constant of proportionality known as modulus of rigidity or shear modulus.

In the following, all materials are assumed to obey Hooke's Law.

1.5.1 Modulus of elasticity (Young's modulus, E)

It is the ratio of normal stress to normal strain in proportional limit.

$$E = \frac{\sigma}{\varepsilon}$$

Unit: N/m^2 or Pa (same unit as that of stress)

Young's modulus (E) is also defined as the slope of stress-strain diagram in the elastic zone. It indicates stiffness of a material and is constant for a given material. It explains the deformation behavior of a material subjected to normal stresses. High Young's modulus (E) value indicates low deformation.

1.5.2 Modulus of rigidity (shear modulus, G)

It is the ratio of shear stress to shear strain in proportional limit.

$$G = \frac{\tau}{\phi}$$

Unit: N/m^2 or Pa

Modulus of rigidity indicates the deformation behavior of a material subjected to shear stresses.

1.6 STRESS-STRAIN DIAGRAM

When a bar (or specimen) is subjected to gradually increasing tensile load, stresses and strains can be found out for a number of loading conditions. The stresses and corresponding strains (upto failure of specimen) when plotted on a graph constitute the stress-strain diagram. These diagrams (curves) differ from material to material.

1.6.1 Stress-strain diagram for a ductile material

A ductile material is the one which shows a significant deformation before fracture (failure). For example, mild steel. Stress-strain diagram for such a material is explained in Fig. 1.7 and the salient points are also described as follows:

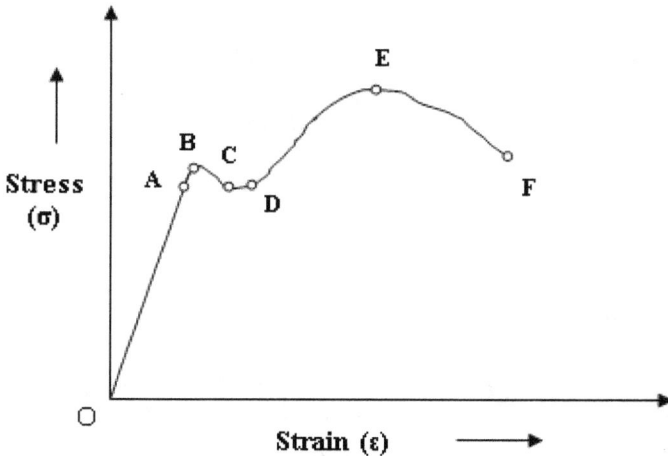

Fig. 1.7 Stress-strain diagram of a ductile material

A =	Proportional limit	OA =	Linear zone
B =	Elastic limit	OB =	Elastic zone
C =	Upper yield point	CD =	Yielding zone
D =	Lower yield point	EF =	Necking zone
E =	Ultimate point (Ultimate strength)		
F=	Fracture point (Fracture strength)		

Proportional limit – Upto which stress is directly proportional to strain.

Elastic limit – Upto which material behaves elastically.

Upto point A (proportional limit) stress and strain are directly proportional to each other. But elasticity property of the material is continued upto point B (elastic limit). Upto elastic limit, material regains its original configurations after removal of load. Beyond the elastic zone, the material enters into plastic zone and removal of load does not return the specimen to its original dimension. On further loading, the diagram reaches the point C (upper yield point) beyond which the load decreases with increase in strain upto point D (lower yield point). After this yielding zone, the stress again increases with increase in strain till the stress reaches the maximum value at point E (ultimate strength). At point E, necking of material starts and the cross sectional area decreases at a rapid rate until the specimen fractures, i.e., upto point F (fracture point).

While drawing the stress strain diagram all the stresses are calculated on the basis of original cross sectional area. Taking the instantaneous area of cross section, the stresses calculated are known to be true stresses and the corresponding strains are called true strains. Considering the original cross sectional area, the stresses calculated are known to be the engineering stress or nominal stress and the corresponding strains are engineering strains or nominal strains.

The stress-strain diagram shown in Fig. 1.7 is the engineering stress-strain curve for mild steel. Fig. 1.8 explains both the engineering and true stress strain diagrams for the ductile material.

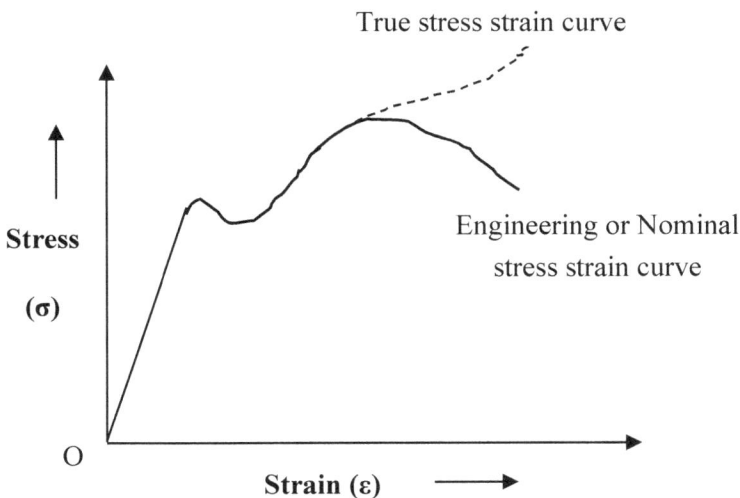

Fig. 1.8 Engineering and true stress-strain diagram for ductile material

1.6.2 Stress-strain diagram for a brittle material

A brittle material is one which shows very small deformation before its fracture. For example, cast iron. The stress-strain diagram for such a material is presented in Fig. 1.9 in which no well-defined yield point is observed.

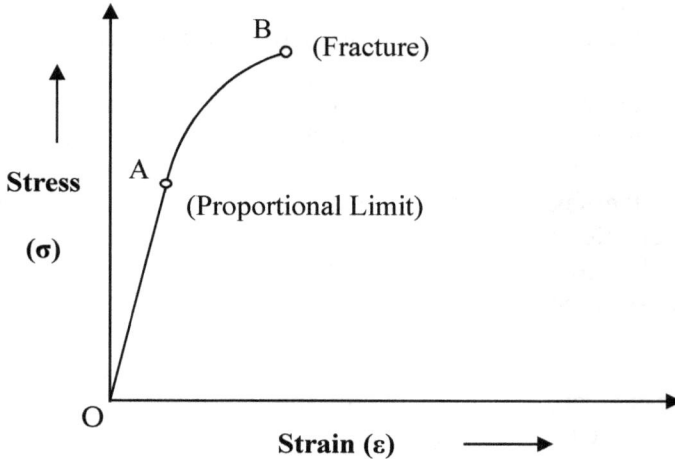

Fig. 1.9 Stress-strain diagram of a brittle material

1.6.3 Stress-strain diagram in compression

For ductile materials, stress-strain diagram in compression follows the same path as that in tensile test at least upto the yield point and even slightly beyond the yield point also. For larger values, the diagrams diverge. In addition, no necking is observed in compression test.

For brittle materials, stress strain diagram in compression is same as that in tensile test but the stresses at various salient points are considerably different.

1.7 ALLOWABLE STRESS AND FACTOR OF SAFETY

While designing a component, it must be ensured that the maximum stress that may be induced during working life do not exceed a certain safe limit. Such a safe limiting stress is known as allowable stress (also called permissible stress or design stress). This is ensured by adopting a suitable factor of safety which is greater than unity.

$$\text{Allowable stress} = \frac{\text{Yield stress}}{\text{Factor of safety}} \quad \text{for ductile material}$$

$$\text{Allowable stress} = \frac{\text{Ultimate stress}}{\text{Factor of safety}} \quad \text{for brittle material}$$

Factor of safety is defined as the ratio of the failure stress of the material to the stress that is allowed. Factor of safety is required for the following reasons.

   i.   To account for variation of material strength.
   ii.   To account for variation of load.
   iii. To account for inaccuracy in construction.
   iv.   To account for error in load estimation and design.
   v.   To limit the deformation to a permissible value.
   vi.   To account for any unforeseen items.

Example 1.1 *A punch of diameter 25 mm is used to punch a hole in a 4 mm thick plate. A force of 80 kN is required. Find the average stress induced in the plate and compressive stress induced in the punch.*

Solution:

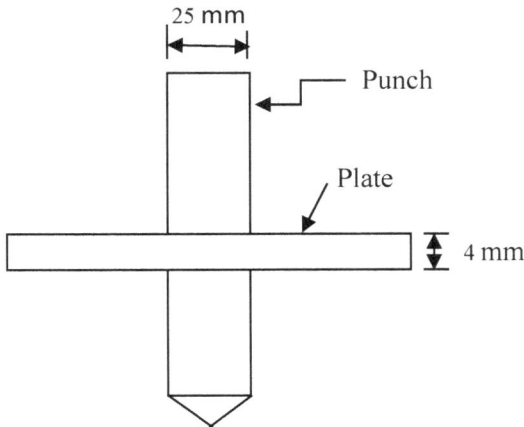

Fig. 1.10

Load, P = 80 x 10³ N
Diameter of punch, d = 25 mm
Plate thickness, t = 4 mm
Refer to Fig. 1.10

Shear stress induced in the plate

$$\tau = \frac{\text{Load}}{\text{Shearing area of plate}} = \frac{P}{\pi d t} = \frac{80 \times 10^3}{\pi \times 25 \times 4} = 254.6 \, ^N/_{mm^2}$$

Compressive stress induced in the punch

$$\sigma = \frac{\text{Load}}{\text{Compressive area of punch}} = \frac{P}{\frac{\pi}{4} d^2} = \frac{80 \times 10^3}{\frac{\pi}{4} \times 25^2}$$

$$= 162.97 \, ^N/_{mm^2}$$

Example 1.2 *Two plates each of cross sectional area 40 mm × 10 mm are connected using two number of 12 mm diameter rivets by a lap joint as shown (Fig. 1.11). Determine the tensile strength of plate and the shearing strength of the rivets.*

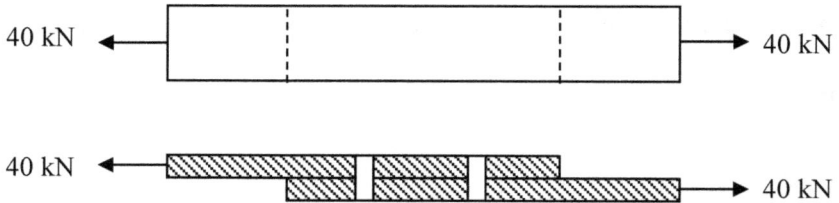

Fig. 1.11

Solution:

Load, P = 40 x 10³ N, plate width, b = 40 mm, plate thickness, t = 10 mm and rivet diameter, d = 12 mm

Shear stress induced in the rivet

$$\tau = \frac{\text{Load}}{\text{Shearing area of two rivets}} = \frac{P}{2\left(\frac{\pi}{4} d^2\right)} = \frac{40 \times 10^3}{2\left(\frac{\pi}{4} \times 12^2\right)}$$

$$= 176.83 \,^N\!/_{mm^2}$$

Tensile stress induced in the plate

$$\sigma = \frac{\text{Load}}{\text{Net area of plate}} = \frac{P}{(b-d)t} = \frac{40 \times 10^3}{(40-12)10} = 142.85 \,^N\!/_{mm^2}$$

Example 1.3 *In a tensile test conducted on a mild steel bar, the following readings were obtained: diameter of steel bar = 30 mm, gauge length of bar = 200 mm, load at elastic limit = 250 kN, elongation at a load of 150 kN = 0.21 mm, maximum load = 380 kN, total elongation = 60 mm and rod diameter at failure = 22.5 mm.*
Calculate (i) Young's modulus, (ii) stress at elastic limit, (iii) the percentage elongation and (iv) the percentage reduction in area.

Solution:
Cross sectional area, $A = \frac{\pi}{4}d^2 = \frac{\pi}{4}(30)^2 = 706.85 \text{ mm}^2$

(i) Young's modulus (E): To calculate the Young's modulus, stress and strain within elastic limit are required.

$$\text{Stress}, \sigma = \frac{\text{Load}}{\text{Area}} = \frac{150 \times 10^3}{706.85} = 212.2 \,^N\!/_{mm^2}$$

Here elastic limit 250 kN cannot be used as the corresponding elongation is not available. For 150 kN load the corresponding elongation is given.

$$\text{Strain}, \varepsilon = \frac{\text{change in length}}{\text{original length}} = \frac{0.21}{200} = 0.00105$$

$$\therefore \text{ Young's modulus}, E = \frac{\sigma}{\varepsilon} = \frac{212.2}{0.00105} = 202095.2 \,^N\!/_{mm^2}$$

$$= 202.095 \text{ GPa}$$

(ii) Stress at elastic limit:

$$\text{Stress at elastic limit} = \frac{\text{Load at elastic limit}}{\text{Area}} = \frac{250 \times 10^3}{706.85}$$

$$= 353.68 \, \text{N}/\text{mm}^2$$

(iii) Percentage elongation:

$$\% \text{ elongation} = \frac{\text{Total change in length}}{\text{Original length (or gauge length)}} \times 100$$

$$= \frac{60}{200} \times 100 = 30\%$$

(iv) Percentage reduction in area:

$$= \frac{\text{reduction in area}}{\text{original area}} \times 100 = \frac{\left[\frac{\pi}{4} \times 30^2 - \frac{\pi}{4} \times 22.5^2\right]}{\frac{\pi}{4} \times 30^2} \times 100$$

$$= 43.75\%$$

1.8 ELONGATION OF A UNIFORM BAR

1.8.1 Elongation of a uniform bar subjected to external load

Consider a uniform bar of length *l* and area of cross section *A* is subjected to external load P, as shown in Fig. 1.12a.

Let, δl = change in length of bar and E = Young's modulus of elasticity.

Normal stress, $\sigma = \dfrac{P}{A}$

Strain, $\epsilon = \dfrac{\delta l}{l}$

As per Hooke's law $\sigma = E\epsilon$

$$\frac{P}{A} = E\,\frac{\delta l}{l}$$

$$\boxed{\delta l = \frac{Pl}{AE}}$$ → *Elongation of uniform bar subjected to external load*

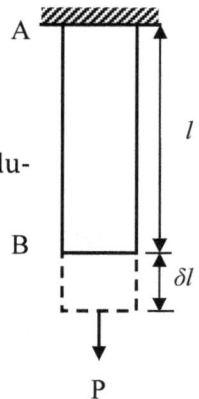

Fig. 1.12a

Also $\quad \delta l = \dfrac{\sigma l}{E}$

Note:

Stiffness of bar $= \dfrac{\text{Force}}{\text{Deformation}} = \dfrac{P}{\delta l} = \dfrac{AE}{l}$

1.8.2 Elongation of uniform bar due to self-weight

Consider a uniform bar AB of length l and area of cross section A, as shown in Fig. 1.12b.

Here no external load is applied.
Elongation of the bar due to its own weight will be computed.

w = specific weight of bar material

Consider elemental length dx of the bar at a distance x from B (Fig. 1.12b)

Weight of bar for a length x is $P = wAx$

Elongation of bar for length dx is

$$\delta l_{dx} = \dfrac{P\, l_{dx}}{AE} = \dfrac{wAx\,dx}{AE} = \dfrac{wx\,dx}{E}$$

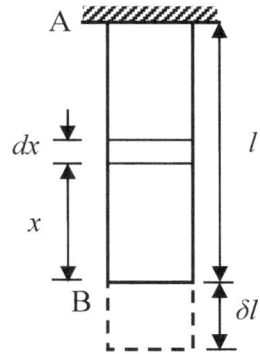

Fig. 1. 12b

Total Elongation of bar, $\quad \delta l = \displaystyle\int_0^l \dfrac{wx\,dx}{E} = \dfrac{w}{E}\left[\dfrac{x^2}{2}\right]_0^l = \dfrac{w\,l^2}{2\,E}$

$$\boxed{\delta l = \dfrac{w\,l^2}{2\,E}} \longrightarrow \quad \textit{Elongation of bar due to self-weight}$$

Note: For a bar of varying cross section (Fig. 1.12c) subjected to axial load P, total elongation is equal to the sum of elongations of various portions of the bar.

Change in length, $\qquad \delta l = (\delta l)_1 + (\delta l)_2 + (\delta l)_3$

$$\delta l = \sum \frac{Pl}{AE} = \frac{Pl_1}{A_1E_1} + \frac{Pl_2}{A_2E_2} + \frac{Pl_3}{A_3E_3}$$

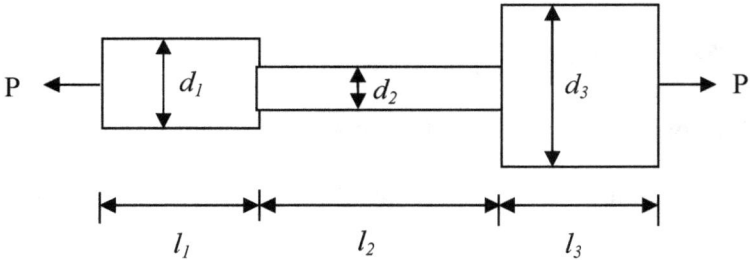

Fig. 1.12c

Example 1.4 *A round bar as shown (Fig. 1.13) is subjected to an axial tensile load of 100 kN. What must be the diameter 'd' if the stress there is to be 100 MN/m². Find also the total elongation. Given E= 290 GPa.*

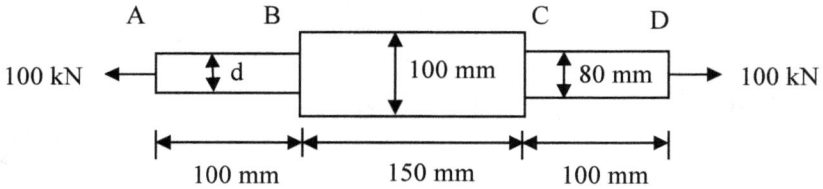

Fig. 1.13

Solution:

Stress, $\sigma = \dfrac{P}{\frac{\pi}{4} \times d^2}$ or $100 \times 10^6 = \dfrac{100 \times 10^3}{\frac{\pi}{4} \times d^2}$

$\therefore$ d = 0.0356 m = 35.6 mm → Required diameter

$$\delta l = \left[\frac{Pl}{AE}\right]_{AB} + \left[\frac{Pl}{AE}\right]_{BC} + \left[\frac{Pl}{AE}\right]_{CD} = \frac{P}{E}\left[\frac{l_{AB}}{A_{AB}} + \frac{l_{BC}}{A_{BC}} + \frac{l_{CD}}{A_{CD}}\right]$$

$$= \frac{100 \times 10^3}{290 \times 10^9}\left[\frac{0.1}{\frac{\pi}{4} \times (0.0356)^2} + \frac{0.15}{\frac{\pi}{4} \times (0.1)^2} + \frac{0.1}{\frac{\pi}{4} \times (0.08)^2}\right]$$

$$= 3.48 \times 10^{-5} \text{ m} = 0.0348 \text{ mm}$$

Example 1.5 *A tie bar on a vertical pressing machine is 2 m long and 40 mm in diameter. What is the stress and extension under a load of 100 kN? Given E = 205000 N/mm².*

Solution:

Given, l = 2 m = 2000 mm, d = 40 mm and load, P =100 × 10³ N

$$\text{Stress, } \sigma = \frac{P}{A} = \frac{P}{\frac{\pi}{4} \times d^2} = \frac{100 \times 10^3}{\frac{\pi}{4} \times (40)^2} = 79.57 \, {}^{N}\!/_{mm^2}$$

But, $\sigma = E.\varepsilon = E \times \dfrac{\delta l}{l}$

$$\therefore \text{ Extension, } \delta l = \frac{\sigma \times l}{E} = \frac{79.57 \times 2000}{205000} = 0.776 \text{ mm}$$

Example 1.6 *A brass tube 50 mm outside diameter, 40 mm bore and 300 mm long is compressed between two end washers by a load of 25kN and the reduction in length measured is 0.2 mm. Assuming Hooke's Law to apply, Calculate Young's modulus.*

Solution:

Given,

d_o = 50 mm, P = 25000 N

d_i = 40 m, δl = 0.2 mm and l = 300 mm

$$\therefore \text{ Stress, } \sigma = \frac{P}{A} = \frac{P}{\frac{\pi}{4} \times (d_o{}^2 - d_i{}^2)} = \frac{25000}{\frac{\pi}{4} \times (50^2 - 40^2)}$$

$$= 35.36 \, {}^{N}\!/_{mm^2}$$

$$\text{Strain, } \quad \varepsilon = \frac{\delta l}{l} = \frac{0.2}{300} = 6.66 \times 10^{-4}$$

$$\therefore \text{ Young's modulus, } E = \frac{\sigma}{\varepsilon} = \frac{33.56}{6.66 \times 10^{-4}} = 53051 \, {}^{N}\!/_{mm^2}$$

Alternately,

$$\delta l = \frac{Pl}{AE}$$

$$\therefore E = \frac{Pl}{A\delta l} = \frac{25000 \times 300}{\frac{\pi}{4} \times (50^2 - 40^2) \times 0.2} = 53051 \, \text{N}/\text{mm}^2$$

Example 1.7 *A compound bar 900 mm long is made of a rod of steel 300 mm long and 30 mm diameter securely fastened to a copper rod of 600 mm long (Fig. 1.14). Under a pull of 50 kN, the extension in each portion are found to be equal. Calculate (a) the diameter 'd' of the copper rod (b) the stress in steel and copper (c) the work done in extending the compound bar.*
Given E_{steel} =205000 N/mm² and E_{copper} =110000 N/mm².

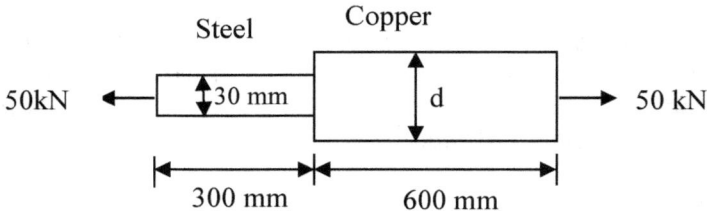

Fig. 1.14

Solution:

For steel portion, diameter, $d_1 = 30$ mm, length, $l_1 = 300$ mm

For copper portion, diameter, $d_2 = d$ (to be calculated), length, $l_2 = 600$ mm

(i) Given, elongation of steel portion = elongation of copper portion

$$\left[\frac{Pl}{AE}\right]_{Steel} = \left[\frac{Pl}{AE}\right]_{Copper}$$

$$\frac{50000 \times 300}{\frac{\pi}{4} \times (30)^2 \times 205000} = \frac{50000 \times 600}{\frac{\pi}{4} \times (d)^2 \times 110000}$$

$$\therefore d = 58 \text{ mm} \rightarrow \textit{Diameter of copper rod}$$

(ii) Stresses:

$$\sigma_{steel} = \frac{P}{A_s} = \frac{P}{\frac{\pi}{4} \times d_1{}^2} = \frac{50000}{\frac{\pi}{4} \times (30)^2} = 70.73\ {}^N/_{mm^2}$$

$$\sigma_{Copper} = \frac{P}{A_c} = \frac{P}{\frac{\pi}{4} \times d_2{}^2} = \frac{50000}{\frac{\pi}{4} \times (58)^2} = 18.92\ {}^N/_{mm^2}$$

(iii) Work done = Average load x deformation = $\frac{P}{2} \times \delta l$

Total deformation = $(\delta l)_{steel} + (\delta l)_{Copper} = 2 \times (\delta l)_{steel}$

$$= 2 \times \left[\frac{Pl}{AE}\right]_{Steel} = 2 \times \frac{50000 \times 300}{\frac{\pi}{4} \times (30)^2 \times 205000} = 0.207\ mm$$

$$\therefore \text{Workdone} = \frac{50000}{2} \times 0.207 = 5175\ N\ mm = 5.175\ Nm$$

Example 1.8 *A steel rod having a constant cross section of 0.003 m² is attached to one end and is subjected to three axial forces as shown in Fig. 1.15. Find the displacement of the free end D, if E=200 GPa.*

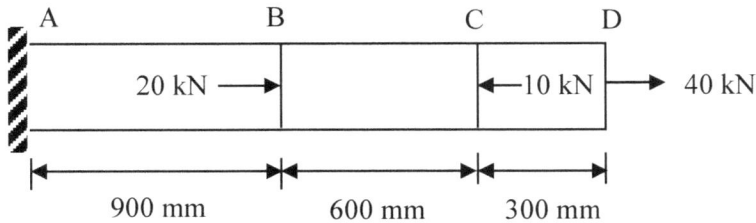

Fig. 1.15

Solution:

Active force = 20-10+40 = 50 kN $\therefore$ Reaction at A = 50 kN

Now the free body diagram is shown (Fig. 1.15a)

$$\delta l = \left[\frac{Pl}{AE}\right]_{AB} + \left[\frac{Pl}{AE}\right]_{BC} + \left[\frac{Pl}{AE}\right]_{CD} = \frac{1}{AE}[(Pl)_{AB} + (Pl)_{BC} + (Pl)_{CD}]$$

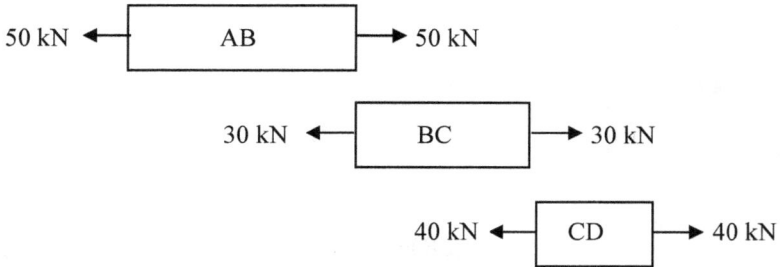

Fig. 1.15a

$$= \frac{1}{0.003 \times 200 \times 10^9} \left[\begin{array}{c} (50 \times 10^3 \times 0.9) + (30 \times 10^3 \times 0.6) \\ +(40 \times 10^3 \times 0.3) \end{array} \right]$$

$$= 1.25 \times 10^{-4} \text{ m} = 0.125 \text{ mm}$$

Example 1.9 *A bar ABCD has its three parts: AB = 2 m, BC = 1 m and CD = 1.5 m. The diameter of AB, BC and CD are 12 mm, 10 mm and 15 mm, respectively. The bar is subjected to forces as shown (Fig. 1.16). Find (i) the force P required for equilibrium, (ii) Change in length of bar. Given E = 200 GPa.*

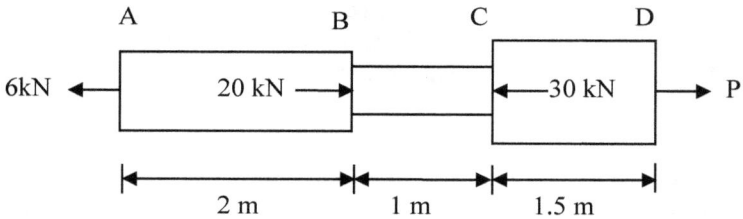

Fig. 1.16

Solution:

Parts AB, BC and CD are shown separately (Fig. 1.16a)

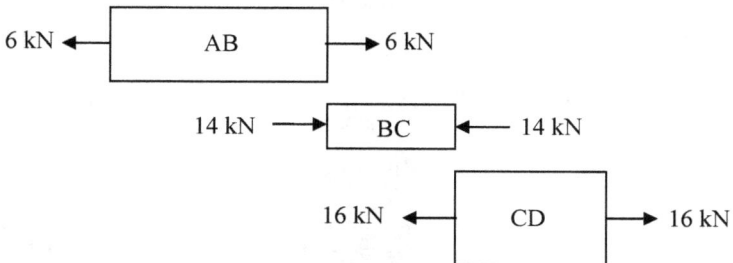

Fig. 1.16a

∴ P = 16 kN (force required for equilibrium)

$$\delta l = \left[\frac{Pl}{AE}\right]_{AB} + \left[\frac{Pl}{AE}\right]_{BC} + \left[\frac{Pl}{AE}\right]_{CD}$$

$$= \left[\frac{6 \times 10^3 \times 2}{\frac{\pi}{4} \times (0.012)^2} - \frac{14 \times 10^3 \times 1}{\frac{\pi}{4} \times (0.010)^2} + \frac{16 \times 10^3 \times 1.5}{\frac{\pi}{4} \times (0.015)^2}\right] \times \frac{1}{200 \times 10^9}$$

$$= 3.2 \times 10^{-4} \text{ m} = 0.32 \text{ mm}$$

Example 1.10 *A steel wire of length 200 m hangs vertically. The specific weight of steel is 78 kN/m³. Find the (i) elongation of wire due to its self-weight (ii) elongation of upper 75 m length of wire. Given E = 200 GPa.*

Solution:

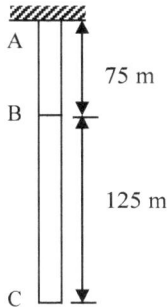

Fig. 1.17

$$w = 78 \times 10^3 \frac{N}{m^3}$$

Length of wire, $l = 200$ m
Refer to Fig. 1.17

(i) Elongation of wire due to self-weight

$$= \frac{wl^2}{2E} = \frac{78 \times 10^3 \times 200^2}{2 \times 200 \times 10^9} = 7.8 \times 10^{-3} \text{m} = 7.8 \text{ mm}$$

(ii) Elongation of AB portion

= Elongation of AB portion due to self-weight + Elongation of AB portion due to external load (weight of BC Portion)

$$= \left[\frac{wl^2}{2E}\right]_{AB} + \left[\frac{Pl}{AE}\right]_{AB}$$

where, P = External load due to weight of BC portion = specific weight x volume

$$= w \times A \times 125 = (78 \times 10^3 \times A \times 125) \text{ N}$$

∴ Elongation of portion AB

$$= \left[\frac{78 \times 10^3 \times 75^2}{2 \times 200 \times 10^9}\right] + \left[\frac{(78 \times 10^3 \times A \times 125) \times 75}{200 \times 10^9 \times A}\right]$$

$$= 4.75 \times 10^{-3}\text{m} = 4.75 \text{ mm}$$

Example 1.11 *A rod of 40 mm diameter and 4 m length is subjected to a tensile load of 40 kN. A 20 mm diameter bore is made centrally on the rod. To what length the rod should be bored so that the total extension will be increased by 30% under the same tensile load. Given E = 2 x 10⁵ N/mm².*

Solution:

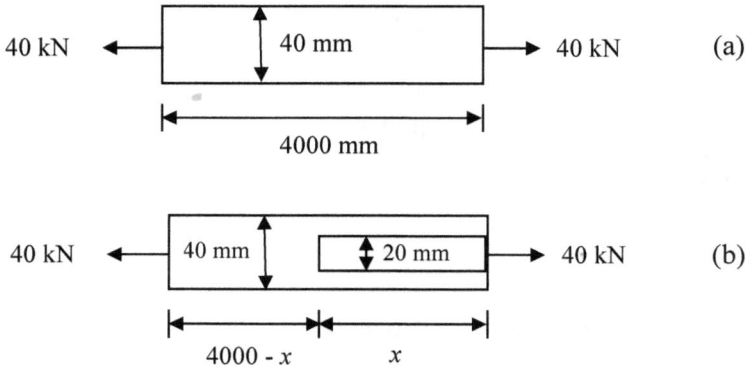

Fig. 1.18

1st Case (no bore) (Fig. 1.18(a))

Elongation $\delta l = \dfrac{Pl}{AE} = \dfrac{40 \times 10^3 \times 4000}{\frac{\pi}{4} \times (40)^2 \times 2 \times 10^5} = 0.6366$ mm

2nd Case: When bore of 20 mm diameter and length x (say) made (Fig. 1.18(b))

Now total elongation = $1.3 \times 0.6366 = 0.8276$ mm

This total elongation is the sum of elongation of portion without bore and elongation of portion with bore.

$$0.8276 = \left[\frac{Pl}{AE}\right]_{Unbored\ portion} + \left[\frac{Pl}{AE}\right]_{bored\ portion}$$

$$= \frac{40 \times 10^3 \times (4000 - x)}{\frac{\pi}{4} \times (40)^2 \times 2 \times 10^5} + \frac{40 \times 10^3 \times x}{\frac{\pi}{4} \times (40^2 - 20^2) \times 2 \times 10^5}$$

$$= \frac{40 \times 10^3}{\frac{\pi}{4} \times 2 \times 10^5} \left[\frac{4000 - x}{1600} + \frac{x}{1200}\right]$$

$\therefore x = 3600$ mm

Therefore, the rod should be bored to a length of 3600 mm or 3.6 m.

Example 1.12 *A member formed by connecting a steel bar to an aluminium bar is shown (Fig. 1.19). Assuming that the bars are prevented from buckling sideways, calculate the magnitude of the force P that will cause the total length of the member to decrease by 0.25 mm. E_s = 210 GPa, E_a = 70 GPa.*
What is the total work done by the force P?

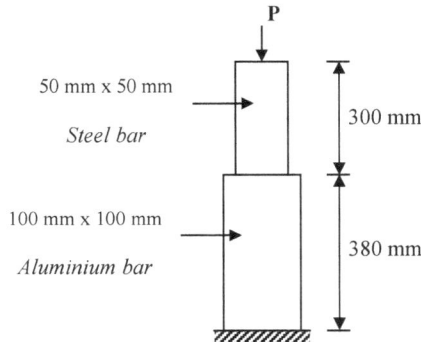

P

50 mm x 50 mm
Steel bar

300 mm

100 mm x 100 mm
Aluminium bar

380 mm

Fig. 1.19

Solution:

For steel: $l = 300$ mm, $A = 2500$ mm^2 and $E = 210 \times 10^3$ N/mm^2

For aluminium: $l = 380$ mm, $A = 10000$ mm^2 and $E = 70 \times 10^3$ N/mm^2

Total change in length = 0.25 mm

$$\delta l = \left[\frac{Pl}{AE}\right]_{Steel} + \left[\frac{Pl}{AE}\right]_{aluminium}$$

$$0.25 = \left[\frac{P \times 300}{2500 \times 210 \times 10^3}\right] + \left[\frac{P \times 380}{10000 \times 70 \times 10^3}\right]$$

$\therefore \ P = 2.243 \times 10^5 N = 224.3$ kN

Work done = Average force × distance $= \dfrac{P}{2} \times \delta l$

$$= \frac{1}{2} \times 2.243 \times 10^5 \times 0.25 \times 10^{-3} = 28.044 \text{ Nm}$$

1.9 ELONGATION OF TAPERED BAR

1.9.1 Elongation of tapered bar of circular cross section subjected to external load

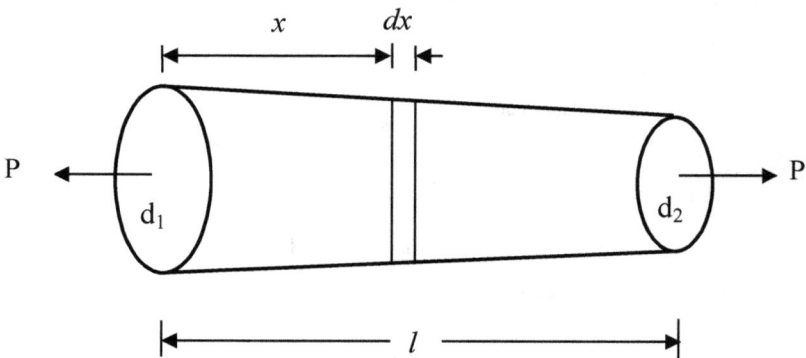

Fig. 1.20a

Consider a circular bar of uniform tapering section as shown in Fig. 1.20a.

Let, P = axial pull on bar
l = length of bar
d_1 = diameter of bigger end
d_2 = diameter of smaller end

Diameter of bar at a distance x from the bigger end (left end)

$$d_x = d_1 - \left(\frac{d_1 - d_2}{l}\right)x = d_1 - kx \quad \text{where } k = \frac{d_1 - d_2}{l}$$

Area of cross section, $A_x = \frac{\pi}{4}(d_1 - kx)^2$

Elongation of elemental length of bar, $\delta l_{dx} = \dfrac{P\,dx}{A_x E}$

Total Elongation $\delta l = \displaystyle\int_0^l \frac{P\,dx}{A_x E} = \int_0^l \frac{P\,dx}{\frac{\pi}{4}(d_1 - kx)^2\,E}$

$$= \frac{4P}{\pi E}\int_0^l \frac{dx}{(d_1 - kx)^2}$$

$$= \frac{4P}{\pi E}\left(\frac{1}{-k}\right)\left[\frac{-1}{d_1 - kx}\right]_0^l = \frac{4P}{\pi E k}\left[\frac{1}{d_1 - kl} - \frac{1}{d_1}\right]$$

$$= \frac{4P}{\pi E\left(\frac{d_1 - d_2}{l}\right)}\left[\frac{1}{d_1 - \left(\frac{d_1 - d_2}{l}\right)l} - \frac{1}{d_1}\right]$$

$$= \frac{4Pl}{\pi E(d_1 - d_2)}\left[\frac{1}{d_2} - \frac{1}{d_1}\right] = \frac{4Pl}{\pi E(d_1 - d_2)}\frac{(d_1 - d_2)}{d_1 d_2}$$

$$\boxed{\delta l = \frac{4Pl}{\pi E d_1 d_2}} \quad\longrightarrow\quad \textit{Elongation of tapered bar}$$

1.9.2 Elongation of Tapered plate subjected to external load

Consider a uniform tapered plate
of thickness t as shown in Fig. 1.20b.

Let, P = axial pull on plate
l = length of plate
B_1 and B_2 = width of plate at smaller
and bigger ends respectively.

Width of plate at a distance x from
bottom end (small end)

$$B_x = B_1 + \left(\frac{B_2 - B_1}{l}\right) x = B_1 + kx$$

where $\quad k = \dfrac{B_2 - B_1}{l}$

Area of cross section, $A_x = B_x\, t = (B_1 + kx)t$

Elongation of elemental length of plate $= \dfrac{P\, dx}{A_x E}$

Total elongation,

$$\delta l = \int_0^l \frac{P\, dx}{A_x E} = \int_0^l \frac{P\, dx}{(B_1 + kx)t\, E} = \frac{P}{tE} \int_0^l \frac{dx}{B_1 + kx}$$

$$= \frac{P}{tE}\left[\log_e(B_1 + kx)\,\frac{1}{k}\right]_0^l = \frac{P}{tEk}\left[\log_e(B_1 + kl) - \log_e B_1\right]$$

$$= \frac{P}{tE\left(\frac{B_2 - B_1}{l}\right)}\left[\log_e\left\{B_1 + \left(\frac{B_2 - B_1}{l}\right)l\right\} - \log_e B_1\right]$$

$$= \frac{Pl}{tE(B_2 - B_1)}\left[\log_e B_2 - \log_e B_1\right]$$

$$\boxed{\delta l = \frac{Pl}{tE(B_2 - B_1)}\log_e\left(\frac{B_2}{B_1}\right)} \longrightarrow \textit{Elongation of tapered plate}$$

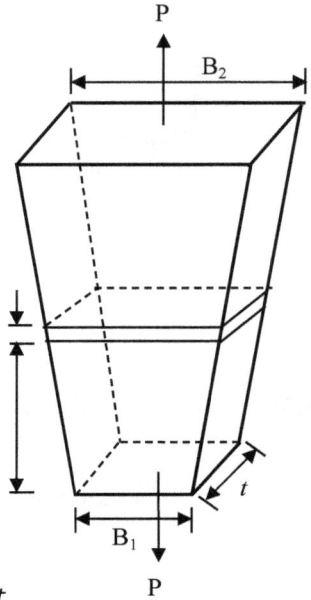

P

B_2

l dx

x

B_1

P

Fig. 1.20b

Example 1.13 *The cross section of a bar is given by* $\left(1+\frac{x^2}{100}\right) cm^2$ *where x cm is the distance from one end. Find the extension under a load of 20 kN on a length of 10 cm. Given E = 200 GPa.*

Solution:

Refer to Fig. 1.21

$$A = \left(1 + \frac{x^2}{100}\right) cm^2$$

Given,

$$P = 20 \times 10^3 \text{ N}, l = 10 \text{ cm}$$

$$E = 200 \times 10^9 \frac{N}{m^2} = 2 \times 10^7 \frac{N}{cm^2}$$

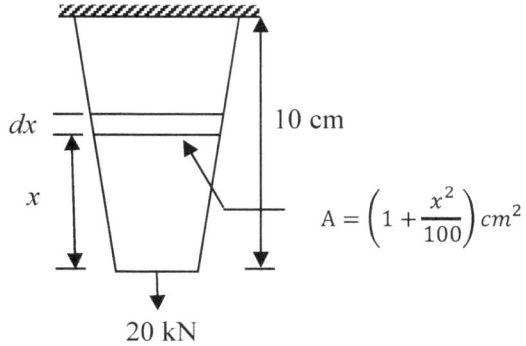

Fig. 1.21

Consider dx length of the bar at a distance x from one end (as shown in Fig. 1.21)

Elongation of bar for length dx is

$$\delta dx = \frac{P\, l_{dx}}{A_{dx}\, E}$$

$$= \frac{20 \times 10^3 \times dx}{\left(1 + \frac{x^2}{100}\right) \times 2 \times 10^7}$$

$\therefore$ Total elongation, $\delta l = \int_0^{10} \frac{1}{1000} \times \frac{dx}{\left(1 + \frac{x^2}{100}\right)}$

$$= \int_0^{10} \frac{1}{1000} \times \frac{100}{(100 + x^2)}\, dx$$

$$= \int_0^{10} \frac{1}{10} \frac{dx}{(100 + x^2)} = \frac{1}{10} \times \frac{1}{10} \times \left[\tan^{-1} \frac{x}{10}\right]_0^{10} = \frac{1}{100}\left[\frac{\pi}{4} - 0\right]$$

$$= 0.00785 \text{ cm}$$

1.10 STATICALLY DETERMINATE AND INDETERMINATE PROBLEMS

Equilibrium equations:

(i) $\sum F_V = 0$ (Algebraic sum of vertical forces is zero)
(ii) $\sum F_H = 0$ (Algebraic sum of horizontal forces is zero)
(iii) $\sum M = 0$ (Algebraic sum of moments is zero)

The problems that can be solved by using equilibrium equations are known to be statically determinate problems. Statically indeterminate problems are those problems which cannot be solved by using equilibrium equations alone. In addition to equilibrium equations, compatibility equations are also needed to solve these problems.

This is explained with reference to a problem as follows (refer to Fig. 1.22):

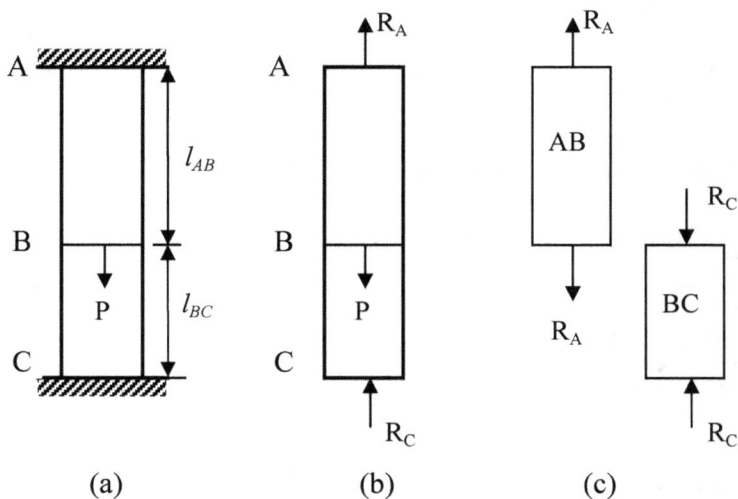

(a)   (b)   (c)

Fig. 1.22

Load P is applied at section B as shown. Free body diagrams of portions AB and BC are shown. Portion AB is subjected to tensile load of R_A (reaction at A) while portion BC is subjected to a compressive load of R_C (reaction at C).

Equilibrium equation:

$$R_A + R_C = P \tag{1}$$

Compatibility equation:

Elongation of AB = Contraction of BC

$$\left[\frac{Pl}{AE}\right]_{AB} = \left[\frac{Pl}{AE}\right]_{BC}$$

$$\frac{R_A \times l_{AB}}{AE} = \frac{R_C \times l_{BC}}{AE} \tag{2}$$

Solving equations (1) and (2), R_A and R_C can be found out and then required stresses can be computed.

Example 1.14 *A uniform bar AB of cross sectional area 10 cm² is fixed horizontally at both ends. It carries an axial load of 30 kN as shown (Fig 1.23). If AB = l and AC = l/3, determine the reactions at supports A and B and the stresses in portions AC and BC. If E=2×10¹¹N/mm², determine the displacement of section C.*

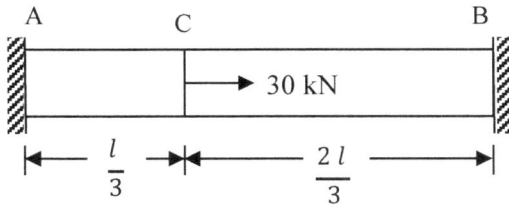

A   C   B

→ 30 kN

$\dfrac{l}{3}$ $\dfrac{2\,l}{3}$

Fig. 1.23

Solution:

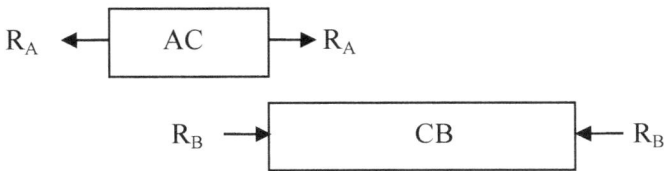

R_A ← AC → R_A

R_B → CB ← R_B

Fig. 1.23a

Free body diagrams for the portions AC and CB are shown (Fig. 1.23a).

Equilibrium equation: $R_A + R_B = P = 30000 \tag{1}$

Compatibility equation:

$$\frac{R_A \times l_{AC}}{AE} = \frac{R_B \times l_{BC}}{AE}$$

$$R_A \times \frac{l}{3} = R_B \times \frac{2l}{3}$$

$$\therefore \ R_A = 2R_B \tag{2}$$

From (1) and (2), $R_A = 20000 \text{ N}$ and $R_B = 10000 \text{ N}$

Stresses:

$$\sigma_{AC} = \frac{R_A}{A} = \frac{20000}{10 \times 10^{-4}} = 20 \times 10^6 \frac{\text{N}}{\text{m}^2} (Tensile)$$

$$\sigma_{CB} = \frac{R_B}{A} = \frac{10000}{10 \times 10^{-4}} = 10 \times 10^6 \frac{\text{N}}{\text{m}^2} (Compressive)$$

Displacement of section C = Elongation of AC or contraction of BC

$$= \left[\frac{Pl}{AE}\right]_{AC} = \frac{20 \times 10^3 \times \frac{l}{3}}{10 \times 10^{-4} \times 2 \times 10^{11}} = 3.33 \times 10^{-6} \ l \text{ m}$$

Example 1.15 *A straight uniform bar clamped at both ends is loaded as shown in Fig 1.24. Initially the bar is stress free. Determine the stresses in all the three portions of the bar of cross sectional area 1000 mm².*

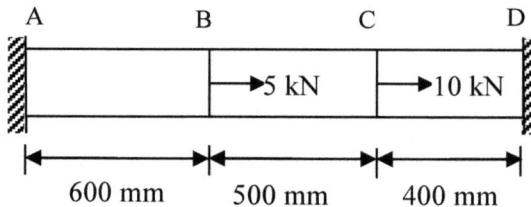

Fig. 1.24

Solution:

Let reactions at A and D be R_A and R_D respectively.

Equilibrium equation:

$$R_A + R_D = 5 + 10 \quad \text{or} \quad R_D = 15 - R_A \tag{1}$$

Free body diagrams for the portions AB, BC and CD are shown (Fig. 1.24a)

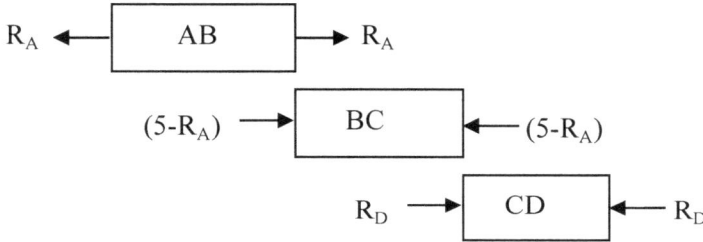

Fig. 1.24a

Compatibility equation:

Elongation of AB = Contraction of BC + Contraction of CD

$$\left[\frac{Pl}{AE}\right]_{AB} = \left[\frac{Pl}{AE}\right]_{BC} + \left[\frac{Pl}{AE}\right]_{CD}$$

$$\frac{R_A \times 600}{AE} = \frac{(5 - R_A) \times 500}{AE} + \frac{R_D \times 400}{AE}$$

$$\therefore \quad 6R_A = 5(5 - R_A) + 4R_D \tag{2}$$

Solving (1) and (2) $R_A = 5.67$ kN and $R_D = 9.33$ kN

$\therefore$ Force in AB portion = 5.67 kN (*Tensile*)

Force in BC portion = 0.67 kN (*Tensile*)

Force in CD portion = 9.33 kN (*Compressive*)

Stresses:

$$\sigma_{AB} = \frac{5.67 \times 10^3}{1000} = 5.67 \frac{N}{mm^2} \ (\textit{Tensile})$$

$$\sigma_{BC} = \frac{0.67 \times 10^3}{1000} = 0.67 \frac{N}{mm^2} \ (\textit{Tensile})$$

$$\sigma_{CD} = \frac{9.33 \times 10^3}{1000} = 9.33 \frac{N}{mm^2} (Compressive)$$

Note: If required, one can find

Displacement of point B = Elongation of AB due to $\sigma_{AB} = \left[\frac{\sigma l}{E}\right]_{AB}$

$$= \frac{5.67 \times 600}{E} mm$$

where E = modulus of elasticity of bar material (N/mm²).

Similarly, displacement of point C = contraction of CD due to

$$\sigma_{CD} = \left[\frac{\sigma l}{E}\right]_{CD} = \frac{9.33 \times 400}{E} mm$$

Example 1.16 *A horizontal steel bar ABC is placed on a smooth table. Its left end is fixed and right end is 0.75 mm away from another support. A load of 60 kN is applied axially at the cross section B and acts as shown (Fig. 1.25). Find the stresses in the two portions AB and BC. Given E = 2×10⁵ N/mm².*

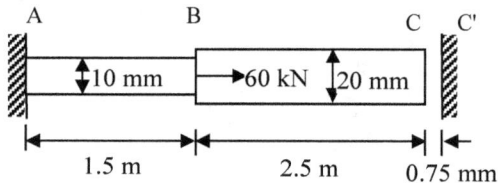

Fig. 1.25

Solution:

Cross sectional area:

$$A_1 = \frac{\pi}{4} \times (10)^2 = 78.5 \; mm^2 \; (Portion \; AB)$$

$$A_2 = \frac{\pi}{4} \times (20)^2 = 314.16 \; mm^2 \; (Portion \; BC)$$

$$l_1 = 1500 \; mm, l_2 = 2500 \; mm \; and \; P = 60 \times 10^3 N$$

Elongation of AB under 60 kN load $= \dfrac{60 \times 10^3 \times 1500}{78.5 \times 2 \times 10^5}$

$$= 5.73 \text{ mm} \quad \left[\delta l = \frac{Pl}{AE} \right]$$

As this elongation is more than the gap, first the gap will be closed and then reaction at C (i.e. R_C) will be developed.

Equilibrium equation:

$$R_A + R_C = 60 \times 10^3 \tag{1}$$

Compatibility equation:

Elongation of AB = Gap + Contraction of BC

$$\frac{R_A \times 1500}{78.5 \times 2 \times 10^5} = 0.75 + \frac{R_C \times 2500}{314.16 \times 2 \times 10^5} \tag{2}$$

Solving (1) and (2)

$$R_A = 23.18 \times 10^3 \text{ N } (Tensile)$$

$$R_C = 36.82 \times 10^3 \text{ N } (Compressive)$$

Stresses:

$$\sigma_{AB} = \frac{R_A}{A_1} = \frac{23.18 \times 10^3}{78.5} = 295.2 \ \frac{\text{N}}{\text{m}^2} \ (Tensile)$$

$$\sigma_{BC} = \frac{R_B}{A_2} = \frac{36.82 \times 10^3}{314.16} = 117.2 \ \frac{\text{N}}{\text{m}^2} \ (Compressive)$$

Example 1.17 *Two rods A and B of equal free length hang vertically 60 cm apart and support a rigid bar horizontally (Fig. 1.26). The bar remains horizontal when carrying a load of 5000 kg at 20 cm from A. If the stress in B is 50 N/mm², find the stress in A and the areas of A and B. Given E_A = 200000 N/mm² and E_B = 90000 N/mm².*

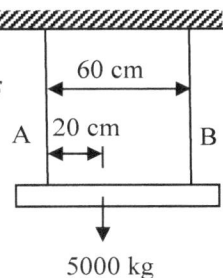

Fig. 1.26

Solution:

$$\sigma_B = 50\ \frac{N}{mm^2},\ E_A = 200000\ \frac{N}{mm^2}\ \text{and}\ E_B = 90000\ \frac{N}{mm^2}$$

Let P_A and P_B = Load carried by rods A and B respectively

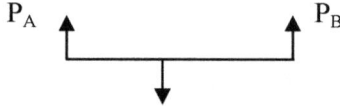

Equilibrium equation: $P_A + P_B = 5000$ (1)

Compatibility equation:

Elongation of rod A = Elongation of rod B (bar remains horizontal)

$$\frac{P_A l_A}{A_A E_A} = \frac{P_B l_B}{A_B E_B} \tag{2}$$

Moment equation:

Taking moments about A

$$5000 \times 20 = P_B \times 60$$

$$\therefore P_B = 1666.67\ \text{kg}$$

$$\therefore P_A = 5000 - 1666.67 = 3333.33\ \text{kg}$$

Given,

$$\sigma_B = 50\ \frac{N}{mm^2}\ \text{i.e.}\ \frac{P_B}{A_B} = 50$$

$$A_B = \frac{1666.67 \times 9.81}{50} = 327\ mm^2$$

From (2)

$$\frac{3333.33 \times l_A}{A_A \times 200000} = \frac{1666.67 \times l_B}{327 \times 90000}$$

$\therefore A_A = 294.29$ mm²

$\therefore$ Cross sectional area of rods A and B are 294.29 mm² and 327 mm².

Stress in rod A, $\sigma_A = \dfrac{P_A}{A_A} = \dfrac{3333.33 \times 9.81}{294.29} = 111 \dfrac{N}{mm^2}$

Example 1.18 *A load of 2 MN is applied on a short concrete column 500 mm × 500 mm. The column is reinforced with four steel bars of 10 mm diameter, one in each corner. Find the stresses in the concrete and the steel bars. Take* E_{steel} = 2.1×10⁵ *N/mm², * $E_{concrete}$ = 1.4×10⁴ *N/mm².*

Solution:

Area of column = 500 × 500 = 250000 mm²

Area of four steel bars, $A_S = 4 \times \dfrac{\pi}{4} \times 10^2 = 314.15$ mm²

Area of concrete, A_C = Column area − steel area

= 250000 − 314.15 = 2.496 × 10⁵ mm²

Deformation of Steel = Deformation of concrete

$$\left[\dfrac{\sigma l}{E}\right]_{Steel} = \left[\dfrac{\sigma l}{E}\right]_{Concrete} \qquad \rightarrow \qquad \dfrac{\sigma_s}{E_s} = \dfrac{\sigma_c}{E_c}$$

$$\sigma_s = \dfrac{E_s}{E_c}\,\sigma_c = \dfrac{2.1 \times 10^5}{1.4 \times 10^4}\,\sigma_c = 15\,\sigma_c$$

Again, total load carried = $\sigma_s A_s + \sigma_c A_c$

or $2 \times 10^6 = (15\,\sigma_c \times 314.15) + (\sigma_c \times 2.496 \times 10^5)$

$\therefore \sigma_c = 7.86 \dfrac{N}{mm^2}$ (*Stress in concrete*)

$\sigma_s = 15 \times 7.86 = 117.9 \dfrac{N}{mm^2}$ (*Stress in steel*)

Example 1.19 *Two vertical rods, one of steel and other of copper are rigidly fixed at the top end. Their lower ends are connected using a cross bar. A load of 5000 N is applied on cross bar such that the cross bar remains horizontal even after loading (Fig. 1.27). Find the stresses in each rod and the position of load on cross bar. Take E_s = 2×10⁵ N/mm², E_c = 1×10⁵ N/mm².*

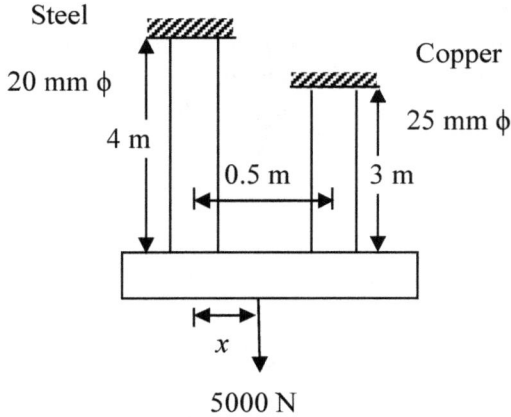

Fig. 1.27

Solution:

l_s = 4000 mm, l_c = 3000 mm

$$A_s = \frac{\pi \times 20^2}{4} = 314.15 \text{ mm}^2$$

$$A_c = \frac{\pi \times 25^2}{4} = 490.87 \text{ mm}^2$$

Let P_s and P_c are the loads carried by the steel and copper rods, respectively.

(i) *Vertical equilibrium equation:* $P_S + P_C = 5000$ (1)

(ii) *Moment equation:* $P_C \times 500 = 5000\,x$ or $P_C = 10\,x$ (2)

(iii) *Compatibility equation:*

Elongation of steel rod = Elongation of copper rod

$$\frac{P_S \times 4000}{314.15 \times 2 \times 10^5} = \frac{P_C \times 3000}{490.87 \times 2 \times 10^5}$$

$$P_C = 1.04 \, P_S \qquad\qquad (3)$$

From (1), (2) and (3) on solving

$P_S = 2450.98$ N, $P_C = 2549.02$ N and $x = 254.9$ mm

Stresses:

$$\sigma_s = \frac{P_s}{A_s} = 7.8 \, \frac{N}{m^2} \, (Stress\ in\ steel\ rod)$$

$$\sigma_c = \frac{P_c}{A_c} = 5.19 \, \frac{N}{m^2} \, (Stress\ in\ copper\ rod)$$

Example 1.20 *A rigid bar AB is hinged at A and supported by a bronze rod 2 m long and steel rod 1 m long as shown in Fig. 1.28. A load of 80 KN is applied at the end B. The area of cross section of steel rod is 900 mm² and that of bronze is 600 mm². Find the stresses in each rod.*

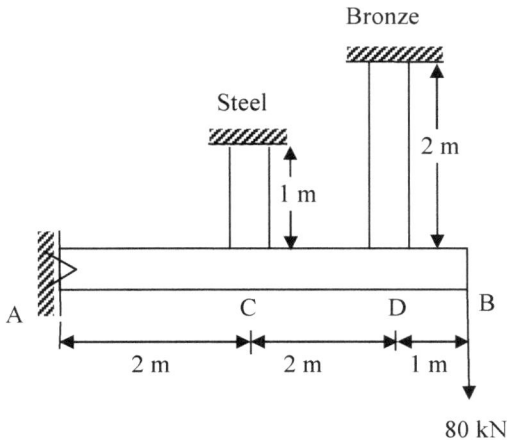

Fig. 1.28

Solution:

Let P_S and P_C are the loads carried by steel and bronze rods, respectively.

Refer to Fig. 1.28a

(i) *Vertical equilibrium equation:*

$$P_S + P_b = R_A + 80 \qquad (1)$$

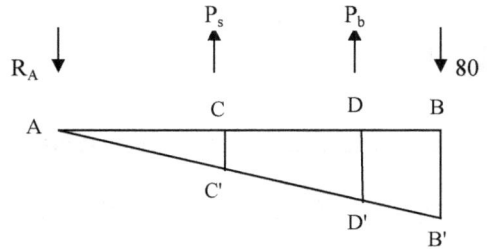

Fig. 1.28a

(ii) *Moment equation (About A):*

$$(P_S \times 2) + (P_b \times 4) = 80 \times 5$$

or $P_S + 2 \times P_b = 200 \qquad (2)$

(iii) Compatibility equation:

$$\frac{CC'}{AC} = \frac{DD'}{AD}$$

$$\frac{CC'}{2} = \frac{DD'}{4} \qquad \text{or} \qquad CC' = \frac{1}{2} \times DD'$$

i.e. Elongation of steel rod $= \frac{1}{2} \times$ Elongation of bronze rod

$$\frac{P_S \times 1000}{900 \times 2 \times 10^5} = \frac{1}{2} \times \frac{P_b \times 2000}{600 \times 0.8 \times 10^5}$$

$$P_S = 3.75\, P_b \qquad\qquad\qquad (3)$$

Solving (1), (2) and (3)

$P_S = 130.4$ kN, $P_b = 34.7$ kN and $R_A = 85.2$ kN

Stresses:

$$\sigma_s = \frac{P_S}{A_s} = 144.88\ \frac{N}{m^2} \ \ (\textit{Stress in steel rod})$$

$$\sigma_b = \frac{P_b}{A_b} = 57.83\ \frac{N}{m^2} \ \ (\textit{Stress in bronze rod})$$

Example 1.21 *A steel bolt and sleeve assembly is shown (Fig. 1.29). The nut is tightened on to the tube through the rigid end blocks until the external force in the bolt is 40 kN. If an external load of 30 kN is then applied to the end block tending to fall them apart, estimate the resulting forces in the bolt and sleeve.*

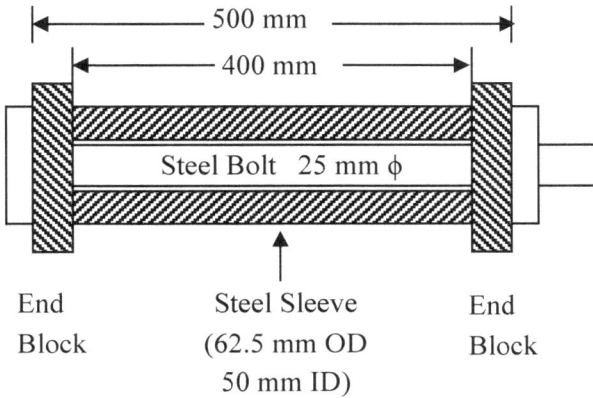

Fig. 1.29

Solution:

Cross sectional area of bolt, $A_b = \frac{\pi}{4} \times 25^2 = 490.87$ mm^2

Cross sectional area of sleeve, $A_S = \frac{\pi}{4} \times (62.5^2 - 50^2) = 1104.46$ mm^2

It is given that nut is tightened until the force in the bolt is 40 kN.

Therefore, sleeve is also subjected to 40 kN force but in opposite direction (Fig. 1.29a).

Fig. 1.29a

(i) Due to nut tightening: Nut tightening force in bolt, $P_b = 40$ kN *(Tensile)* given.

Equal and opposite force will be developed in sleeve. Force in sleeve, $P_S = 40$ kN *(Compressive)*

(ii) Due to external load:

Let load carried by bolt = P_b'

Load carried by sleeve = P_S'

$$P_S' + P_b' = 30 \text{ kN} \tag{1}$$

(iii) *Compatibility equation:*

Elongation of bolt = Elongation of sleeve

$$\left[\frac{Pl}{AE}\right]_{bolt} = \left[\frac{Pl}{AE}\right]_{sleeve}$$

$$\frac{P_b' \times 500}{490.87 \times E} = \frac{P_S' \times 400}{1104.46 \times E} \quad \text{or} \quad P_b' = 0.355\, P_S' \tag{2}$$

Solving (1) and (2)

$P_S' = 22.14$ kN *(Tensile)*

$P_b' = 7.86$ kN *(Tensile)*

Due to	Force in bolt	Force in sleeve
Nut tightening	40 kN *(Tensile)*	40 kN *(Compressive)*
External load	7.86 kN *(Tensile)*	22. 14 kN *(Tensile)*
Combined effect	47.86 kN *(Tensile)*	17.86 kN *(Compressive)*

Example 1.22 *A sleeve and bolt assembly is made by inserting a steel bolt 25 mm diameter inside a copper sleeve 26 mm inner diameter and 46 mm outer diameter and 60 mm length. The bolt is tightened with a nut between two washers each of 4 mm thick first by hand pressure and then with spanner to 3/4th of a turn. Calculate the resultant stress in bolt and sleeve which are of thread 2 mm. Take $E_s = 2.1 \times 10^5$ N/mm² and $E_c = 0.9 \times 10^5$ N/mm².*

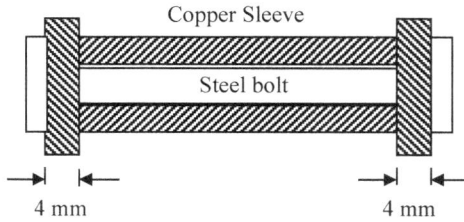

Fig. 1.30

Solution:

Refer to Fig. 1.30

Cross sectional area of steel bolt,

$$A_S = \frac{\pi}{4} \times 25^2 = 490.87 \text{ mm}^2$$

Cross sectional area of copper sleeve,

$$A_C = \frac{\pi}{4} \times (46^2 - 26^2) = 1130.9 \text{ mm}^2$$

Pitch of thread, p = 2 mm

Nut is rotated to 3/4th of turn $\therefore n = 3/4$

Since, there is no external load applied, internal forces shall be equal and opposite.

When nut is tightened, bolt is subjected to tensile force and sleeve to compressive force.

Tensile force in bolt = compressive force in sleeve

Compatibility equation:

Elongation in steel bolt + contraction of copper sleeve

= relative movement between bolt and nut = np

$$\left[\frac{Pl}{AE}\right]_{steel\ bolt} + \left[\frac{Pl}{AE}\right]_{copper\ sleeve} = np$$

$$\frac{P \times 608}{490.87 \times 2.1 \times 10^5} + \frac{P \times 600}{1130.9 \times 0.9 \times 10^5} = \frac{3}{4} \times 2$$

$$P_S' = 127.19 * 10^3 \text{ N}$$

Stresses:

$$\sigma_{\text{steel bolt}} = \frac{P}{A_s} = \frac{127.19 \times 10^3}{490.87} = 259.1 \frac{\text{N}}{\text{m}^2} \ (Tensile)$$

$$\sigma_{\text{copper sleeve}} = \frac{P}{A_b} = \frac{127.19 \times 10^3}{1130.9}$$

$$= 112.46 \frac{\text{N}}{\text{m}^2} \ (Compressive)$$

Example 1.23 *Three identical wires support a rigid bar ABC as shown (Fig. 1.31). Determine the forces developed in each wire when a load of 15 kN is applied at D.*

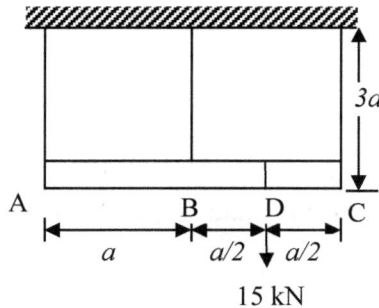

Fig. 1.31

Solution:

Let P_A, P_B and P_C be the forces developed in wires A, B and C respectively (Fig. 1.31a).

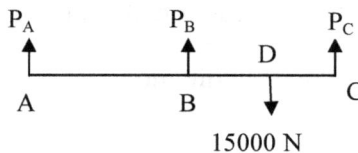

Fig. 1.31a

Equilibrium equation: $P_A + P_B + P_C = 5000$ (1)

Moment equation: Taking moments about A

$$15000\left(a + \frac{a}{2}\right) = (P_B \times a) + (P_C \times 2a)$$

$$P_B + 2 \times P_C = 15000 \times \frac{3}{2} = 22500 \qquad\qquad (2)$$

Compatibility equation:

As the bar is rigid, the final position will be as shown (Fig. 1.31b).

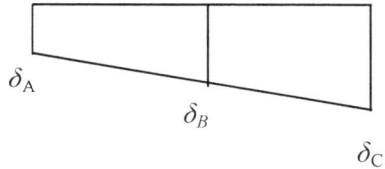

Fig. 1.31b

Let δ_A, δ_B and δ_C be the elongations of the wires A, B and C respectively.

If $\delta \rightarrow$ Increase in elongations of wire B over that of wire A

Then, $\delta_B = \delta_A + \delta$ and $\delta_C = \delta_A + 2 \times \delta$

If P' = Force required for elongation δ of the wire, then

$P_B = P_A + P'$ and $P_C = P_A + 2P'$ (3)

Using (3) in (1) we get, $P_A + P_A + P' + P_A + 2P' = 15000$

or $P_A + P' = 5000$ (4)

Using (3) in (2) we get, $P_A + P' + 2(P_a + 2P') = 22500$

or $3P_A + 5P' = 22500$ (5)

Solving (4) and (5) $P' = 3750$ N and $P_A = 1250$ N

Then $P_B = P_A + P' = 5000$ N

$P_C = P_A + 2P' = 8750$ N

Example 1.24 *Three bars AO, BO and CO are connected to ceiling at points A, B and C, respectively on the same line (Fig. 1.32). The lower ends of all the bars are connected together at point O as shown. If a load of 60 kN is applied vertically downward at O, find the stresses developed in the bars. Area of cross section of each bar = 40 cm².*

Solution:

Let P_1, P_2, P_3 = load carried by the bars 1, 2 and 3 respectively

By symmetry $P_1 = P_3$

Equilibrium equation:

$P_1 \sin 60 + P_2 + P_3 \sin 60 = 60 \times 10^3$

$$P_1 \times \frac{\sqrt{3}}{2} + P_2 + P_3 \times \frac{\sqrt{3}}{2} = 60 \times 10^3$$

$$\sqrt{3}P_1 + P_2 = 60 \times 10^3 \tag{1}$$

Compatibility equation:

Elongation of bar AO = Elongation of bar BO in the same direction

Elongation of bar AO $= \left[\dfrac{Pl}{AE}\right]_{AO} = \dfrac{P_1 \times \dfrac{150}{\cos 30}}{40 \times E} = \dfrac{4.33\ P_1}{E}$ cm

Elongation of bar BO $= \left[\dfrac{Pl}{AE}\right]_{BO} = \dfrac{P_2 \times 150}{40 \times E} = \dfrac{3.75\ P_2}{E}$ cm

Elongation of bar AO along OB = Elongation of bar BO

$$\frac{\left(\dfrac{4.33P_1}{E}\right)}{\cos 30} = \frac{3.75\ P_2}{E}$$

$$P_1 = 0.75\ P_2$$

Refer to Fig. 1.32a

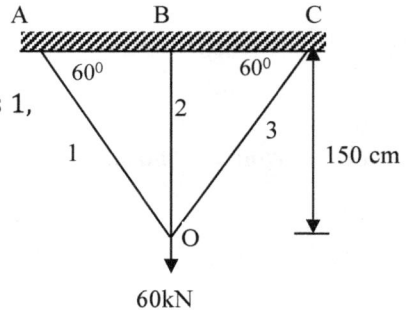

O' = final position of rod,

A'O'=Elongation of wire AO

$$OO' = \frac{A'O'}{\cos 30}$$

$$\text{Elongation of BO} = \frac{\text{Elongation of AO}}{\cos 30}$$

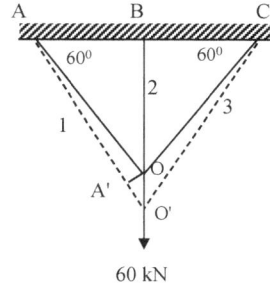

Fig. 1.32a

Solving (1) and (2), $P_1 = 19.57$ kN and $P_2 = 26.09$ kN

Stresses:

$$\sigma_{AO} = \frac{P_1}{A} = \frac{19.57 \times 10^3}{40 \times 10^2} = 4.892 \ \frac{N}{mm^2} \ (Tensile)$$

$$\sigma_{BO} = \frac{P_2}{A} = \frac{26.09 \times 10^3}{40 \times 10^2} = 6.522 \ \frac{N}{m^2} \ (Tensile)$$

$$\sigma_{CO} = \sigma_{AO} = 4.892 \ \frac{N}{mm^2} \ (Tensile)$$

1.11 ELASTIC CONSTANTS

(i) Young's modulus (E)

$$E = \frac{\text{Normal stress } (\sigma)}{\text{Normal strain } (\varepsilon)} \qquad \text{within proportional limit}$$

(ii) Shear modulus (G)

$$G = \frac{\text{Shear stress } (\tau)}{\text{Shear strain } (\phi)} \qquad \text{within proportional limit}$$

(iii) Bulk modulus (K)

Consider a rectangular block of a material subjected to equal normal stress σ in three mutually perpendicular directions as shown (Fig. 1.33).

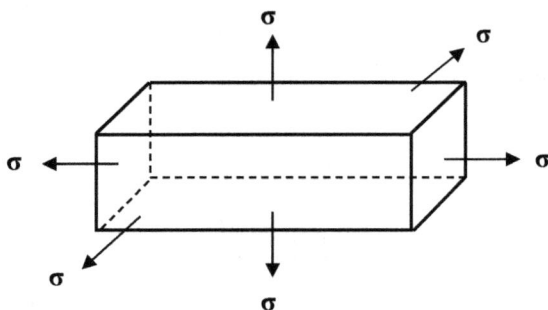

Fig. 1.33

Bulk modulus (K) can be defined as the ratio of normal stress to volumetric strain.

$$K = \frac{\text{Normal stress } (\sigma)}{\text{Volumetric strain } (\varepsilon_v)} = \frac{\sigma}{\left(\frac{\delta V}{V}\right)}$$

In more general terms,

$$\text{Bulk modulus, } K = \frac{\text{Volumetric stress}}{\text{Volumetric strain}}$$

If σ_x, σ_y and σ_z are normal stresses along three mutually perpendicular directions, then

$$\text{Volmetric stress} = \frac{\sigma_x + \sigma_y + \sigma_z}{3}$$

$$\text{Therefore, } K = \frac{\sigma_x + \sigma_y + \sigma_z}{3\varepsilon_v}$$

(iv) Poisson's ratio (v)

Poisson's Ratio is the ratio of lateral strain to longitudinal strain. It is denoted by **v** (nu).

$$v = \frac{\text{Lateral strain}}{\text{Longitudinal strain}}$$

$$= \frac{\text{Strain measured in a direction normal to the direction of stress}}{\text{Strain measured in the direction of stress}}$$

$$= \frac{\delta d/d}{\delta l/l}$$

1.12 RELATIONSHIP BETWEEN STRESSES AND STRAINS

1.12.1 Uniaxial stress condition

Stress exists only in one direction (Fig. 1.34a)

σ_x exists, $\sigma_y = 0$ and $\sigma_z = 0$

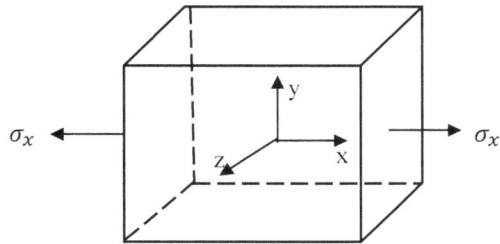

Fig. 1.34a

Strains:

$$\varepsilon_x = \frac{\sigma_x}{E}, \qquad \varepsilon_y = -v\frac{\sigma_x}{E}, \qquad \varepsilon_z = -v\frac{\sigma_x}{E}$$

$$(v = \frac{\text{Strain in y or z direction}}{\text{Strain in x direction}}$$

−ve sign indicates decrease in dimension in y and z directions)

∴ Volumetric strain,

$$\varepsilon_V = \varepsilon_x + \varepsilon_y + \varepsilon_z = \frac{\sigma_x}{E} - v\frac{\sigma_x}{E} - v\frac{\sigma_x}{E} = \frac{\sigma_x}{E}(1 - 2v)$$

1.12.2 Biaxial stress condition

Stresses exist in two mutually perpendicular directions (Fig. 1.34b).

σ_x, σ_y exist and $\sigma_z = 0$

$$\varepsilon_x = \frac{\sigma_x}{E} - v\frac{\sigma_y}{E},$$

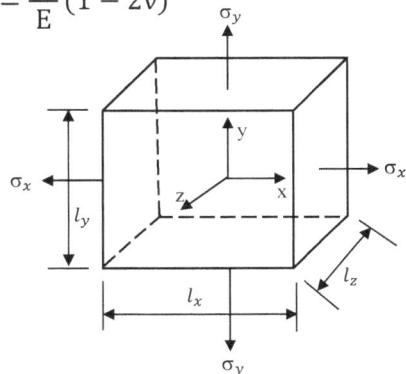

Fig. 1.34b

$$\varepsilon_y = \frac{\sigma_y}{E} - v\frac{\sigma_x}{E} \quad \text{and} \quad \varepsilon_z = -v\frac{\sigma_x}{E} - v\frac{\sigma_y}{E}$$

Now volumetric strain,

$$\varepsilon_V = \varepsilon_x + \varepsilon_y + \varepsilon_z = \left[\frac{\sigma_x}{E} - v\frac{\sigma_y}{E}\right] + \left[\frac{\sigma_y}{E} - v\frac{\sigma_x}{E}\right] + \left[-v\frac{\sigma_x}{E} - v\frac{\sigma_y}{E}\right]$$

$$= \frac{\sigma_x + \sigma_y}{E} - 2v\frac{\sigma_x}{E} - 2v\frac{\sigma_y}{E} = \frac{\sigma_x + \sigma_y}{E}(1 - 2v)$$

1.12.3 Triaxial stress condition

σ_x, σ_y, σ_z all three exist (Fig. 1.34c)

$$\varepsilon_x = \frac{\sigma_x}{E} - v\frac{\sigma_y}{E} - v\frac{\sigma_z}{E}$$

$$\varepsilon_y = \frac{\sigma_y}{E} - v\frac{\sigma_x}{E} - v\frac{\sigma_z}{E}$$

$$\varepsilon_z = \frac{\sigma_z}{E} - v\frac{\sigma_x}{E} - v\frac{\sigma_y}{E}$$

$$\varepsilon_V = \varepsilon_x + \varepsilon_y + \varepsilon_z$$

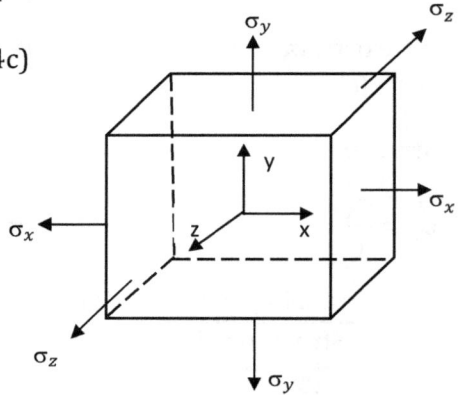

Fig. 1.34c

$$= \left[\frac{\sigma_x}{E} - v\frac{\sigma_y}{E} - v\frac{\sigma_z}{E}\right] + \left[\frac{\sigma_y}{E} - v\frac{\sigma_x}{E} - v\frac{\sigma_z}{E}\right] + \left[\frac{\sigma_z}{E} - v\frac{\sigma_x}{E} - v\frac{\sigma_y}{E}\right]$$

$$= \frac{\sigma_x + \sigma_y + \sigma_z}{E}(1 - 2v)$$

Note:
Condition for constancy of volume:
For volume to remain constant, change in volume, $\delta V = 0$ and volumetric strain, $\varepsilon_V = 0$

$$\varepsilon_V = \frac{\sigma_x + \sigma_y + \sigma_z}{E}(1 - 2v) = 0$$

For steel, $v = 0.25$ (say) and $1 - 2v = 0.5 \neq 0$

$\therefore$ Condition for constancy of volume: $\sigma_x + \sigma_y + \sigma_z = 0$

For rubber, $v = 0.5$ (say) and $1 - 2v = 0$

$\therefore$ Condition for constancy of volume: $\sigma_x + \sigma_y + \sigma_z$ need not be zero.

1.13 RELATIONSHIP BETWEEN ELASTIC CONSTANTS

1.13.1 Relation between K, E and v

Consider an element subjected to equal stresses in three mutually perpendicular directions (Fig. 1.35a).

$$\sigma_x = \sigma, \quad \sigma_y = \sigma, \quad \sigma_z = \sigma$$

$\therefore$ Volumetric strain

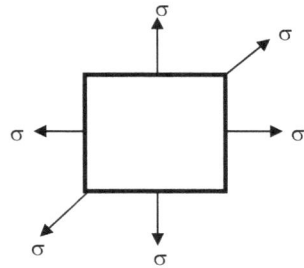

Fig. 1.35a

$$\varepsilon_V = \frac{\sigma_x + \sigma_y + \sigma_z}{E}(1 - 2v) = \frac{3\sigma}{E}(1 - 2v)$$

$$\text{Bulk modulus}, K = \frac{\text{Normal stress}}{\text{Vometric strain}} = \frac{\sigma}{\varepsilon_V} = \frac{\sigma}{\frac{3\sigma}{E}(1 - 2v)}$$

$$\boxed{K = \frac{E}{3(1 - 2v)}}$$

For $v = 0$, $\quad K = \dfrac{E}{3}$ and for $v = 0.5$, $K = \infty$

$$\therefore \quad \frac{E}{3} \le K \le \infty$$

1.13.2 Relation between G, E and v

Consider a square element ABCD subjected to shear stress τ (Fig. 1.35b)

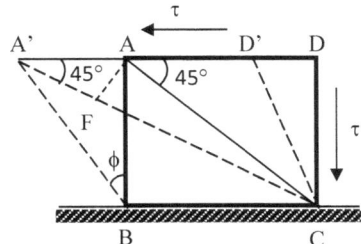

Fig. 1.35b

Shear strain $\phi = \dfrac{AA'}{AB}$ (for small ϕ, $\tan\phi \approx \phi$)

AF is perpendicular to A'C

As $\angle$ACF is very small $\therefore$ AC = FC

$\angle$AA'F $\cong$ $\angle$ DAC

Normal strain along diagonal AC,

$$\varepsilon = \frac{\text{increase in length of diagonl AC}}{\text{original lenth of diagonal AC}} = \frac{\text{A'F}}{\text{AC}} = \frac{\text{AA'cos45}}{\sqrt{2}\text{AB}} = \frac{\text{AA'}}{2\text{AB}} = \frac{\phi}{2}$$

$\therefore \phi = 2\varepsilon$

From Hooke's law for shear stress, $\tau = G\phi = G\times 2\varepsilon$

$$\varepsilon = \frac{\tau}{2G} \tag{1}$$

Now normal strain along diagonal AC is

$$\varepsilon = \frac{\sigma_1}{E} - v\frac{\sigma_2}{E} = \frac{\tau}{E} - v\frac{(-\tau)}{E} = \frac{\tau}{E}(1 + v) \tag{2}$$

From (1) and (2)

$$\boxed{G = \frac{E}{2(1 + v)}}$$

$$\boxed{\begin{array}{cc} K = \dfrac{E}{3(1 - 2v)} & G = \dfrac{E}{2(1 + v)} \\[3mm] v = \dfrac{3K - 2G}{6K + 2G} & E = \dfrac{9KG}{3K + G} \end{array}}$$

(Usually E and v are found from experiments in laboratory and other two elastic constants are calculated using formula).

Example 1.25 *Derive an expression for volumetric strain of a cylindrical bar of length l and diameter d subjected to an axial load P. A cylindrical bar of diameter 5 cm is subjected to an axial pull of 40 kN/cm². The extension in a gauge length of 200 mm measured by the dial type extensometer is 0.3 mm and decrease in diameter is 0.02 mm. Calculate the Young's modulus, Poisson's ratio, shear modulus and bulk modulus for the bar material.*

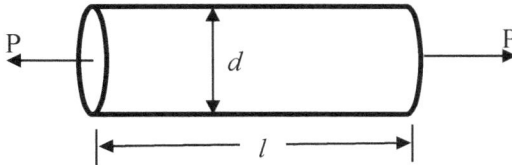

Fig. 1.36

Solution:

Refer to Fig. 1.36

Axial load = P

Stress, $\sigma_x = \dfrac{P}{A} = \dfrac{P}{\frac{\pi}{4}d^2} = \dfrac{4P}{\pi d^2}$

Volumetric strain,

$$\varepsilon_V = \frac{\sigma_x + \sigma_y + \sigma_z}{E}(1 - 2v) = \frac{(4P/\pi d^2)}{E}(1 - 2v)$$

$(\sigma_x = \dfrac{4P}{\pi d^2}, \qquad \sigma_y = 0, \qquad \sigma_z = 0)$

$$\boxed{\varepsilon_V = \frac{4P}{\pi d^2 E}(1 - 2v)}$$

Change in volume, $\delta V = \varepsilon_V V = \dfrac{4P}{\pi d^2 E}(1 - 2v)\dfrac{\pi}{4}d^2 l$

$$\boxed{\therefore \text{ Change in volume, } \delta V = \frac{Pl}{E}(1 - 2v)}$$

Axial pull of 40 kN/cm^2 given. This is axial stress (or normal stress) $\sigma = 400$ N/mm^2

Given $\delta l = 0.3$ mm, $l = 200$ mm, $\delta d = 0.02$ mm, $d = 50$ mm

$\therefore$ Lateral strain $= \dfrac{\delta d}{d} = \dfrac{0.02}{50} = 4 \times 10^{-4}$

Axial strain $= \dfrac{\delta l}{l} = \dfrac{0.3}{200} = 1.5 \times 10^{-3}$

Poisson's ratio, $\nu = \dfrac{\text{lateral strain}}{\text{axial strain}} = \dfrac{4 \times 10^{-4}}{1.5 \times 10^{-3}} = 0.266$

Young's modulus, $E = \dfrac{\text{normal stress}}{\text{normal strain}} = \dfrac{400}{1.5 \times 10^{-3}}$

$= 2.67 \times 10^5$ N/mm^2

Bulk modulus, $K = \dfrac{E}{3(1 - 2\nu)} = \dfrac{2.67 \times 10^5}{3(1 - 2 \times 0.266)}$

$= 1.89 \times 10^5$ N/mm^2

Shear modulus, $G = \dfrac{E}{2(1 + \nu)} = \dfrac{2.67 * 10^5}{2(1 + 0.266)} = 1.05 \times 10^5$ N/mm^2

Example 1.26 *Consider a carefully conducted test where an aluminium bar of 500 mm diameter is stressed in a testing machine as shown (Fig. 1.37). At certain distance the applied force P is 100 kN while the measured elongation of rod is 0.219 mm in a 300 mm gauge length and diameter is decreased by 0.01215 mm. Calculate the Poisson's ratio and elastic modulus of material.*

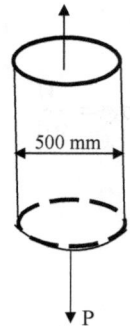

500 mm

P

Fig. 1.37

Solution:

$P = 100$ kN, $\delta l = 0.219$ mm, $l = 300$ mm, $\delta d = 0.01215$ mm, $d = 50$ mm

Lateral strain $= \dfrac{\delta d}{d} = \dfrac{0.01215}{50} = 2.43 \times 10^{-4}$

Axial strain or longitudinal strain $= \dfrac{\delta l}{l} = \dfrac{0.219}{300} = 7.3 \times 10^{-4}$

Poisson's ratio, $v = \dfrac{\text{lateral strain}}{\text{axial strain}} = \dfrac{2.43 \times 10^{-4}}{7.3 \times 10^{-4}} = 0.332$

Normal stress, $\sigma = \dfrac{P}{A} = \dfrac{100 \times 10^3}{\frac{\pi}{4} \times 50^2} = 50.92 \ \dfrac{N}{mm^2}$

$\therefore$ Modulus of elasticity $E = \dfrac{\sigma}{\varepsilon} = \dfrac{50.92}{7.3 \times 10^{-4}} = 6.97 \times 10^4 \ N/mm^2$

Example 1.27 *A steel bar of 200 mm length and 30 mm × 20 mm cross section is subjected to tensile force of 60 kN in the direction of its length. If v = 0.3 and E = 2 × 10^5 N/mm^2, find (i) change in length (ii) change in lateral dimension and (iii) change in volume.*

Solution:

Given $l = 200 \ mm, \ b = 30 \ mm, \ h = 20 \ mm$

$\sigma_x = \dfrac{60 \times 10^3}{20 \times 30} = 100 \ N/mm^2$

$\sigma_y = 0, \ \sigma_z = 0$

$\varepsilon_x = \dfrac{\sigma_x}{E} - v\dfrac{\sigma_y}{E} - v\dfrac{\sigma_z}{E} = \dfrac{\sigma_x}{E} = \dfrac{100}{2 \times 10^5} = 0.0005$

$\varepsilon_x = \dfrac{\delta l}{l}$

$\therefore \delta l = \varepsilon_x \, l = 0.0005 \times 200 = 0.1 \ mm$ (increase in length)

$\varepsilon_y = \dfrac{\sigma_y}{E} - v\dfrac{\sigma_x}{E} - v\dfrac{\sigma_z}{E} = -v\dfrac{\sigma_x}{E} = -0.3 \times \dfrac{100}{2 \times 10^5} = -0.00015$

$$\varepsilon_y = \frac{\delta b}{b}$$

$$\therefore \delta b = \varepsilon_y\, b = -0.00015 \times 30 = -0.0045 \text{ mm (decrease)}$$

$$\varepsilon_z = \frac{\sigma_z}{E} - \nu \frac{\sigma_x}{E} - \nu \frac{\sigma_y}{E} = -\nu \frac{\sigma_x}{E} = -0.3 \times \frac{100}{2 \times 10^5} = -0.00015$$

$$\varepsilon_z = \frac{\delta h}{h}$$

$$\therefore \delta h = \varepsilon_z\, h = -0.00015 \times 20 = -0.003 \text{ mm (decrease)}$$

Now, volumetric strain, $\varepsilon_V = \varepsilon_x + \varepsilon_y + \varepsilon_z = 0.0005 - 0.00015 - 0.00015 = 0.0002$

$$\therefore \delta V = \varepsilon_V\, V = 0.0002(200 \times 30 \times 20) = 24 \text{ mm}^3 \text{ (increase)}$$
$\rightarrow$ Change in volume

(Also $\varepsilon_V = \dfrac{\sigma_x + \sigma_y + \sigma_z}{E}(1 - 2\nu)$ and then $\delta V = \varepsilon_V\, V$)

Example 1.28 *An element is subjected to biaxial stress condition as shown (Fig. 1.38). Derive the expression for stresses σ_x and σ_y in terms of strains ε_x and ε_y.*

Solution:

$$\varepsilon_x = \frac{\sigma_x}{E} - \nu \frac{\sigma_y}{E} \qquad (1)$$

$$\varepsilon_y = \frac{\sigma_y}{E} - \nu \frac{\sigma_x}{E} \qquad (2)$$

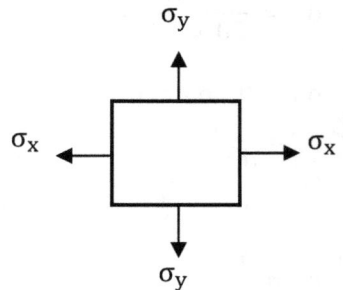

Fig. 1.38

Multiplying (1) with ν and then adding (2) we get

$$\nu\varepsilon_x + \varepsilon_y = \left(\frac{\nu\sigma_x}{E} - \nu^2\frac{\sigma_y}{E}\right) + \left(\frac{\sigma_y}{E} - \nu\frac{\sigma_x}{E}\right) = \frac{\sigma_y}{E}(1 - \nu^2)$$

$$\therefore \sigma_y = \frac{E}{1 - \nu^2}\left(\nu\varepsilon_x + \varepsilon_y\right)$$

Similarly, now multiplying (2) with v and adding (1) we get

$$\varepsilon_x + v\varepsilon_y = \left(\frac{\sigma_x}{E} - v\frac{\sigma_y}{E}\right) + \left(v\frac{\sigma_y}{E} - v^2\frac{\sigma_x}{E}\right) = \frac{\sigma_x}{E}(1 - v^2)$$

$$\therefore \sigma_x = \frac{E}{1 - v^2}\left(\varepsilon_x + v\varepsilon_y\right)$$

$$\sigma_x = \frac{E}{1 - v^2}\left(\varepsilon_x + v\varepsilon_y\right)$$

$$\sigma_y = \frac{E}{1 - v^2}\left(v\varepsilon_x + \varepsilon_y\right)$$

$\longrightarrow$ *Stresses in terms of strains*

Example 1.29 *A concrete cube of dimension 1 m is dropped in sea. If specific weight of sea water is 1025 kg/m³ and the bulk modulus of concrete is 1×10^4 N/mm², find the change in volume of cube. Depth of sea is 600 m.*

Solution:

Given specific weight of sea water, $w = 1025$ kg/m³
$= 1025 \times 9.81$ N/m³ $= 10055$ N/m³

Water pressure on cube $= w\, h$ (h = depth of sea)
$= 10055 \times 600 = 6.03 \times 10^6$ N/m² $= 6.03$ N/mm²

Bulk modulus, $\quad K = \dfrac{\sigma}{\varepsilon_v}$

(stress on the cube is due to water pressure. It is compressive and therefore negative)

$$1 \times 10^4 = \frac{-6.03}{\varepsilon_V}$$

$$\therefore \varepsilon_V = -6.03 \times 10^{-4}$$

$\therefore$ Change in volume $\delta V = \varepsilon_V\, V = -6.03 \times 10^{-4}\, (1000 \times 1000 \times 1000) = -6.03 \times 10^5$ mm³

$\therefore$ Volume decreases by 6.03×10^5 mm^3

Example 1.30 *If the tension test bar is found to be tapered from (D + a) cm to (D - a) cm diameter (Fig. 1.39), prove that the error involved in using the mean diameter to calculate the Young's modulus is* $\left(\frac{10a}{D}\right)^2$ *percent.*

Solution:

Actual condition: $d_1 = D + a, \ \ d_2 = D - a$

$$\therefore \ \delta l = \frac{4Pl}{\pi E d_1 d_2} = \frac{4Pl}{\pi E(D+a)(D-a)} = \frac{4Pl}{\pi E(D^2 - d^2)}$$

$$\therefore \ E = \frac{4Pl}{\pi(D^2 - d^2)\delta l}$$

Ideal condition: (mean diameter is used)

Mean diameter, $d = \dfrac{D+a+D-a}{2} = D$

$$\delta l = \frac{Pl}{AE} = \frac{Pl}{\frac{\pi}{4}D^2 E}$$

$$\therefore \ E = \frac{4Pl}{\pi D^2 \ \delta l}$$

$$\therefore \ \% \text{ error} = \frac{(E)_{\text{actual}} - (E)_{\text{ideal}}}{(E)_{\text{actual}}} \times 100$$

$$= \frac{\frac{4pl}{\pi(D^2 - a^2)\delta l} - \frac{4Pl}{\pi D^2 \ \delta l}}{\frac{4pl}{\pi(D^2 - a^2)\delta l}} \times 100 = \frac{\left(\frac{1}{D^2 - a^2}\right) - \frac{1}{D^2}}{\left(\frac{1}{D^2 - a^2}\right)} \times 100$$

$$= \frac{D^2 - D^2 + a^2}{D^2(D^2 - a^2)} \times (D^2 - a^2) \times 100$$

$$= \left(\frac{10 \ a}{D}\right)^2 \text{ percent.}$$

Fig. 1.39

Example 1.31 *Two pieces of material A and B have the same bulk modulus. Young's modulus (E) for material B is 1% greater than that for material A. Find the value of shear modulus (G) for the material B in terms of E and G for the material A.*

Solution:

$$E = \frac{9KG}{3K + G}$$

Given $K_A = K_B$ and $E_B = 1.01\, E_A$

$$K = \frac{EG}{9G - 3E}$$

As $K_A = K_B$

$$\frac{E_A G_A}{9G_A - 3E_A} = \frac{E_B G_B}{9G_B - 3E_B}$$

$$\frac{E_A G_A}{9G_A - 3E_A} = \frac{1.01\, E_A G_B}{9G_B - 3\,(1.01\, E_A)}$$

$$9G_B E_A G_A - 3.03 E_A^2 G_A = 9.09 E_A G_A G_B - 3.03\, E_A^2 G_B$$

$$G_B(3.03E_A - 0.09G_A) = 3.03E_A G_A$$

$$\therefore G_B = \frac{(1.01)E_A G_A}{(1.01)E_A - (0.03)G_A}$$

Example 1.32 *A determination of Young's modulus (E) and shear modulus (G) gives value of 205000 N/mm² and 80700 N/mm², respectively. Calculate the Poisson's ratio and the bulk modulus. Find the change in diameter produced in a bar of this material 5 cm in diameter acted on by an axial tensile load of 150 kN.*

Solution:

$$E = 205000 \text{ N/mm}^2\,, \quad G = 80700 \text{ N/mm}^2$$

$$G = \frac{E}{2(1 + v)} \quad \therefore\ v = \frac{E}{2G} - 1 = \frac{205000}{2 \times 80700} - 1 = 0.27$$

$$K = \frac{E}{3(1 - 2v)} = \frac{205000}{3(1 - 2 \times 0.27)} = 148550 \text{ N/mm}^2$$

$$d = 5 \text{ cm} = 50 \text{ mm}, \quad P = 150 \times 10^3 \text{ N}$$

$$\sigma = \frac{P}{A} = \frac{P}{\frac{\pi}{4} \times d^2} = \frac{150 \times 10^3}{\frac{\pi}{4} \times 50^2} = 76.39 \text{ N/mm}^2$$

$$\varepsilon = \frac{\sigma}{E} = \frac{76.39}{205000} = 3.72 \times 10^{-4}$$

Poisson's ratio, $\quad v = \dfrac{\text{lateral strain}}{\text{longitudinsl strain}} = \dfrac{\left(\frac{\delta d}{d}\right)}{\varepsilon}$

$$0.27 = \frac{(\delta d / 50)}{3.72 \times 10^{-4}}$$

$\therefore$ Change in diameter, $\delta d = 0.00503$ mm

1.14 COMPOSITE BAR

A composite bar (also known as compound bar) is a bar made of two or more bars (usually of different materials) put in parallel and rigidly fixed with each other so that they behave as one unit for extension or contraction when subjected to axial tensile or compressive loads.

A composite bar subjected to axial compressive load P is shown (Fig. 1.40) and this load P will be shared by individual bars.

Let P_1 and P_2 = Load shared by bar 1 and bar 2, respectively.

It may be noted that the strains produced in bars (1) and (2) will be same.

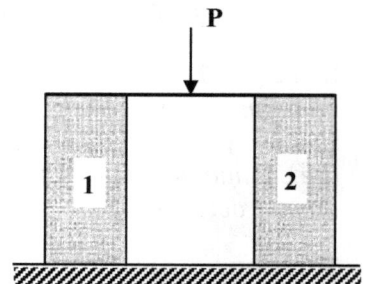

Fig. 1.40

Now, *equilibrium equation:*

$$P_1 + P_2 = P \tag{1}$$

Compatibility equation:

Contraction of bar 1 = Contraction of bar 2

i. e. $\left[\dfrac{Pl}{AE}\right]_1 = \left[\dfrac{Pl}{AE}\right]_2$ (2)

in which the symbols have their usual meaning for the bars 1 and 2. From the equations (1) and (2) presented here, the values of P_1 and P_2 can be calculated and the stresses in the individual bars can be computed considering the respective cross sectional areas. This procedure can be extended for compound bars made of more than two bars.

Example 1.33 *A steel rod of 20 mm diameter is enclosed centrally in a hollow copper tube of external diameter 40 mm and internal diameter 35 mm. The composite bar is then subjected to an axial pull of 50 kN. If the length of each is equal to 200 mm, determine the stresses in the rod and the tube. Also calculate the change in length. Given E_{steel} = 2×10^5 N/mm² and E_{copper} = 1×10^5 N/mm².*

Solution:

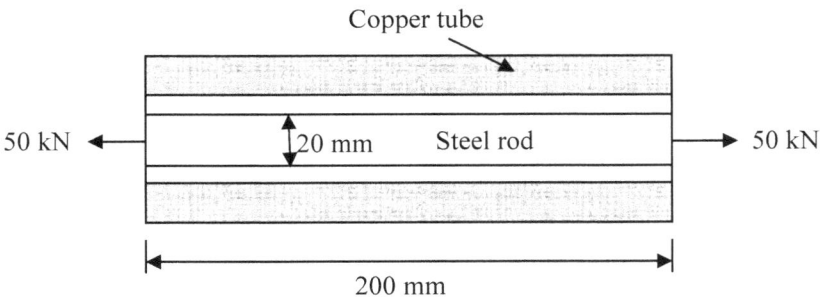

Fig. 1.41

Refer to Fig. 1.41.

Axial load, P = 50000 N

Cross sectional areas:

For steel rod $-$ $A_s = \dfrac{\pi}{4}(20)^2 = 314.15$ mm²

For copper tube $-$ $A_c = \dfrac{\pi}{4}(40^2 - 35^2) = 294.5$ mm²

Let P_s, P_c = Load shared by steel rod and copper tube, respectively

Equilibrium equation:

$$P_s + P_c = 50000 \qquad\qquad (1)$$

Compatibility equation:

Elongation of steel rod = Elongation of copper tube

i.e. $\left[\dfrac{Pl}{AE}\right]_{steel} = \left[\dfrac{Pl}{AE}\right]_{copper}$

$$\frac{P_s \times 200}{314.15 \times 2 \times 10^5} = \frac{P_c \times 200}{294.5 \times 1 \times 10^5}$$

$$P_s = 2.136\ P_c \qquad\qquad (2)$$

Solving (1) and (2)

$$P_s = 34043\ \text{N} \quad \text{and} \quad P_c = 15957\ \text{N}$$

Stresses:

$$\sigma_{steel\,rod} = \frac{P_s}{A_s} = \frac{34043}{314.15} = 108.36\ \text{N}/\text{mm}^2\ (Tensile)$$

$$\sigma_{copper\,tube} = \frac{P_c}{A_c} = \frac{15957}{294.5} = 54.18\ \text{N}/\text{mm}^2\ (Tensile)$$

Change in length,

$$\delta l = \left[\frac{Pl}{AE}\right]_{steel} = \frac{34043 \times 200}{314.15 \times 2 \times 10^5} = 0.1083\ \text{mm}$$

1.15 BAR OF UNIFORM STRENGTH

Elongation of tapered bar under the action of external load is already discussed earlier. A bar under the action of external axial load can be designed to have equal normal stress developed throughout its length by varying the cross section along its length. This is being discussed in the present section. Figure 1.42 shows an axially loaded bar of length l.

Let
A_1 = cross sectional area of small end of bar

Consider elemental length dx of the bar at a distance x from small end (Fig. 1.42).

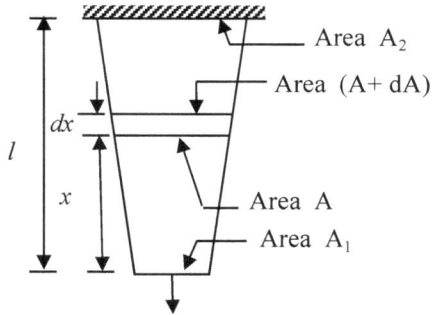

A = cross sectional area of bar at a distance x from small end.

$A + dA$ = cross sectional area of bar at a distance $x + dx$ from small end as shown.

Fig. 1.42

w = specific weight

For equilibrium,

Elemental weight + $\sigma A = \sigma (A + dA)$

$w\,A\,dx + \sigma A = \sigma (A + dA)$

$w\,A\,dx = \sigma\,dA$

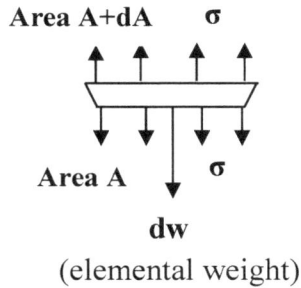

$$\frac{dA}{A} = \frac{w}{\sigma}\,dx$$

On integration,

$$\int_{A_1}^{A} \frac{dA}{A} = \int_{0}^{x} \frac{w}{\sigma}\,dx$$

$$[\log_e A]_{A_1}^{A} = \frac{w}{\sigma}\,[x]_0^x$$

$$\log_e \frac{A}{A_1} = \frac{w}{\sigma}\,x \qquad or \qquad \frac{A}{A_1} = e^{\frac{wx}{\sigma}}$$

$$A = A_1 \, e^{\frac{wx}{\sigma}}$$ ⟶   *Area of any cross section at a distance x from small end*

$$A_2 = A_1 \, e^{\frac{wl}{\sigma}}$$ ⟶   *Cross sectional area at big end*

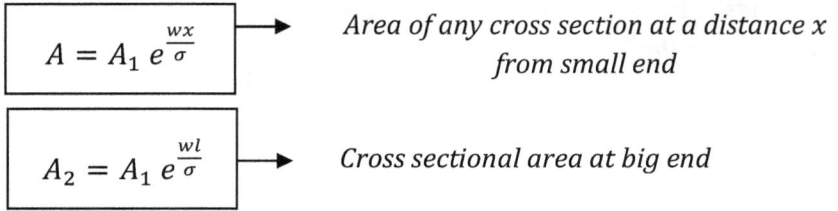

Example 1.34 *To a vertically hanging bar of varying cross section, load of 16 kN is applied at its lower end (1000 mm² cross section) and the cross section of the bar varies in such a way that same stress is developed at all sections throughout its length under the action of external load and its own weight. Length of the bar is 8 m. Calculate (i) area at the bigger end (ii) area at the mid-length (iii) extension of the bar. Take E=200 GPa and density=7800 kg/m³.*

Solution:

$$\sigma = \frac{\text{Load}}{\text{Area}} = \frac{16000}{1000} = 16 \text{ N/mm}^2$$

Specific weight $w = \rho \, g = 7800 \times 9.81 = 76518 \text{ N/m}^3$

(i) Area at bigger end:

$$A_2 = A_1 \, e^{\frac{wl}{\sigma}} = 1000 \times e^{\frac{76518 \times 8}{16 \times 10^6}} = 1000 \times 1.039 = 1039 \text{ mm}^2$$

(ii) Area at mid-length:

$$A = A_1 \, e^{\frac{wx}{\sigma}} = 1000 \times e^{\frac{76518 \times 4}{16 \times 10^6}} = 1000 \times 1.0193 = 1019.3 \text{ mm}^2$$

(iii) Extension of bar

$$\delta l = \frac{\sigma}{E} l = \frac{16 \times 10^6}{200 \times 10^9} \times 8 = 0.00064 \text{ m} = 0.64 \text{ mm}$$

1.16 THERMAL STRESSES

Consider a bar of length *l* subjected to t^0 C increase in temperature. The free elongation of the bar is *lαt* where α is the coefficient of thermal expansion. If the bar is prevented from elongation by fixing between two supports, then compressive stresses will be developed.

This compressive stress is due to temperature change and is called temperature stress or thermal stress.

If the bar is allowed to elongate freely, then temperature stresses will not be developed. Similarly, if a bar is subjected to $t^0\ C$ decrease in temperature, the free contraction of the bar is $l\alpha t$. When the bar is allowed to contract freely then thermal stresses will not be developed but if the bar is prevented from contraction then tensile stresses will be developed (This tensile stress is due to temperature change).

As discussed earlier, composite bars are made of two different materials (value of α is different for the materials).

Case I: Consider a composite bar subjected to rise in temperature
(a) In low α material, elongation is imposed – tensile stress develops.
(b) In high α material elongation is prevented – compressive stress develops.

Case II: Consider a composite bar subjected to fall in temperature
(a) In low α material, contraction is imposed – compressive stress develops
(b) In high α materials, contraction is prevented – tensile stress develops.

Note:
The prevented elongation (or contraction) may be considered as δl, the formula $\delta l = \frac{Pl}{AE}$ is to be used and P to be found out. This P is the load to which the body is subjected due to the temperature change (tensile or compressive). Then the stresses can be calculated.

Example 1.35 *A steel bar 25 mm diameter is rigidly fixed to two parallel supports 8 m apart. Find the pull exerted by the bar on the supports when the temperature of bar is increased by 100°C (i) if the supports do not yield (ii) if yielding of both supports is 2.5 mm. Take* $E_{steel} = 210 \times 10^9 \frac{N}{m^2}$ *and* $\alpha_{steel} = 12 \times 10^{-6}/°C.$

Solution:

Case (i): Supports do not yield (move)

Free elongation $= l\alpha t = 8000 \times 12 \times 10^{-6} \times 100 = 9.6$ mm

Permitted elongation = 0 (No move)

Prevented elongation= 9.6 mm (free elongation – permitted elongation)

Prevented elongation $= \dfrac{Pl}{AE}$

$9.6 = \dfrac{P \times 8000}{490.87 \times 2.1 \times 10^5}$ $\left(A = \dfrac{\pi}{4}(25)^2 = 490.87 \text{ mm}^2\right)$

$\therefore P = 123.699 \times 10^3 \text{N} = 123.699$ kN $\rightarrow$ Pull exerted

$\therefore$ stress $= \dfrac{123.699}{490.87} = 0.2519 \dfrac{\text{kN}}{\text{mm}^2}$

Case (ii):   Supports yield by 2.5mm

Free elongation $= l\alpha t = 9.6$ mm,

Permitted elongation $= 2.5$mm

$\therefore$ Prevented elongation $= 9.6 - 2.5 = 7.1$ mm

Prevented elongation $= \dfrac{Pl}{AE}$ or $7.1 = \dfrac{P \times 8000}{490.87 \times 2.1 \times 10^5}$

$\therefore P = 91.48 \times 10^3 = 91.48$ kN $\rightarrow$ Pull exerted

$\therefore$ Stress $= \dfrac{91.48}{490.87} = 0.186 \dfrac{\text{kN}}{\text{mm}^2}$

Example 1.36 *A composite bar made of aluminium and steel is firmly held between two unyielding supports as shown (Fig. 1.43). An axial load of 200 kN is applied at B at 50°C. Find the stresses in each material when temperature is 100°C.*

Given $E_s = 2.1 \times 10^5 \dfrac{N}{mm^2}$, $\quad E_{al} = 0.7 \times 10^5 \dfrac{N}{mm^2}$,

$\alpha_s = 11.8 \times 10^{-6} \ /°C$, $\quad \alpha_{al} = 29 \times 10^{-6} \ /°C$

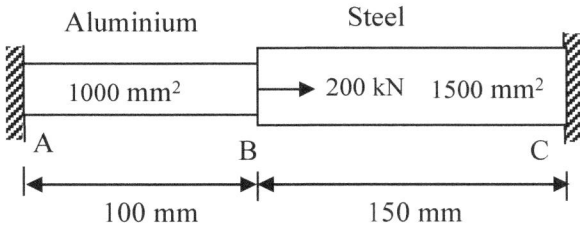

Fig. 1.43

Solution:

Equilibrium equation: $R_A + R_C = 200000$ $\hspace{2cm}$ (1)

Compatibility equation:

Elongation of AB = Contraction of BC

$$\frac{R_A \times 100}{1000 \times 0.7 \times 10^5} = \frac{R_c \times 150}{1500 \times 2.1 \times 10^5} \hspace{2cm} (2)$$

Solving (1) and (2), $R_A = 50000$ N and $R_c = 150000$ N

Stresses:

$$\sigma_{al} = \frac{R_A}{A_{al}} = \frac{50000}{1000} = 50 \frac{N}{mm^2} \ (Tensile)$$

$$\sigma_{st} = \frac{R_C}{A_{Steel}} = \frac{150000}{1500} = 100 \ \frac{N}{mm^2} \ (Compreesive)$$

These stresses are due to applied axial load of 200 kN.

Due to temperature change:

Rise in temperature, $t = 100 - 50 = 50°C$

Free elongation= $(l\alpha t)_{al} + (l\alpha t)_{st}$

$$= (100 \times 29 \times 10^{-6} \times 50) + (150 \times 11.8 \times 10^{-6} \times 50)$$

$$= 0.2335 \text{ mm}$$

Permitted elongation = 0 (both ends are fixed)

∴ Prevented elongation = 0.2335 mm

Now prevented elongation $= \left[\dfrac{Pl}{AE}\right]_{al} + \left[\dfrac{Pl}{AE}\right]_{st}$

$$0.2335 = \frac{P \times 100}{1000 \times 0.7 \times 10^5} + \frac{P \times 150}{1500 \times 2.1 \times 10^5}$$

∴ $P = 1.23 \times 10^5$ N (*Compressive*)

$$\sigma_{al} = \frac{p}{A_{al}} = \frac{1.23 \times 10^5}{1000} = 123 \frac{N}{mm^2} \ (Compressive)$$

$$\sigma_{st} = \frac{P}{A_{st}} = \frac{1.23 \times 10^5}{1500} = 82 \frac{N}{mm^2} \ (Compressive)$$

Due to	Aluminium bar	Steel bar
External load	50 (tensile)	100 (compressive)
Temperature change	123 (compressive)	82 (compressive)
Final stress (N/mm²)	**73 (compressive)**	**182 (compressive)**

Example 1.37 *A steel tape is 100 m long at normal temperature. What will be the correction for 100 m on a day when the temperature is 15°C more than the normal and the pull on the tape is 150 N. The cross section of the tape is 10 mm × 1 mm. Given $\alpha_{st} = 12 \times 10^{-6}/$°C and $E_{st} = 210$ GPa.*

Solution:

$l = 100$ m, t = 15°C, $\alpha = 12 \times 10^{-6}$ /°C,

$A = 10 \text{ x } 1 = 10 \text{ mm}^2$, $E = 210 \ GPa = 210 \times 10^3 N/mm^2$

Free elongation $= l\alpha t = 100 \times 12 \times 10^{-6} \times 15 = 0.018$ m

This free elongation is due to temperature rise of 15°C

Again, change in length of tape due to applied load of 150 N is

$$\delta l = \frac{Pl}{AE} = \frac{150 \times 100 \times 10^3}{10 \times 210 \times 10^3} = 7.14 \text{ mm}$$

Total elongation= 7.14+18=25.14 mm (due to applied load and temperature rise)

∴ The correction for 100 m will be (−25.14 mm)

Example 1.38 *The composite bar shown (Fig. 1.44) is 0.2 mm short of distance between the rigid supports at room temperature. What is the maximum temperature rise which will not produce stresses in the bar? Find the stresses induced (if any) when the temperature rise is 40°C. Given that* $A_{steel} : A_{copper} = 4 : 3$
$\alpha_{steel} = 12 \times 10^{-6} \ /°C$, $\alpha_{copper} = 17.5 \times 10^{-6} \ /°C$,
$E_{steel} = 2 \times 10^5 \frac{N}{mm^2}$ $E_{copper} = 1.2 \times 10^5 \frac{N}{mm^2}$

Solution:

Let t = Temperature rise to produce 0.2 mm elongation

$\delta l = (l\alpha t)_{steel} + (l\alpha t)_{Copper}$

0.2 = (300× 12 × 10^{-6} × t) + (200× 17.5 × 10^{-6} × t)

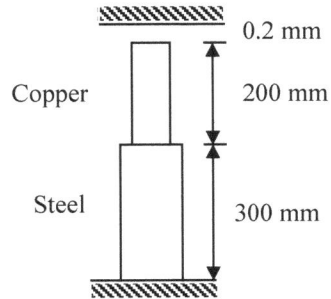

Fig. 1.44

∴ $t = 28.16$ °C
→ Maximum temperature rise which will not produce stress in the bar

Now increase in temperature t = 40°C

Free elongation = $(l\alpha t)_{steel} + (l\alpha t)_{Copper}$

$= (300 \times 12 \times 10^{-6} \times 40) + (200 \times 17.5 \times 10^{-6} \times 40)$

$= 0.284$ mm

Permitted elongation=0.2 mm

∴ Prevented elongation $= 0.284 - 0.2 = 0.084 \; mm$

Let P → Compressive force on bar due to prevented elongation

prevented elongation $= \left[\dfrac{Pl}{AE}\right]_{steel} + \left[\dfrac{Pl}{AE}\right]_{copper}$

$0.084 = \dfrac{P \times 300}{\frac{4}{3} \times A_c \times 2 \times 10^5} + \dfrac{P \times 200}{A_c \times 1.2 \times 10^5}$

$\therefore \dfrac{P}{A_c} = 30.107 \; N/mm^2$

Stresses:

$\sigma_{copper} = \dfrac{P}{A_c} = 30.107 \dfrac{N}{mm^2} (Compressive)$

$\sigma_{steel} = \dfrac{P}{A_s} = \dfrac{P}{\frac{4}{3} A_c} = \dfrac{3}{4} \times 30.107 = 22.58 \dfrac{N}{mm^2} (Compressive)$

Example 1.39 *A steel tube 24 mm external diameter and 18 mm internal diameter encloses a copper rod 15 mm diameter to which it is rigidly connected at its ends. If at temperature 10°C there is no longitudinal stress, calculate the stresses in the rod and the tube when temperature rises to 200°C.*
Given $E_s = 2.1 \times 10^5 \dfrac{N}{mm^2}$, $E_c = 1 \times 10^5 \dfrac{N}{mm^2}$
$\alpha_s = 11 \times 10^{-6} \; /°C$, $\alpha_c = 18 \times 10^{-6} \; /°C$,

Solution:

$A_s = \dfrac{\pi}{4} \times (24^2 - 18^2) = 197.9 \; mm^2$

→ cross sectional area of steel tube

$A_c = \dfrac{\pi}{4} (15^2) = 176.7 \; mm^2$ → cross sectional area of copper rod

Due to temperature rise, elongation is imposed on steel tube and the same is prevented in copper rod

Let P= Tensile force in steel tube (also the compressive force in copper rod)
Steel tube:

Free elongation $= l\alpha t = l \times 11 \times 10^{-6} \times 190 = 2.09 \times 10^{-3} \times l$

Permitted elongation $= \delta$

$\therefore$ Imposed elongation $= \delta - 2.09 \times 10^{-3} \times l$

Now $\delta - 2.09 \times 10^{-3} \times l = \left[\dfrac{Pl}{AE}\right]_{Steel\ tube} = \dfrac{P \times l}{197.9 \times 2.1 \times 10^5}$

$$\therefore P = \frac{41.5 \times 10^6 \times (\delta - 2.09 \times 10^{-3} \times l)}{l} \tag{1}$$

Copper rod:

Free elongation $= l\alpha t = l \times 18 \times 10^{-6} \times 190 = 3.42 \times 10^{-3} \times l$

Permitted elongation $= \delta$ (same elongation as that of tube)

Prevented elongation $= l\alpha t - \delta = 3.42 \times 10^{-3} \times l - \delta$

Now $3.42 \times 10^{-3} \times l - \delta = \left[\dfrac{pl}{AE}\right]_{Copper\ rod} = \dfrac{P \times l}{176.7 \times 1 \times 10^5}$

$$\therefore P = \frac{176.7 \times 1 \times 10^5 \times (3.42 \times 10^{-3} \times l - \delta)}{l} \tag{2}$$

From (1) and (2), $\delta = 2.487 \times 10^{-3} \times l$

$\therefore P = 16.47 \times 10^3 \text{ N} = 16.47 \text{ kN}$

Stresses:

In steel tube, $\sigma_s = \dfrac{P}{A_s} = \dfrac{16.47 \times 10^3}{197.9} = 83.2 \ \dfrac{\text{N}}{\text{mm}^2} \ (Tensile)$

In Copper rod, $\sigma_c = \dfrac{P}{A_c} = \dfrac{16.47 \times 10^3}{176.7}$

$= 93.2 \dfrac{N}{mm^2}$ (*Compressive*)

Example 1.40 *A composite bar is constructed from a steel rod of 25 mm diameter surrounded by a copper tube of 50 mm outside diameter and 25 mm inside diameter. The rod and tube are joined by 20 mm diameter pins at the ends. Find the shear stress set up in the pins if after pinning the temperature is raised by 50°C. $E_s = 210$ GPa, $\alpha_s = 11 \times 10^{-6}$ /°C, $E_c = 105$ GPa, $\alpha_c = 17 \times 10^{-6}$ /°C.*

Solution:

$A_s = \dfrac{\pi}{4} \times 25^2 = 490.87 \text{ mm}^2$

$A_c = \dfrac{\pi}{4} \times (50^2 - 25^2) = 1472.6 \text{ mm}^2$

Due to temperature rise, elongation is imposed on steel rod and is prevented on copper tube.

Let P= Tensile force in steel rod due to temperature rise (also equal to compressive force in copper tube)

Steel rod:

Free elongation $= l\alpha t = l \times 11 \times 10^{-6} \times 50 = 5.5 \times 10^{-4} \times l$

Permitted elongation $= \delta$ (say)

Imposed elongation $= \delta - l\alpha t = \delta - 5.5 \times 10^{-4} \times l$

$\therefore \delta - 5.5 \times 10^{-4} \times l = \dfrac{P \times l}{490.87 \times 210 \times 10^3}$

$\therefore P = \dfrac{1.03 \times 10^8 (\delta - 5.5 \times 10^{-4} \times l)}{l}$ (1)

Copper tube:

Free elongation $= l\alpha t = l \times 17 \times 10^{-6} \times 50 = 8.5 \times 10^{-4} \times l$

Permitted elongation $= \delta$

Prevented elongation $= l\alpha t - \delta = 8.5l \times 10^{-4} - \delta$

$$\therefore\; 8.5 \times 10^{-4} \times l - \delta = \frac{P \times l}{1472.6 \times 105 \times 10^3}$$

$$\therefore\; P = \frac{1.54 \times 10^8 (8.5 \times 10^{-4} \times l - \delta)}{l} \tag{2}$$

From (1) and (2)

$$\delta = \frac{1.875 \times 10^{-3} \times l}{2.54} = 7.29 \times 10^{-4} \times l$$

$$\therefore\; P = \frac{1.54 \times 10^8 (8.5 \times 10^{-4} \times l - 7.29 \times 10^{-4} \times l)}{l} = 18545.9\ \text{N}$$

Shear stress in pin, $\tau_{pin} = \dfrac{P}{2A_{pin}} = \dfrac{18545.9}{2 \times \frac{\pi}{4} \times 20^2} = 29.51\ \dfrac{\text{N}}{\text{mm}^2}$

(Pin is subjected to double shear)

Example 1.41 *A circular bar whose cross section varies uniformly from d_1 and d_2 is fixed at its ends. It is subjected to $t°C$ rise in temperature. If E be the Young's modulus and α be the coefficient of thermal expansion for the bar material, what will be the maximum stress induced in the bar.*

Solution:

$d_1 =$ Diameter at bigger end
$d_2 =$ Diameter at smaller end
$l =$ length of bar

Prevented elongation $= l\alpha t$

Contraction of bar under the action of external load P (if acting) is

$$\delta l = \frac{4Pl}{\pi E d_1 d_2}$$

$$\therefore l\alpha t = \frac{4Pl}{\pi E d_1 d_2} \qquad or \qquad P = \frac{\pi E d_1 d_2 \alpha t}{4}$$

$$\text{Maximum stress} = \frac{\frac{\pi E d_1 d_2 \alpha t}{4}}{\frac{\pi}{4} \times d_2{}^2}$$

(Maximum stress corresponds to minimum area)

$$\therefore \sigma_{max} = E\alpha t \frac{d_1}{d_2} \qquad \rightarrow \quad \text{Maximum stress induced in the tapered bar}$$

Example 1.42 *A wooden wheel of diameter d_1 is fitted with a steel ring by increasing the temperature of ring through $t°C$. Find the original diameter of steel ring, if the coefficient of thermal expansion and Young's modulus for the ring material are given as α and E, respectively. Also calculate the stress developed in the ring when it cools back to the normal temperature.*

Solution:

d_1 = diameter of ring after heating

Let d = diameter of ring before heating

After heating, circumference of ring $= \pi d_1$

Before heating, circumference of ring $= \pi d$

Rise in temperature = t°C

$$\therefore \pi d + \pi d\alpha t = \pi d_1 \qquad or \qquad d(1 + \alpha t) = d_1$$

$$\therefore d = \frac{d_1}{1 + \alpha t} \quad \rightarrow \quad \text{Original diameter of steel ring}$$

When it cools back to the normal temperature

Prevented contraction $= \pi d_1 - \pi d = \pi(d_1 - d)$

Prevented contraction will cause tensile stresses

Prevented contraction $= \dfrac{Pl}{AE}$

$$\pi(d_1 - d) = \dfrac{P\pi d_1}{AE}$$

$\therefore$ Stress developed in the ring $= \dfrac{P}{A} = \dfrac{(d_1 - d)E}{d_1} = E\left(1 - \dfrac{d}{d_1}\right)$

HIGHLIGHTS

Definitions

1. *Stress* - It is defined as the internal resistance per unit area offered by the material of the body against external loading (unit: N/m^2).

2. *Strain* - Normal strain or linear strain is the ratio of change in dimension to original dimension. Shear strain is the angular deformation caused due to shear stress.

3. *Hooke's Law* - Within elastic limit, stress is directly proportional to strain.

4. *Volumetric strain* - It is the ratio of change in volume to the original volume. It is also equal to the sum of strains in three mutually perpendicular directions.

5. *Young's modulus* (modulus of elasticity, E) - It is the ratio of normal stress to normal strain within proportional limit.

6. *Shear modulus* (modulus of rigidity, G) - It is the ratio of shear stress to shear strain within proportional limit.

7. *Factor of safety* - It is the ratio of the failure stress of the material to the stress that is allowed.

$$\text{F. O. S.} = \frac{\text{Yield stress}}{\text{Allowable stress}} \qquad \text{(for ductile material)}$$

$$\text{F. O. S.} = \frac{\text{Ultimate stress}}{\text{Allowable stress}} \qquad \text{(for brittle material)}$$

8. *Nominal stress* – It is the ratio of load to original cross sectional area (also known as engineering stress). True Stress is the ratio of load to instantaneous cross sectional area.

9. *Bulk modulus* - It is the ratio of normal Stress to volumetric strain.

10. *Poisson's ratio* - It is the ratio of lateral strain to the longitudinal strain.

11. *Temperature stress* - When the elongation or contraction of a bar caused due to change in temperature is prevented by making its ends rigidly fixed, then the stress induced in the bar is known as temperature stress. The corresponding strain is known as temperature strain.

Concepts and Formulae

1. Elongation of uniform bar, $\delta l = \dfrac{Pl}{AE} = \dfrac{\sigma l}{E}$

2. Stiffness of bar $= \dfrac{\text{Force}}{\text{Deformation}} = \dfrac{P}{\delta l} = \dfrac{AE}{l}$

3. For bar of cross section varying in steps subjected to axial load P

$$\delta l = \frac{Pl_1}{A_1 E_1} + \frac{Pl_2}{A_2 E_2} + \frac{Pl_3}{A_3 E_3} + \ldots$$

4. For bars of cross section varying in steps and subjected to different loads at different cross sections,

$$\delta l = \frac{P_1 l_1}{A_1 E_1} + \frac{P_2 l_2}{A_2 E_2} + \frac{P_3 l_3}{A_3 E_3} + \ldots$$

(Tensile load causes elongation and compressive load causes contraction)

5. Elongation of uniform bar due to self weight,

$$\delta l = \frac{wl^2}{2E}, w = \text{specific weight}$$

6. Elongation of tapered bar, $\delta l = \dfrac{4Pl}{\pi E d_1 d_2}$

7. Elongation of tapered plate, $\delta l = \dfrac{Pl}{tE(B_2 - B_1)} \log_e \left(\dfrac{B_2}{B_1}\right)$

8. For solving statically indeterminate problems, compatibility equations are also required in addition to the equilibrium equations.

9. Volumetric Strain

$$\varepsilon_v = \frac{\sigma_x}{E}(1 - 2v) \qquad \text{for uniaxial stress condition}$$

$$= \frac{\sigma_x + \sigma_y}{E}(1 - 2v) \quad \text{for biaxial stress condition}$$

$$= \frac{\sigma_x + \sigma_y + \sigma_z}{E}(1 - 2v) \quad \text{for triaxial stress condition}$$

Also $\varepsilon_v = \varepsilon_x + \varepsilon_y + \varepsilon_z$

10. Relation between elastic constants

$$K = \frac{E}{3(1 - 2v)}, \qquad G = \frac{E}{2(1 + v)}, \qquad v = \frac{3k - 2G}{6K + 2G},$$

$$E = \frac{9KG}{3K + G}$$

11. For composite bars made of two or more bars of equal lengths and subjected to tensile or compressive load, the elongation or contraction in each bar will be equal and the total load will be equal to the sum of the loads carried by each member.

12. For a bar of uniform strength the cross sectional area at a distance x from smaller end (having area of cross section A_1) is

$A = A_1 e^{\frac{wx}{\sigma}}$

w = specific weight of bar element, σ = stress induced in the bar.

13. Thermal stresses:

Free elongation $= l\alpha t$

Prevented elongation $= \dfrac{Pl}{AE}$

Thermal stress $= \dfrac{P}{A}$

Similarly, for contraction also, thermal stress can be computed.

14. Maximum temperature stress in bar of tapering section is

$\sigma_{max} = E\alpha t \dfrac{d_1}{d_2}$,

d_1, d_2 = diameters at bigger and smaller ends.

15. The deformation due to self-weight of the body is half of that produced by external force of intensity equal to self-weight of body.

SHORT TYPE QUESTIONS

1. Hooke's law is valid for _____ type of material?
                        [Ans. Linearly elastic material]

2. What is the ratio of deformation of a bar due to its own weight to the deformation due to axial load equal to self-weight of the body?
                        [Ans. $\frac{1}{2}$]

[Explanation: Self weight $- \;\; \delta l = \dfrac{wl^2}{2E}$,

External load $- \;\; \delta l = \dfrac{Pl}{AE} = \dfrac{(wAl)l}{AE} = \dfrac{wl^2}{E}$]

3. If the Poisson's ratio of a material is 0.4 then ratio of shear modulus to Young's modulus is _____ . [Ans. 0.357]

[Explanation: $\dfrac{G}{E} = \dfrac{1}{2(1+v)} = \dfrac{1}{2(1+0.4)}$]

4. A square plate of thickness t is subjected to a tensile stress σ_x in x-direction and compressive stress $\sigma_y = -\sigma_x$ in y- direction. If E is the modulus of elasticity and v is the Poisson's ratio, then the change in plate thickness δt in the z- direction is

(a) zero   (b) $\dfrac{\sigma_x(1-v)t}{E}$ (c) $\dfrac{\sigma_y(1-v)t}{E}$ (d) $\dfrac{\sigma_x(1+v)t}{E}$

[Ans. (a)]

[Explanation: $\varepsilon_z = -v\dfrac{\sigma_x}{E} - v\dfrac{\sigma_y}{E} = -v\dfrac{\sigma_x}{E} - v\left(\dfrac{-\sigma_x}{E}\right) = 0,\ \varepsilon_z = \dfrac{\delta t}{t}$ ∴ $\delta t = 0$]

5. In question No. 4 the strain in x- direction is

(a) zero   (b) $\dfrac{\sigma_x(1+v)}{E}$ (c) $\dfrac{-\sigma_x(1+v)}{E}$ (d) None of these

[Ans. (b)]

[Explanation: $\varepsilon_x = \dfrac{\sigma_x}{E} - v\dfrac{\sigma_y}{E} = \dfrac{\sigma_x}{E} - v\left(\dfrac{-\sigma_x}{E}\right) = \dfrac{\sigma_x}{E}(1+v)$,

$\varepsilon_y = \dfrac{\sigma_y}{E} - v\dfrac{\sigma_x}{E} = -\dfrac{\sigma_x}{E}(1+v)$]

6. A mild steel plate is subjected to tri-axial stresses. If the stresses in the two mutually perpendicular directions are 40 N/mm² (compressive) and 15 N/mm² (tensile), what will be the stress in the third direction for no change of volume?

[Ans. 25 N/mm² (tensile)]

[Explanation : $\sigma_x + \sigma_y + \sigma_z = 0$ or $-40 + 15 + \sigma_z = 0$

∴ $\sigma_z = 25$ N/mm²]

7. The length of a wire is increased by 2 mm on applying an axial tensile load. For the wire of same material, having half the length and twice the radius, the deflection under the same load will be

(a) 2 mm   (b) 1 mm   (c) 0.5 mm   (d) 0.25 mm

[Ans. (d)]

[Explanation: 1st case, $\delta l_1 = \left(\dfrac{Pl}{AE}\right)_1 = \dfrac{P.l}{\pi r^2 E} = 2$ given,

2nd case , $\delta l_2 = \left(\dfrac{Pl}{AE}\right)_2 = \dfrac{Pl/2}{4\pi r^2 E} = \dfrac{1}{8}\dfrac{Pl}{\pi r^2 E} = \dfrac{1}{8} \times 2 = 0.25$ mm]

8. A free bar of length l is uniformly heated from 0^0C to a temperature t^0C. If α is the coefficient of linear expansion and E is the modulus of elasticity, the stress in the bar is
(a) $\alpha t E$ (b) $\alpha t E/2$ (c) zero   (d) None of these

[Ans. (c)]

9. An aluminium bar 1.8 m long has a 25 mm square cross section over 0.6 m of its length and a 25 mm diameter circular cross section over other 1.2 m. How much will the bar elongate under a tensile load of 17.5 kN if E = 75 GPa?

[Ans. 0.794 mm]

[Explanation: $\delta l = \left(\frac{Pl}{AE}\right)_{sq} + \left(\frac{Pl}{AE}\right)_{cir} = \frac{P}{E}\left(\frac{l_1}{A_1} + \frac{l_2}{A_2}\right) = \frac{17500}{75\times10^3}\left(\frac{600}{25^2} + \frac{1200}{\frac{\pi}{4}\times25^2}\right) = 0.794$ mm]

10. The cross section of a wire is circular. If its radius decreases to half of the original value due to stretch of wire by load, then the modulus of elasticity of wire will be
(a) One fourth   (b) Halved   (c) Doubled (d) Unaffected

[Ans. (d)]

[Explanation – E is material property]

11. Match list I (elastic properties) with list II (nature of strain produced) and select the correct answer using the codes given below the lists.

List – I	List – II
A. Young's modulus	1. Shear strain
B. Modulus of Rigidity	2. Normal strain
C. Bulk modulus	3. Transverse strain
D. Poisson's Ratio	4. Volumetric strain

(a) A B C D   (b) A B C D   (c) A B C D   (d) A B C D
    1 2 3 4   2 1 3 4   2 1 4 3   1 2 4 3

[Ans. (c)]

12. In terms of Poisson's ratio (v), the ratio of Young's modulus (E) and shear modulus (G) of elastic materials is
(a) $2(1+v)$ (b) $2(1-v)$ (c) $\frac{1}{2}(1+v)$ (d) $\frac{1}{2}(1-v)$

[Ans. (a)]

13. A tapered bar (diameters of end sections being d_1 and d_2) and a bar of uniform cross section of diameter d have the same length and are subjected to the same axial pull. Both the bars will have the same extensions if d is equal to

(a) $\frac{d_1+d_2}{2}$ (b) $\sqrt{d_1 d_2}$ (c) $\frac{\sqrt{d_1 d_2}}{2}$ (d) $\frac{\sqrt{d_1+d_2}}{2}$

[Ans. (b)]

[Explanation $-\ \delta l_1 = \frac{4Pl}{\pi E d_1 d_2}$, $\delta l_2 = \frac{Pl}{AE} = \frac{Pl}{\frac{\pi}{4}d^2 E}$]

EXERCISE PROBLEMS

1. A steel wire 2 m long and 3 mm in diameter is extended by 0.75 mm when a weight W is suspended from the wire. If the same weight is suspended from a brass wire, 2.5 m long and 2 mm in diameter, it is elongated 4.64 mm. Determine the modulus of elasticity of brass if that of steel is 2×10^5 N /mm^2.

[Ans. 90986 N/mm^2]

2. Two straight rods one made of steel and other of brass hang vertically. Each rod is 1 m long and they support a rigid bar horizontally. When a load of 25 kN is placed at 40 cm from the steel rod on the horizontal bar, the deflection of the two vertical rods are found to be equal. If the area of steel rod is 3 cm^2, Find (i) the area of the other rod (ii) the stresses in the rods (iii) strains in the rods.

[Ans. (i) 4.706 cm^2 (ii) σ_s = 50 mN/m^2, σ_b = 21.25 MN/m^2
(iii) ε_s = 0.25×10^{-3}, ε_b = 0.25×10^{-3}]

3. The composite bar shown (Fig. 1.45) is subjected to a tensile force of 30 kN. The extension observed is 0.372 mm. Find the Young's modulus of brass if that of steel is 2×10^5 N/mm^2.

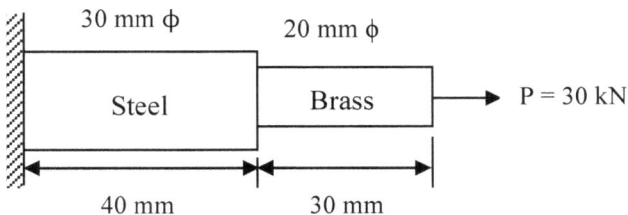

30 mm φ	20 mm φ
Steel	Brass	P = 30 kN
40 mm	30 mm

Fig. 1.45   [Ans. 99777.6 N/mm^2]

4. Two rods one of brass and other of steel are rigidly connected at ends. Both the rods have a cross section 10 mm × 10 mm and have a length of 1m. This combined rod is pulled by a tension = 10 kN. Find the (i) elongation (ii) stresses in two rods (iii) strains in two rods. Take E_{steel} = 2×10^5 N/mm^2, E_{brass} = 1×10^5 N/mm^2

[Ans. (i) 3.33 mm (ii) σ_{steel} = 66.67 N/mm^2, σ_{brass} = 33.33 N/mm^2 (iii) ε_s = ε_b = 33.33×10^{-5}]

5. A stepped bar shown (Fig. 1.46) is subjected to an axially applied compressive load of 35 kN. Find the maximum and minimum stresses produced.

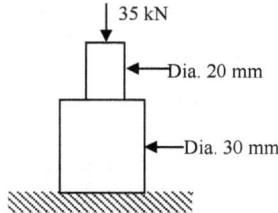

Fig. 1.46

[Ans. 111.4 N/mm^2, 49.51 N/mm^2]

6. Determine the value of Poisson's ratio and Young's modulus if modulus of rigidity of material is 0.5×10^5 N/mm^2 and bulk modulus is 1.12×10^5 N/mm^2.

[Ans. 0.305, 1.3×10^5 N/mm^2]

7. A steel bolt 12 mm diameter passes through a brass tube of 16 mm internal diameter, 25 cm long and 20 mm external diameter. The bolt is tightened by a nut at 15°C so as to exert a compressive force of 1500 kg on the tube. Calculate (i) the stresses in each (ii) the stresses in each when the temperature of tube and bolt is raised to 50°C.
Use E_s = 2×10^6 kg/cm^2, α_s = 12×10^{-6}/°C, E_b = 1×10^6 kg/cm^2, α_b = 19×10^{-6}/°C

[Ans. (i) σ_s = 1327.76 kg/cm^2 (Tensile), σ_b = 1327.4 kg/cm^2 (Compressive), (ii) σ_s = 1490.76 kg/cm^2 (Tensile), σ_b = 1490.76 kg/cm^2 (Compressive)]

8. A steel ring is fitted on a wooden wheel of diameter 1.5 m by raising the temperature of steel ring through 40°C. What should be the original internal diameter of the ring? Also calculate the stress de-

veloped in the ring when it cools back to the normal temperature. Take $E_s = 2.1 \times 10^5$ N/mm^2, $\alpha_s = 12 \times 10^{-6}/^0$C

[Ans. d = 1.499 m, σ = 100.8 N/mm^2]

Chapter 2

Compound Stresses and Strains

Learning Objectives

After going through this chapter, the reader will be able to
- determine the stresses on oblique planes in members sub-jected to biaxial stresses and shear stress.
- compute the principal stresses and maximum in-plane shear stress and locate the associated planes.
- draw the Mohr's circle for plane stress condition and inter-pret it.
- determine the principal strains analytically and graphically for plane strain condition.
- express the principal stresses in terms of principal strains.

2.1 INTRODUCTION

The effects of simple stresses which are totally normal (tensile or compressive) or totally tangential (shear) and acting on a particular plane are discussed in the previous Chapter. Though the stress is simple on that particular plane, it may be complex in other planes. Also, actual engineering problems involve both the normal stresses as well as shear stresses acting simultaneously. For example, a beam is always under bending and shear, a shaft may be considered subjected to torque, bending and direct stresses. In this Chapter, we will see the combined effect of these stresses.

2.2 THREE-DIMENSIONAL STRESS SYSTEM

There are 18 stress components on the elemental cuboid as shown in Fig. 2.1.

σ_{xx} = Stress on x plane acting along x direction (commonly written as σ_x)

Similarly, σ_{yy} (or σ_y) and σ_{zz} (or σ_z) denote the stresses on y and z planes, respectively along y and z directions.

τ_{xy} = Stress on x plane acting along y direction
τ_{yx} = Stress on y plane acting along x direction
The stresses τ_{xz}, τ_{zx}, τ_{yz} and τ_{zy} are defined in a similar way.

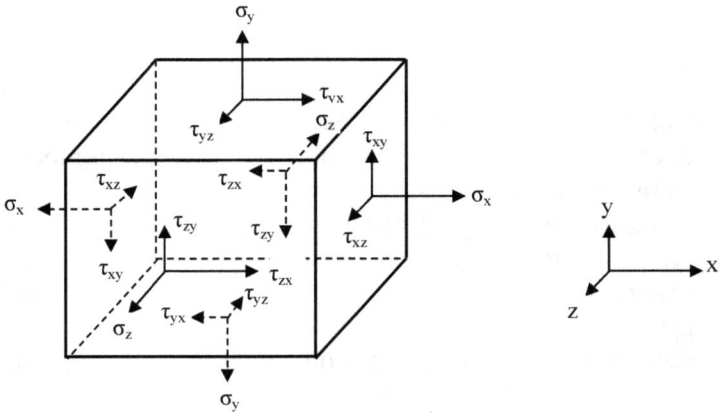

Fig. 2.1

2.3 PLANE STRESS CONDITION

In a cuboidal element when stresses are acting only on two pairs of parallel planes and no stress is present on the third pair of parallel planes, the element is said to be under plane stress condition. The plane stress condition is a two-dimensional stress condition (Fig. 2.2a).

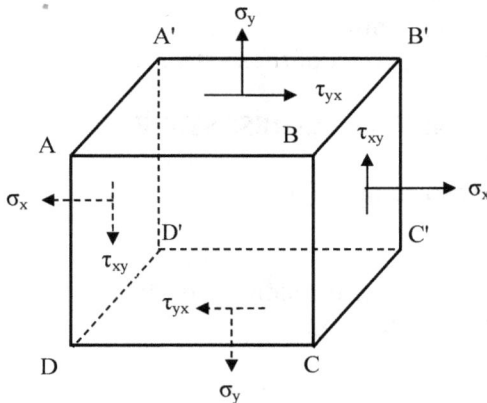

Fig. 2.2a

As shown in Fig. 2.2a, one pair of parallel planes AA'D'D and BB'C'C is subjected to tensile stress σ_x and shear stress τ_{xy} and another pair of parallel planes AA'B'B and DD'C'C is subjected to tensile stress σ_y and shear stress τ_{yx}. But no stress is present on the other pair of parallel planes ABCD and A'B'C'D'. Therefore, the shown cuboidal element is said to be under plane stress condition. It can also be represented as shown in Fig. 2.2b.

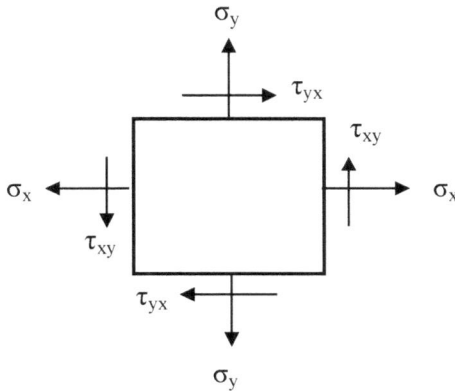

Fig. 2.2b

2.4 COMPLEMENTARY SHEAR STRESS

Consider a rectangular block of length dx, height dy and of unit thickness as shown (Fig. 2.3a). A shear stress τ_{xy} acts on faces AA'D'D and BB'C'C as shown.

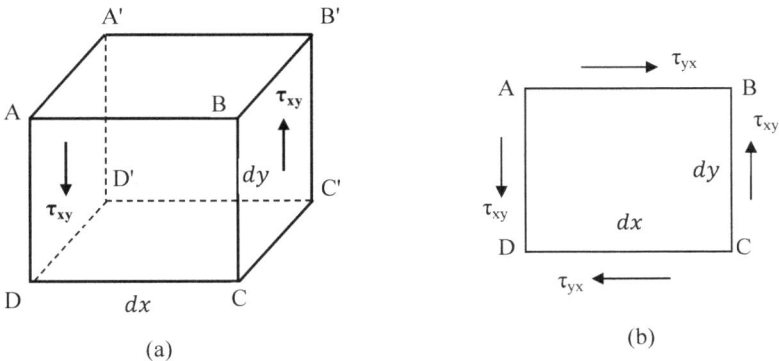

(a)   (b)

Fig. 2.3

Forces caused by the stresses form a couple and for equilibrium, there should be an equal and opposite couple.

Let τ_{yx} = Magnitude of shear stress forming the balancing couple
Taking moments of forces about DD'

$$\tau_{xy} \times (dy \times 1) \times dx = \tau_{yx} \times (dx \times 1) \times dy$$

$$\Rightarrow \qquad \tau_{xy} = \tau_{yx}$$

Therefore, a shear stress is automatically accompanied by a shear stress of equal intensity but of opposite turning moment. That is shear stress is complementary in nature.

2.5 ELEMENT SUBJECTED TO BIAXIAL STRESSES AND A SHEAR STRESS

Consider an elemental rectangular block (unit thickness) subjected to stresses σ_x, σ_y and τ as shown (Fig. 2.4a).

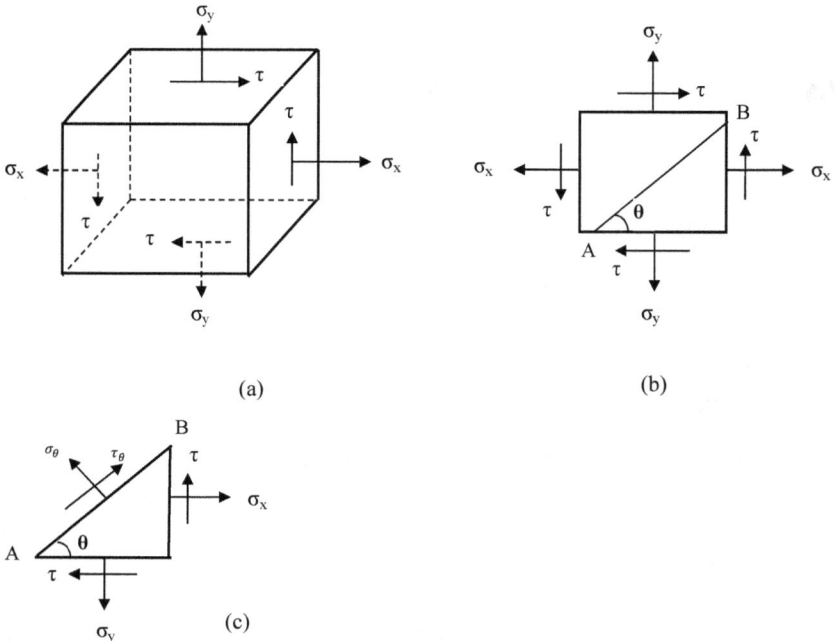

(a)

(b)

(c)

Fig. 2.4

It is of interest to find out the values of the normal stress and shear stress on a plane AB which is inclined at an angle θ as shown (Figs. 2.4 b and c).

σ_θ = Normal stress on plane AB
τ_θ = Shear stress on plane AB

Sign convention

(i) Normal stress – Tensile stress is positive and compressive stress is negative

(ii) Shear stress – Shear stress τ is positive if acting upward in positive y-direction on the right face of element otherwise is considered negative (Fig. 2.5).

Fig. 2.5

(iii) θ measured anticlockwise with respect to horizontal (σ_x direction) is considered positive and if measured in clockwise direction with respect to horizontal is considered negative.

The portion ABC of the elemental block is shown along with the forces acting on the various planes (Fig. 2.6).

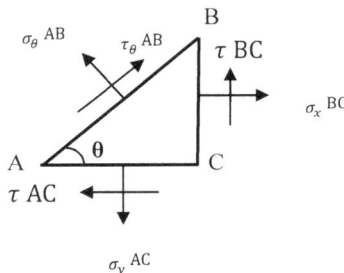

Fig. 2.6

Now the equilibrium equation yields

σ_θ AB $+ \tau$ AC sin $\theta + \tau$ BC cos $\theta = \sigma_x$ BC sin $\theta + \sigma_y$ AC cos θ
(AC $=$ AB cos θ and BC $=$ AB sin θ)

σ_θ AB $+ \tau$ AB cos θ sin $\theta + \tau$ AB sin θ cos θ
$$= \sigma_x \text{ AB sin } \theta \text{ sin } \theta + \sigma_y \text{ AB cos } \theta \text{ cos } \theta$$

$\sigma_\theta = \sigma_x \sin^2 \theta + \sigma_y \cos^2 \theta - 2 \tau \sin \theta \cos \theta$

$$= \sigma_x \left(\frac{1 - \cos 2\theta}{2}\right) + \sigma_y \left(\frac{1 + \cos 2\theta}{2}\right) - \tau \sin 2\theta$$

$$\boxed{\sigma_\theta = \frac{\sigma_y + \sigma_x}{2} + \left(\frac{\sigma_y - \sigma_x}{2}\right) \cos 2\theta - \tau \sin 2\theta}$$

Similarly, another equilibrium equation yields:

τ_θ AB $+ \sigma_x$ BC cos $\theta + \tau$ BC sin $\theta = \sigma_y$ AC sin $\theta + \tau$ AC cos θ

τ_θ AB $+ \sigma_x$ AB sin θ cos $\theta + \tau$ AB sin θ sin θ
$$= \sigma_y \text{ AB cos } \theta \text{ sin } \theta + \tau \text{ AB cos } \theta \text{ cos } \theta$$

$\tau_\theta + \sigma_x \sin \theta \cos \theta + \tau \sin^2 \theta = \sigma_y \cos \theta \sin \theta + \tau \cos^2 \theta$

$\tau_\theta = (\sigma_y - \sigma_x) \sin \theta \cos \theta + \tau (\cos^2 \theta - \sin^2 \theta)$

$$\boxed{\tau_\theta = \left(\frac{\sigma_y - \sigma_x}{2}\right) \sin 2\theta + \tau \cos 2\theta}$$

Resultant stress, $\sigma_r = \sqrt{\sigma_\theta^2 + \tau_\theta^2}$

Obliquity

The angle between the resultant stress and the normal to oblique plane is known as obliquity (φ).

$\tan \varphi = \dfrac{\tau_\theta}{\sigma_\theta}$

In case, element is subjected to the biaxial stresses (σ_x and σ_y only exist and $\tau = 0$) (Fig. 2.7)

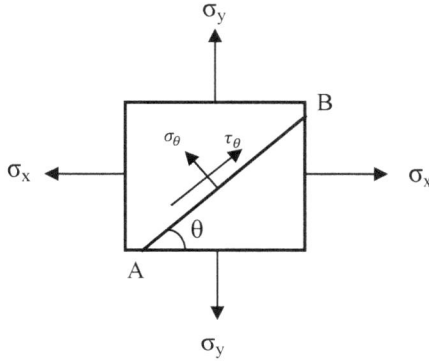

Fig. 2.7

$$\sigma_\theta = \frac{\sigma_y + \sigma_x}{2} + \left(\frac{\sigma_y - \sigma_x}{2}\right) \cos 2\theta$$

$$= \frac{\sigma_y}{2}(1 + \cos 2\theta) + \frac{\sigma_x}{2}(1 - \cos 2\theta)$$

$$= \sigma_y \cos^2 \theta + \sigma_x \sin^2 \theta$$

$$\tau_\theta = \left(\frac{\sigma_y - \sigma_x}{2}\right) \sin 2\theta$$

Maximum shear stress occurs when $\theta = 45^0$

$$\tau_{max} = \frac{\sigma_y - \sigma_x}{2}$$

$$\tan \varphi = \frac{\tau_\theta}{\sigma_\theta} = \frac{\left(\frac{\sigma_y - \sigma_x}{2}\right) \sin 2\theta}{\sigma_y \cos^2 \theta + \sigma_x \sin^2 \theta} = \frac{(\sigma_y - \sigma_x)\sin \theta \cos \theta}{\sigma_y \cos^2 \theta + \sigma_x \sin^2 \theta}$$

$$= \frac{(\sigma_y - \sigma_x)}{\sigma_y \cot \theta + \sigma_x \tan \theta}$$

For maximum obliquity, $\dfrac{d}{d\varphi}(\tan \varphi) = 0$

$$\boxed{\tan \varphi = \sqrt{\frac{\sigma_y}{\sigma_x}}} \longrightarrow \text{\textit{Condition for Maximum Obliquity}}$$

For element subjected to *uniaxial* stress condition (only σ_x present) (Fig. 2.8a).

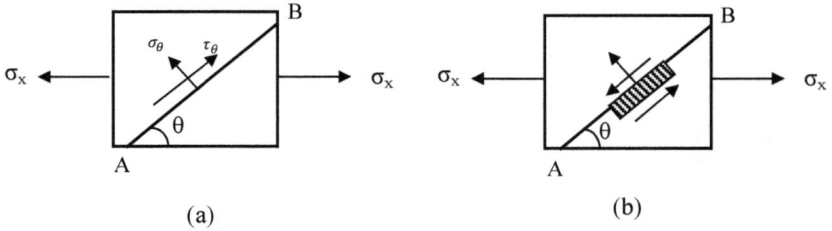

(a)   (b)

Fig. 2.8

$$\sigma_\theta = \sigma_x \sin^2 \theta$$

$$\tau_\theta = \left(\frac{-\sigma_x}{2}\right) \sin 2\theta = -\sigma_x \sin \theta \cos \theta$$

Therefore, the nature of shear stress will be such that it tends to rotate the hatched element anticlockwise (Fig. 2.8b).

Also σ_θ is maximum when $\theta = 90^0$ (which corresponds to the vertical plane).

τ_θ is maximum when $\theta = 45^0, \ 135^0$
and $\tau_{max} = \dfrac{\sigma_x}{2}$

For element subjected to only shear stress (case of simple shear or pure shear) (Fig. 2.9)

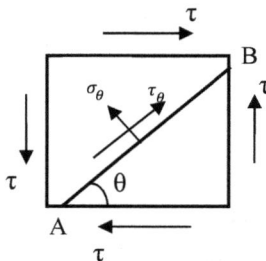

Fig. 2.9

$$\sigma_\theta = \frac{\sigma_y + \sigma_x}{2} + \left(\frac{\sigma_y - \sigma_x}{2}\right) \cos 2\theta - \tau \sin 2\theta = -\tau \sin 2\theta$$

$$\tau_\theta = \left(\frac{\sigma_y - \sigma_x}{2}\right) \sin 2\theta + \tau \cos 2\theta = \tau \cos 2\theta$$

2.6 PRINCIPAL PLANES AND PRINCIPAL STRESSES

Figure 2.10 shows an element subjected to plane stress condition. For this element, through a point A (say point A) any number of planes can be considered and the normal stress as well as shear stress can be computed on these planes using the procedure explained in the previous section.

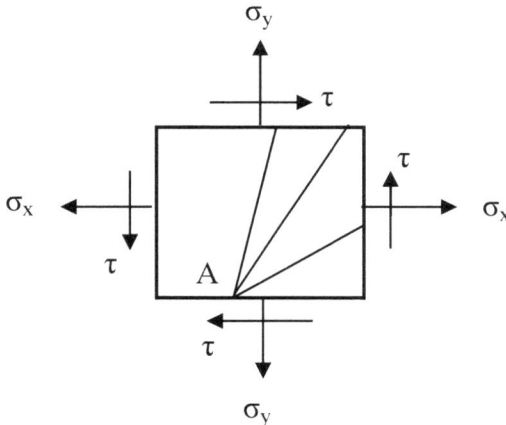

Fig. 2.10

Of these planes, there will be two planes (perpendicular to each other) which are free of shear stresses and only normal stress acts on these planes. These planes are known as principal planes and the corresponding stresses are principal stresses. Principal planes are the planes on which the shear stress is zero, and only normal stress acts on these planes. Principal stresses are the maximum and minimum values of normal stresses. The maximum principal stress is called major principal stress and the minimum principal stress is known as minor principal stress. The plane on which the major principal stress acts is known to be the major principal plane and the plane on which the minor principal stress acts is known as the minor principal plane.

To locate the principal planes, condition is

$$\tau_\theta = 0$$

$$\therefore \left(\frac{\sigma_y - \sigma_x}{2}\right) \sin 2\theta + \tau \cos 2\theta = 0$$

$$\rightarrow \quad \tan 2\theta = \frac{-2\tau}{\sigma_y - \sigma_x}$$

Inclination of principal (normal) planes can be found out from expression

$$\boxed{\tan 2\theta_n = \frac{-2\tau}{\sigma_y - \sigma_x}}$$

Two values of θ_n will be obtained which differ by 90^0

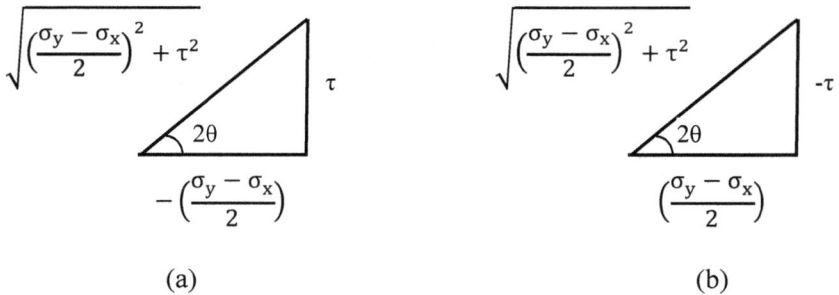

(a)   (b)

Fig. 2.11

With reference to Fig. 2.11

$$\sin 2\theta = \pm \frac{\tau}{\sqrt{\left(\frac{\sigma_y - \sigma_x}{2}\right)^2 + \tau^2}} \quad \text{and} \quad \cos 2\theta = \mp \frac{(\sigma_y - \sigma_x)/2}{\sqrt{\left(\frac{\sigma_y - \sigma_x}{2}\right)^2 + \tau^2}}$$

(when sin 2θ is ±, cos 2θ is ∓)

Now $\sigma_\theta = \dfrac{\sigma_y + \sigma_x}{2} + \left(\dfrac{\sigma_y - \sigma_x}{2}\right) \cos 2\theta - \tau \sin 2\theta$

$$= \frac{\sigma_y + \sigma_x}{2} + \left(\frac{\sigma_y - \sigma_x}{2}\right) \left(\mp \frac{(\sigma_y - \sigma_x)/2}{\sqrt{\left(\frac{\sigma_y - \sigma_x}{2}\right)^2 + \tau^2}}\right)$$

$$- \tau \left(\pm \frac{\tau}{\sqrt{\left(\frac{\sigma_y - \sigma_x}{2}\right)^2 + \tau^2}}\right)$$

$$= \frac{\sigma_y + \sigma_x}{2} \pm \frac{\left(\frac{\sigma_y - \sigma_x}{2}\right)^2 + \tau^2}{\sqrt{\left(\frac{\sigma_y - \sigma_x}{2}\right)^2 + \tau^2}} = \frac{\sigma_y + \sigma_x}{2} \pm \sqrt{\left(\frac{\sigma_y - \sigma_x}{2}\right)^2 + \tau^2}$$

$$\sigma_{\theta_1} = \frac{\sigma_y + \sigma_x}{2} + \sqrt{\left(\frac{\sigma_y - \sigma_x}{2}\right)^2 + \tau^2}$$

$\rightarrow$ Major principal stress (σ_1)

$$\sigma_{\theta_1} = \frac{\sigma_y + \sigma_x}{2} - \sqrt{\left(\frac{\sigma_y - \sigma_x}{2}\right)^2 + \tau^2}$$

$\rightarrow$ Minor principal stress (σ_2)

Principal stress expression

$$\boxed{\sigma_1, \sigma_2 = \frac{\sigma_y + \sigma_x}{2} \pm \sqrt{\left(\frac{\sigma_y - \sigma_x}{2}\right)^2 + \tau^2}}$$

The expressions for σ_1 and σ_2 can be written considering $+$ and $-$ sign, respectively.

2.7 PRINCIPAL SHEAR PLANES AND PRINCIPAL SHEAR STRESSES

Principal shear planes are the planes on which the shear stress is maximum or minimum and these planes are also perpendicular to each other. Principal shear stresses are the maximum and minimum values of shear stresses. These are numerically equal but opposite in sign. These stresses are acting on the principal shear planes. The maximum and minimum shear stresses are known as major principal shear stress and minor principal shear stress, respectively. The corresponding planes respectively are known as major principal shear plane and minor principal shear plane.

To locate principal shear planes

$$\frac{d}{d\theta}(\tau_\theta) = 0$$

$$\therefore \frac{d}{d\theta}\left[\left(\frac{\sigma_y - \sigma_x}{2}\right)\sin 2\theta + \tau \cos 2\theta\right] = 0$$

$$\rightarrow \left(\frac{\sigma_y - \sigma_x}{2}\right) \times 2\cos 2\theta + \tau(-2\sin 2\theta) = 0$$

$$\rightarrow \quad \tan 2\theta = \frac{\sigma_y - \sigma_x}{2\tau}$$

Inclination of principal shear planes can be found out from expression

$$\boxed{\tan 2\theta_s = \frac{\sigma_y - \sigma_x}{2\tau}}$$

Here also two values of θ_s will be obtained which differ from each other by 90^0.

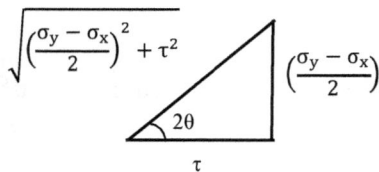

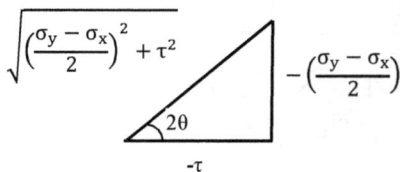

(a)   (b)

Fig. 2.12

With reference to Fig. 2.12

$$\sin 2\theta = \pm \frac{(\sigma_y - \sigma_x)/2}{\sqrt{\left(\frac{\sigma_y - \sigma_x}{2}\right)^2 + \tau^2}} \quad \text{and} \quad \cos 2\theta = \pm \frac{\tau}{\sqrt{\left(\frac{\sigma_y - \sigma_x}{2}\right)^2 + \tau^2}}$$

$$\therefore \quad \tau_\theta = \left(\frac{\sigma_y - \sigma_x}{2}\right) \sin 2\theta + \tau \cos 2\theta$$

$$= \left(\frac{\sigma_y - \sigma_x}{2}\right) \left(\pm \frac{(\sigma_y - \sigma_x)/2}{\sqrt{\left(\frac{\sigma_y - \sigma_x}{2}\right)^2 + \tau^2}} \right) + \tau \left(\pm \frac{\tau}{\sqrt{\left(\frac{\sigma_y - \sigma_x}{2}\right)^2 + \tau^2}} \right)$$

$$= \pm \frac{\left(\frac{\sigma_y - \sigma_x}{2}\right)^2 + \tau^2}{\sqrt{\left(\frac{\sigma_y - \sigma_x}{2}\right)^2 + \tau^2}} \qquad = \pm \sqrt{\left(\frac{\sigma_y - \sigma_x}{2}\right)^2 + \tau^2}$$

$$\tau_{max} = \sqrt{\left(\frac{\sigma_y - \sigma_x}{2}\right)^2 + \tau^2} \quad \rightarrow \quad \text{Major principal shear stress}$$

$$\tau_{min} = -\sqrt{\left(\frac{\sigma_y - \sigma_x}{2}\right)^2 + \tau^2} \quad \rightarrow \quad \text{Minor principal shear stress}$$

Also,

$$\tau_{max} = \frac{\sigma_1 - \sigma_2}{2}, \qquad \tau_{min} = -\left(\frac{\sigma_1 - \sigma_2}{2}\right)$$

Normal stress on principal shear planes

On the principal shear planes

$$\sin 2\theta = \pm \frac{(\sigma_y - \sigma_x)/2}{\sqrt{\left(\frac{\sigma_y - \sigma_x}{2}\right)^2 + \tau^2}} \quad \text{and} \quad \cos 2\theta = \pm \frac{\tau}{\sqrt{\left(\frac{\sigma_y - \sigma_x}{2}\right)^2 + \tau^2}}$$

$$\therefore \quad \sigma_\theta = \frac{\sigma_y + \sigma_x}{2} + \left(\frac{\sigma_y - \sigma_x}{2}\right) \cos 2\theta - \tau \sin 2\theta$$

$$= \frac{\sigma_y + \sigma_x}{2} \pm \left(\frac{\sigma_y - \sigma_x}{2}\right) \frac{\tau}{\sqrt{\left(\frac{\sigma_y - \sigma_x}{2}\right)^2 + \tau^2}} \mp \tau \frac{(\sigma_y - \sigma_x)/2}{\sqrt{\left(\frac{\sigma_y - \sigma_x}{2}\right)^2 + \tau^2}}$$

$$= \frac{\sigma_y + \sigma_x}{2}$$

Note: For an element subjected to simple shear (no normal stress and only shear stress), principal stresses are numerically equal to the shear stresses i.e.

$$\sigma_1 = +\tau \quad \text{and} \quad \sigma_2 = -\tau$$

Angle between principal normal plane and principal shear plane

Inclination of principal normal planes, $\tan 2\theta_n = \dfrac{-2\tau}{\sigma_y - \sigma_x}$

Inclination of principal shear planes, $\tan 2\theta_s = \dfrac{\sigma_y - \sigma_x}{2\tau}$

It may be noted that the product of $\tan 2\theta_n$ and $\tan 2\theta_s$ is -1.

Therefore, $2\theta_n$ and $2\theta_s$ will differ by 90^0 and so θ_n and θ_s will differ by 45^0.

Therefore, *principal shear planes are at an angle of 45^0 with principal normal planes.*

Sum of normal stresses on mutually perpendicular planes

Let the two mutually perpendicular planes be inclined at θ and $(90 + \theta)$ with horizontal.

$$\sigma_\theta = \frac{\sigma_y + \sigma_x}{2} + \left(\frac{\sigma_y - \sigma_x}{2}\right) \cos 2\theta - \tau \sin 2\theta$$

$$\therefore \ \sigma_{90+\theta} = \frac{\sigma_y + \sigma_x}{2} + \left(\frac{\sigma_y - \sigma_x}{2}\right) \cos 2(90 + \theta) - \tau \sin 2(90 + \theta)$$

$$= \frac{\sigma_y + \sigma_x}{2} + \left(\frac{\sigma_y - \sigma_x}{2}\right) \cos (180 + 2\theta) - \tau \sin (180 + 2\theta)$$

$$= \frac{\sigma_y + \sigma_x}{2} - \left(\frac{\sigma_y - \sigma_x}{2}\right) \cos 2\theta \ - \tau \sin 2\theta$$

Now $\quad \sigma_\theta + \sigma_{90+\theta} = \dfrac{\sigma_y + \sigma_x}{2} + \dfrac{\sigma_y + \sigma_x}{2} = \sigma_y + \sigma_x$

The sum of normal stresses on two mutually perpendicular planes at a point is invariant (remains constant).

$$\sigma_\theta + \sigma_{90+\theta} = \sigma_y + \sigma_x = \sigma_1 + \sigma_2$$

Example 2.1 *For the state of stress shown (Fig. 2.13), find (i) principal stresses (ii) principal shear stresses with the associated normal stress. In each case show the results on a properly oriented element.*

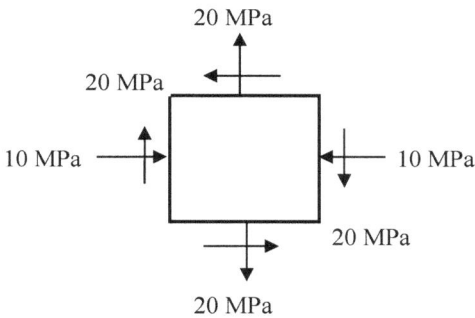

Fig. 2.13

Solution:

Given,

$\sigma_x = -10$ MPa (compressive), $\sigma_y = 20$ MPa (tensile), $\tau = -20$ MPa

Principal stresses:

$$\sigma_{1,2} = \frac{\sigma_y + \sigma_x}{2} \pm \sqrt{\left(\frac{\sigma_y - \sigma_x}{2}\right)^2 + \tau^2}$$

$$= \frac{20 - 10}{2} \pm \sqrt{\left(\frac{20 + 10}{2}\right)^2 + (-20)^2} = 5 \pm 25$$

$\sigma_1 = 5 + 25 = 30\text{MPa} \rightarrow$ Major principal stress

$\sigma_2 = 5 - 25 = -20\text{MPa} \rightarrow$ Minor principal stress

Inclination of Principal planes, $\quad \tan 2\theta_n = \dfrac{-2\tau}{\sigma_y - \sigma_x} = \dfrac{-2(-20)}{20 + 10}$

$2\theta_n = 53.13^0, \qquad 233.13^0$

$\therefore \theta_n = 26.56^0, \qquad 116.56^0$

$$\sigma_{26.56} = \frac{\sigma_y + \sigma_x}{2} + \frac{\sigma_y - \sigma_x}{2} \cos 2\theta - \tau \sin 2\theta$$

$$= \frac{20 - 10}{2} + \frac{20 + 10}{2} \cos(53.13) - (-20) \sin(53.13) = 30 \text{ MPa}$$

$\therefore \sigma_{116.56} = -20\text{MPa}$

The results are shown in Fig. 2.14.

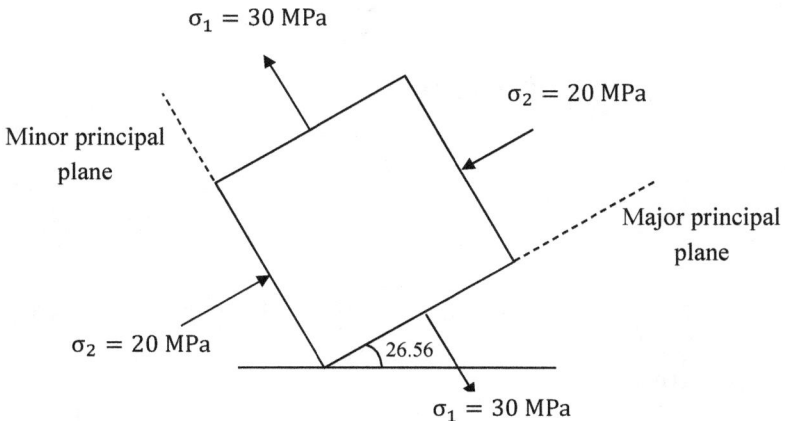

Fig. 2.14

Principal shear stresses:

$$\tau_{max} = \frac{\sigma_1 - \sigma_2}{2} = \frac{30 - (-20)}{2} = 25 \text{ MPa}$$
$$\rightarrow \text{ Major principal shear stress}$$

$\tau_{min} = -25$ MPa $\quad \rightarrow$ Minor principal shear stress

Inclination of principal shear planes:

$$\tan 2\theta_s = \frac{\sigma_y - \sigma_x}{2\tau} = \frac{20 + 10}{2 \times (-20)}$$

$\therefore 2\theta_s = -36.86 = 36.86 + 180 \quad$ and $\quad -36.6 + 180 + 180$

$2\theta_s = 143.1, 323.1$

(When 2θ value is negative, $180°$ may be added twice to obtain two positive values)

$\therefore \theta_s = 71.5°, 161.5°$

$$\tau_{71.5} = \frac{\sigma_y - \sigma_x}{2} \sin(2\theta) + \tau \cos(2\theta)$$

$$= \frac{20 - (-10)}{2} \sin(143.1) + (-20) \cos(143.1) = 25 \text{ MPa}$$

$\therefore \tau_{161.57} = -25$ MPa

Normal stresses on principal shear planes:

Considering $\theta = 71.5°$ in σ_θ expression, we get

$$\sigma_{71.57} = \frac{\sigma_y + \sigma_x}{2} + \frac{\sigma_y - \sigma_x}{2} \cos 2\theta - \tau \sin 2\theta$$

$$= \frac{20 - 10}{2} + \left(\frac{20 - (-10)}{2}\right) \cos(143.14) - (-20) \sin(143.14)$$

$= 5$ MPa

Also, $\sigma_{161.57} = \dfrac{20 - 10}{2} + \left[\dfrac{20 - (-10)}{2}\right] \cos(323.14)$
$$- (-20) \sin(323.14) = 5 \text{ MPa}$$

Normal stresses on principal shear planes can also be calculated as

$$\sigma = \frac{\sigma_1 + \sigma_2}{2} = \frac{30 - 20}{2} = 5 \text{ MPa}$$

The results are shown in Fig. 2.15.

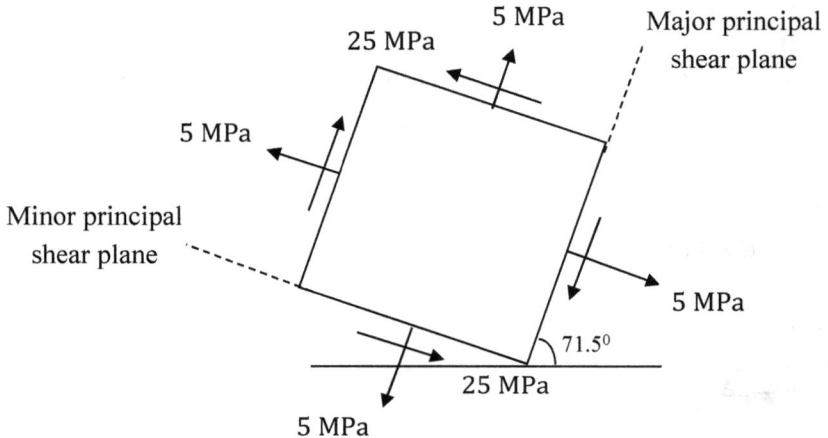

Fig. 2.15

Example 2.2 *For the state of stress specified as shown (Fig. 2.16), determine (i) the normal and shear stresses on a plane whose normal makes 35⁰ with horizontal (ii) principal plane and principal stresses (iii) maximum shear stresses and the corresponding planes.*

Solution:

$\sigma_x = 15$ MPa,
$\sigma_y = -30$ MPa
$\tau = 20$ MPa

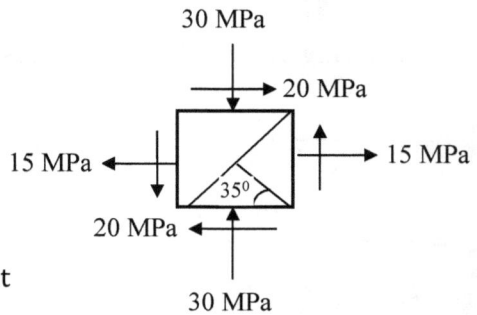

Fig. 2.16

(i) We have to find out σ_θ and τ_θ where $\theta = 90 - 35 = 55°$

$$\sigma_\theta = \frac{\sigma_y + \sigma_x}{2} + \frac{\sigma_y - \sigma_x}{2} \cos 2\theta - \tau \sin 2\theta$$

$$= \frac{-30 + 15}{2} + \frac{-30 - 15}{2} \cos (2 \times 55) - 20 \sin (2 \times 55)$$

$= -18.59$ MPa (−ve sign indicates compressive stress)

$$\tau_\theta = \left(\frac{\sigma_y - \sigma_x}{2}\right) \sin 2\theta + \tau \cos 2\theta$$

$$= \frac{-30 - 15}{2} \sin(2 \times 55) + 20 \cos\,(2 \times 55)$$

$$= -27.98 \text{ MPa}$$

The results are shown in Fig. 2.17.

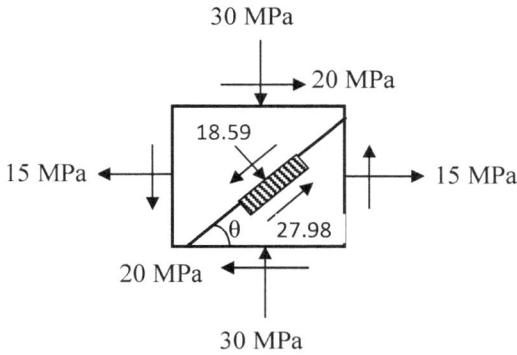

Fig. 2.17

(ii)

$$\sigma_{1,2} = \frac{\sigma_y + \sigma_x}{2} \pm \sqrt{\left(\frac{\sigma_y - \sigma_x}{2}\right)^2 + \tau^2}$$

$$= \frac{-30 + 15}{2} \pm \sqrt{\left(\frac{-30 - 15}{2}\right)^2 + (20)^2}$$

$$= -7.5 \pm 30.1$$

$\therefore \sigma_1 = -7.5 + 30.1 = 22.6$ MPa   (tensile)
$\qquad\qquad \rightarrow$ Major principal stress

$\therefore \sigma_2 = -7.5 - 30.101 = -37.6$ MPa (compressive)
$\qquad\qquad \rightarrow$ Minor principal stress

Inclination of principal planes:

$$\tan 2\theta_n = \frac{-2\tau}{\sigma_y - \sigma_x} = \frac{-2(20)}{-30 - 15} \quad \rightarrow \quad 2\theta_n = 41.6 \text{ or } 221.6$$

$$\therefore \theta_n = 20.8^0, \qquad 110.8^0$$

$$\sigma_{20.8} = \frac{\sigma_y + \sigma_x}{2} + \left(\frac{\sigma_y - \sigma_x}{2}\right) \cos 2\theta - \tau \sin 2\theta$$

$$= \frac{-30 + 15}{2} + \frac{-30 - 15}{2} \cos (41.6) - 20 \sin (41.6)$$

$$= -37.6 \text{ MPa}$$

$$\therefore \sigma_{110.8} = 22.6 \text{ MPa}$$

These results are shown in Fig. 2.18.

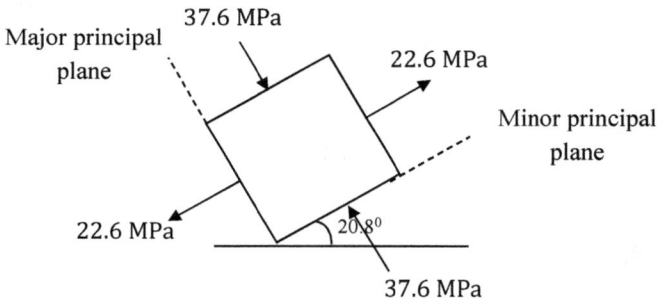

Fig. 2.18

(iii) Maximum shear stress (principal shear stresses)

$$\tau_{max} = \frac{\sigma_1 - \sigma_2}{2} = \frac{22.6 - (-37.6)}{2} = 30.1 \text{ MPa}$$

$$\therefore \tau_{min} = -30.1 \text{ MPa}$$

Inclination of principal shear planes:

$$\tan 2\theta_s = \frac{\sigma_y - \sigma_x}{2\tau} = \frac{-30 - 15}{2 \times 20}$$

$2\theta_s = (-48.36 + 180), (-48.36 + 180 + 180)$

$\therefore \theta_s = 65.8^0, 155.8^0$

$\tau_{65.8} = \left(\frac{\sigma_y - \sigma_x}{2}\right) \sin 2\theta + \tau \cos 2\theta$

$= \frac{-30 + 15}{2} \sin(131.6) + 20 \cos(131.6) = -30.1 \text{ MPa} \quad \rightarrow \quad \tau_{min}$

$\therefore \tau_{155.8} = 30.1 \text{ MPa} \rightarrow \tau_{max}$

Example 2.3 *The stress condition of a plane element is shown in (Fig. 2.19). Determine (a) normal and shearing stress on a plane whose normal is inclined at 30° to the x direction as shown (b) the maximum and minimum value of normal stresses and the inclination of the as- sociated planes (c) magnitude and direction of maximum shear stresses.*

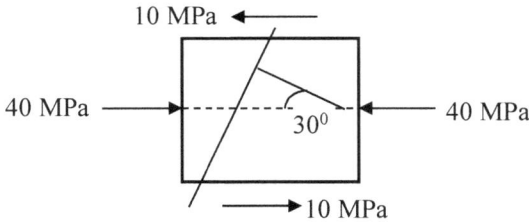

Fig. 2.19

Solution:

$\sigma_x = -40 \text{ MPa}, \qquad \sigma_y = 0, \qquad \tau = -10 \text{ MPa}, \qquad \theta = 60°$

(a) Normal stress on the plane:

$\sigma_\theta = \frac{\sigma_y + \sigma_x}{2} - \frac{\sigma_y - \sigma_x}{2} \cos(2\theta) - \tau \sin(2\theta)$

$= \frac{0 - 40}{2} + \frac{0 + 40}{2} \cos(120) - (-10) \sin(120)$

$= -21.33 \text{ MPa} \quad (\text{compresssive})$

Shear stress on the plane:

$$\tau_\theta = \left(\frac{\sigma_y - \sigma_x}{2}\right) \sin 2\theta + \tau \cos 2\theta$$

$$= \frac{0 + 40}{2} \sin(120)$$
$$+ (-10) \cos(120)$$

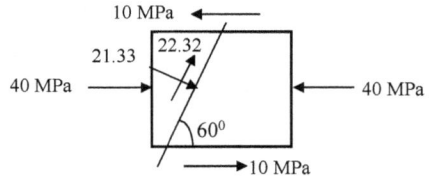

Fig. 2.20

$= 22.32$ MPa

The results are shown in Fig. 2.20.

(a) Principal stresses:

$$\sigma_{1,2} = \frac{\sigma_y + \sigma_x}{2} \pm \sqrt{\left(\frac{\sigma_y - \sigma_x}{2}\right)^2 + \tau^2}$$

$$= \frac{0 - 40}{2} \pm \sqrt{\left(\frac{0 + 40}{2}\right)^2 + (-10)^2}$$

$= -20 \pm 22.36$

$\sigma_1 = -20 + 22.36 = 2.36$ MPa $\rightarrow$ Maximum normal stress

$\sigma_2 = -20 - 22.36 = -42.36$ MPa $\rightarrow$ Minimum normal stress

Inclination of principal planes, $\tan 2\theta_n = \dfrac{-2\tau}{\sigma_y - \sigma_x} = \dfrac{-2(-10)}{0 - (-40)}$

$\therefore 2\theta_n = 26.56^0, 206.56^0$ $\therefore \theta_n = 13.28^0, 103.28^0$

Putting $\theta = 13.28^0$ in σ_θ expression

$$\sigma_{13.28} = \frac{\sigma_y + \sigma_x}{2} + \left(\frac{\sigma_y - \sigma_x}{2}\right) \cos 2\theta - \tau \sin 2\theta$$

$$= \frac{0 - 40}{2} + \frac{0 - (-40)}{2} \cos(26.56) - (-10) \sin(26.56)$$

$= 2.36$ MPa

$\therefore \sigma_{103.28} = -42.36$ MPa

(b) Maximum shear stress

$$\tau_{max} = \frac{\sigma_1 - \sigma_2}{2} = \frac{2.36 - (-43.36)}{2} = 22.36 \text{ MPa}$$

Minimum shear stress $\tau_{min} = -22.36$ MPa

Inclination of principal shear planes:

$$\tan 2\theta_s = \frac{\sigma_y - \sigma_x}{2\tau} = \frac{0 - (-40)}{2 \times (-10)}$$

$\therefore 2\theta_s = 116.57, 296.57$ and $\theta_s = 58.28^0, 148.28^0$

Now, $\theta = 58.28°$ in τ_θ expression yields

$$\tau_{58.28} = \left(\frac{\sigma_y - \sigma_x}{2}\right) \sin 2\theta + \tau \cos\, 2\theta$$

$$= \left(\frac{0 + 40}{2}\right) \sin(2 \times 58.28) + (-10) \cos(2 \times 58.28)$$

$$= 22.36 \text{ MPa}$$

$\therefore \tau_{148.28} = -22.36$ MPa

Normal stresses on principal shear planes:

$$\sigma = \frac{\sigma_1 + \sigma_2}{2} = \frac{2.36 - 42.36}{2} = -20 \text{ MPa}$$

Example 2.4 *The principal stresses at a point in a bar are 200 N/mm² (tensile) and 100 N/mm² (compressive). Determine the magnitude and direction of resultant stress on a plane inclined at 60⁰ to the axis of major principal stress as shown (Fig. 2.21). Also determine the maximum intensity of shear stress in the material at this point.*

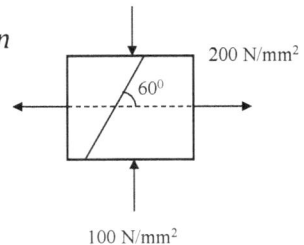

200 N/mm²

60⁰

100 N/mm²

Fig. 2.21

Solution:

$\sigma_x = 200$ MPa, $\sigma_y = -100$ MPa, $\theta = 60°$

When shear stress is absent, normal stresses will be the principal stresses. The greater value out of σ_x and σ_y is the major principal stress (σ_1) and the other one is minor principal stress (σ_2).

Normal stress, $\sigma_\theta = \dfrac{\sigma_y + \sigma_x}{2} + \left(\dfrac{\sigma_y - \sigma_x}{2}\right) \cos 2\theta - \tau \sin 2\theta$

$= \dfrac{-100 + 200}{2} + \left(\dfrac{-100 - 200}{2}\right) \cos(2 \times 60) = 125$ MPa

Shear stress, $\tau_\theta = \left(\dfrac{\sigma_y - \sigma_x}{2}\right) \sin 2\theta + \tau \cos 2\theta$

$= \left(\dfrac{-100 - 200}{2}\right) \sin(2 \times 60) = -129.9$ MPa

∴ Resultant stress, $\sigma_r = \sqrt{\sigma_\theta^2 + \tau_\theta^2} = \sqrt{(125)^2 + (-129.9)^2}$

$= 180.27$ MPa

Let φ = Inclination of resultant stress with normal to inclind plane

∴ $\tan \varphi = \dfrac{129.9}{125}$ ∴ $\varphi = 46.1°$

Maximum shear stress, $\tau_{max} = \dfrac{\sigma_1 - \sigma_2}{2} = \dfrac{200 - (-100)}{2}$

$= 150$ MPa

Example 2.5 *A rectangular block of material is subjected to a tensile stress of 110 N/mm² on one plane and a tensile stress of 47 N/mm² on the plane at right angle to the former. Each of the above stress is accompanied by a shear stress of 63 N/mm² and that associated with the former tensile stress tends to rotate the block anticlockwise. Find (i) the direction and magnitude of principal stress (ii) magnitude of the greatest shear stress.*

Solution:

$\sigma_x = 110$ MPa, $\sigma_y = 47$ MPa,
$\tau = 63$ MPa (refer Fig. 2.22)

(i) Principal stresses:

$$\sigma_{1,2} = \frac{\sigma_x + \sigma_y}{2} \pm \sqrt{\left(\frac{\sigma_y - \sigma_x}{2}\right)^2 + \tau^2}$$

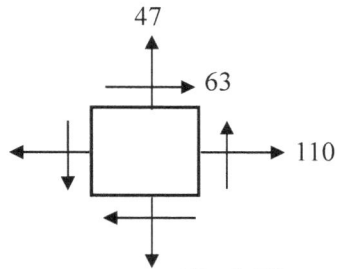

Fig. 2.22

$$= \frac{47 + 110}{2} \pm \sqrt{\left(\frac{47 - 110}{2}\right)^2 + (63)^2}$$

$$= 78.5 \pm 70.43$$

$\therefore \sigma_1 = 78.5 + 70.43 = 148.93$ MPa $\quad \rightarrow \quad$ Major principal stress

$\therefore \sigma_2 = 78.5 - 70.43 = 8.07$ MPa $\quad \rightarrow \quad$ Minor principal stress

Inclination of principal planes, $\tan 2\theta_n = \dfrac{-2\tau}{\sigma_y - \sigma_x} = \dfrac{-2 \times 63}{47 - 110}$

$\therefore 2\theta_n = 63.43, 243.43$

$\theta_n = 31.71°, 121.71°$

$$\sigma_\theta = \frac{\sigma_y + \sigma_x}{2} + \left(\frac{\sigma_y - \sigma_x}{2}\right) \cos 2\theta - \tau \sin 2\theta$$

$$\sigma_{31.71} = \frac{47 + 110}{2} + \frac{47 - 110}{2} \cos(2 \times 31.71) - 63 \sin(2 \times 31.71)$$

$$= 8.07 \text{ MPa}$$

$$\sigma_{121.71} = 148.93 \text{ MPa}$$

(i) Greatest shear stress

$$\tau_{max} = \frac{\sigma_1 - \sigma_2}{2} = \frac{148.93 - 8.07}{2} = 70.42 \text{ MPa}$$

Example 2.6 *A bar of circular cross section 50 mm diameter is subjected to a tensile force of 10 kN. Find the maximum shear stress and its inclination.*

Solution:

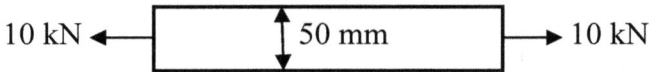

10 kN ← | 50 mm | → 10 kN

Fig. 2.23

Cross Sectional area of bar

$$A = \frac{\pi}{4} \times 50^2 = 1963.49 \text{ mm}^2$$

Normal stress, $\sigma_x = \frac{P}{A} = \frac{10000}{1963.49} = 5.09 \text{ MPa}$

$\sigma_x = 5.09 \text{ MPa}, \sigma_y = 0, \tau = 0$

Principal stresses,

$$\sigma_1, \sigma_2 = \frac{\sigma_y + \sigma_x}{2} \pm \sqrt{\left(\frac{\sigma_y - \sigma_x}{2}\right)^2 + \tau^2} = \frac{\sigma_x}{2} \pm \frac{\sigma_x}{2}$$

$$\sigma_1 = \frac{\sigma_x}{2} + \frac{\sigma_x}{2} = \sigma_x = 5.09 \text{ MPa}$$

$$\sigma_2 = \frac{\sigma_x}{2} - \frac{\sigma_x}{2} = 0$$

Principal shear stresses:

$$\tau_{max} = \frac{\sigma_1 - \sigma_2}{2} = \frac{\sigma_1}{2} = \frac{5.09}{2} = 2.545 \text{ MPa}$$

$$\tau_{min} = -2.545 \text{ MPa}$$

Inclination of principle shear planes, $\tan 2\theta_s = \frac{\sigma_y - \sigma_x}{2\tau} = \infty$ and $2\theta_s = 90, 270$

$$\therefore \theta_s = 45°, 135°$$

Note: When an element is subjected to uniaxial stress condition, principal shear planes are inclined at 45° and 135°.

Example 2.7 *A bolt of 20 mm diameter is subjected to axial tensile force of 20 kN together with shear force of 10 kN. Find the principal stresses and their inclination. Also find the principal shear stresses.*

Solution:

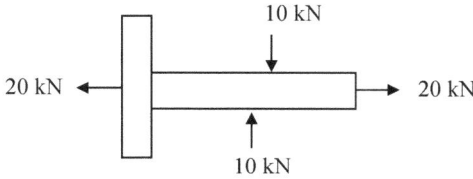

Fig. 2.24

$$A = \frac{\pi}{4} \times 20^2 = 314.15 \text{ mm}^2$$

Normal stress, $\quad \sigma_x = \dfrac{20000}{314.15} = 63.66 \text{ MPa}$

Shear stress, $\quad \tau = \dfrac{\text{Shear force}}{A} = \dfrac{10000}{314.15} = 31.83 \text{ MPa}$

Refer Fig. 2.24

$\sigma_x = 63.66 \text{ MPa}, \quad \sigma_y = 0, \quad \tau = -31.83 \text{ MPa}$

Principal stresses:

$$\sigma_1, \sigma_2 = \frac{\sigma_y + \sigma_x}{2} \pm \sqrt{\left(\frac{\sigma_y - \sigma_x}{2}\right)^2 + \tau^2}$$

$$= \frac{0 + 63.66}{2} \pm \sqrt{\left(\frac{0 - 63.66}{2}\right)^2 + (-31.83)^2}$$

$\sigma_1 = 76.8 \text{ MPa}$

$\sigma_2 = -31.18 \text{ MPa}$

Inclination of principal planes:

$$\tan 2\theta_n = \frac{-2\tau}{\sigma_y - \sigma_x} = \frac{-2 \times (-31.83)}{0 - 63.66}$$

$\therefore 2\theta_n = -45, \quad \therefore 2\theta_n = -45 + 180, -45 + 180 + 180$

$\therefore \theta_n = 67.5^0, 157.5^0$

$$\sigma_{67.5} = \frac{\sigma_y + \sigma_x}{2} + \left(\frac{\sigma_y - \sigma_x}{2}\right) \cos 2\theta - \tau \sin 2\theta$$

$$= \frac{0 + 63.66}{2} + \frac{0 - 63.66}{2} \cos(135) - (-31.83) \sin(135)$$

$$= 76.8 \text{ MPa}$$

$\therefore \sigma_{157.5} = -13.18 \text{ MPa}$

Maximum shear stress,

$$\tau_{max} = \frac{\sigma_1 - \sigma_2}{2} = \frac{76.8 - (-13.18)}{2} = 45.03 \text{ MPa}$$

Minimum shear stress, $\tau_{min} = -45.03 \text{ MPa}$

Example 2.8 *For the rectangular element shown (Fig. 2.25), the following data are given:*
$\sigma_x = 360 \text{ N/mm}^2$, $\sigma_y = 200 \text{ N/mm}^2$, $\tau = 60 \text{ N/mm}^2$.
Determine the magnitude of principal stresses and the inclination of principal planes.

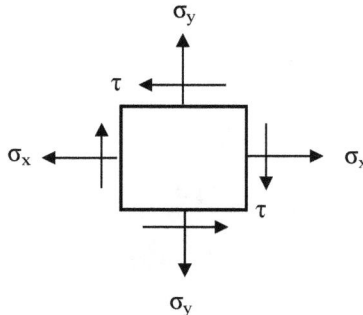

Fig. 2.25

Solution:

$$\sigma_x = 360 \text{ MPa}, \quad \sigma_y = 200 \text{ MPa}, \quad \tau = -60 \text{ MPa}$$

Principal stresses:

$$\sigma_1, \sigma_2 = \frac{\sigma_y + \sigma_x}{2} \pm \sqrt{\left(\frac{\sigma_y - \sigma_x}{2}\right)^2 + \tau^2}$$

$$= \frac{200 + 360}{2} \pm \sqrt{\left(\frac{200 - 360}{2}\right)^2 + (-60)^2} = 280 \pm 100$$

$$\therefore \sigma_1 = 280 + 100 = 380 \text{ MPa} \rightarrow \text{Major principal stress}$$

$$\sigma_2 = 280 - 100 = 180 \text{ MPa} \rightarrow \text{Minor principal stress}$$

Inclination of principal planes:

$$\tan 2\theta_n = \frac{-2\tau}{\sigma_y - \sigma_x} = \frac{-2 \times (-60)}{200 - 360}$$

$$\therefore 2\theta_n = -36.86$$

$$\therefore 2\theta_n = 143.14, 323.14$$

$$\theta_n = 71.56^0, 161.56^0$$

$$\sigma_{71.56} = \frac{\sigma_y + \sigma_x}{2} + \left(\frac{\sigma_y - \sigma_x}{2}\right) \cos 2\theta - \tau \sin 2\theta$$

$$= \frac{200 + 360}{2} + \frac{200 - 360}{2} \cos(2 \times 71.56) - (-60) \sin(2 \times 71.56)$$

$$= 380 \text{ MPa}$$

$$\therefore \sigma_{161.56} = 180 \text{ MPa}$$

Example 2.9 *For the rectangular element shown (Fig. 2.26), the following numerical data are given:*

$\sigma_x = 840 \ N/mm^2$, $\sigma_y = 280 \ N/mm^2$, $\tau = 210 N/mm^2$.
Calculate the normal stress and shear stress on the plane defined by
the angle $\varphi = 30°$.

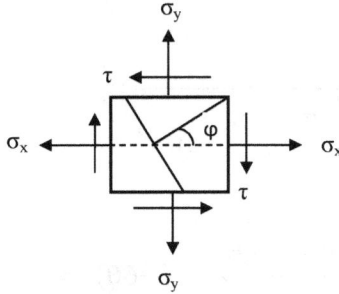

Fig. 2.26

Solution:

$\sigma_x = 840$ MPa, $\sigma_y = 280$ MPa, $\tau = -210$ MPa, $\theta = 120^0$

Normal stress on the plane:

$$\sigma_\theta = \frac{\sigma_y + \sigma_x}{2} + \frac{\sigma_y - \sigma_x}{2} \cos 2\theta - \tau \sin 2\theta$$

$$= \frac{280 + 840}{2} + \frac{280 - 840}{2} \cos(240) - (-210) \sin(240)$$

$= 518.13$ MPa

Shear stress on the plane:

$$\tau_\theta = \frac{\sigma_y - \sigma_x}{2} \sin 2\theta + \tau \cos 2\theta$$

$$= \frac{280 - 840}{2} \sin(240) + (-210)\cos(240)$$

$= 347.48$ MPa

Example 2.10 *For the square element shown (Fig. 2.27) both the
normal stress and shear stress on the diagonal plane AB are zero. If
$\sigma_x = \sigma_y = -350 \ MPa$, what is the value of shear stress τ ? What are
the magnitudes of the principal stresses?*

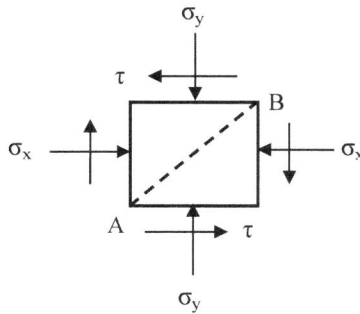

Fig. 2.27

Solution:

$\sigma_x = \sigma_y = -350$ MPa, $\qquad \theta = 45°$

Given that $\sigma_\theta = 0$ and $\tau_\theta = 0$

$$\sigma_\theta = \frac{\sigma_y + \sigma_x}{2} + \left(\frac{\sigma_y - \sigma_x}{2}\right) \cos 2\theta - \tau \sin 2\theta = 0$$

$$\frac{-350 - 350}{2} + \frac{-350 + 350}{2} \cos(90) - \tau \sin(90) = 0$$

$\therefore \tau = -350$ MPa

Principal stresses:

$$\sigma_1, \sigma_2 = \frac{\sigma_y + \sigma_x}{2} \pm \sqrt{\left(\frac{\sigma_y - \sigma_x}{2}\right)^2 + \tau^2}$$

$$= \frac{-350 - 350}{2} \pm \sqrt{\left(\frac{-350 + 350}{2}\right)^2 + (-350)^2}$$

$\therefore \sigma_1 = -350 + 350 = 0 \qquad \rightarrow \qquad$ Major principal stress

$\sigma_2 = -350 - 350 = -700$ MPa $\qquad \rightarrow \qquad$ Minor principal stress

Example 2.11 *A column rests on a foundation block, the top of the latter being horizontal. The column transmits to the block a compressive stress of 174 MPa together with a shear stress of 46.6 MPa . Find*

the magnitude and direction of the principal stresses at a point just below the top of the block.

Solution:

Refer Fig. 2.28

$\sigma_x = 0$
$\sigma_y = -174$ MPa, $\tau = 46.6$ MPa

Principal stress:

$$\sigma_1, \sigma_2 = \frac{\sigma_y + \sigma_x}{2} \pm \sqrt{\left(\frac{\sigma_y - \sigma_x}{2}\right)^2 + \tau^2}$$

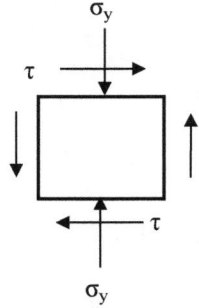

Fig. 2.28

$$= \frac{-174 + 0}{2} \pm \sqrt{\left(\frac{-174 - 0}{2}\right)^2 + 46.6^2} = -87 \pm 98.69$$

$\therefore \sigma_1 = -87 + 98.69 = 11.69$ MPa (tensile)
$\rightarrow$ Major principal stress

$\sigma_2 = -87 - 98.69 = -185.6$ MPa (compressive)
$\rightarrow$ Minor principal stress

$$\tan 2\theta_n = \frac{-2\tau}{\sigma_y - \sigma_x} = \frac{-2 \times 46.6}{-174 - 0}$$

$2\theta_n = 28.17, 208.17$

$\theta_n = 14.08°, 104.08°$

Putting $\theta = 14.08°$ in σ_θ expresssion

$$\sigma_{14.08} = \frac{\sigma_y + \sigma_x}{2} + \left(\frac{\sigma_y - \sigma_x}{2}\right) \cos 2\theta - \tau \sin 2\theta$$

$$= \frac{-174 + 0}{2} + \left(\frac{-174 - 0}{2}\right) \cos (28.17) - 46.6 \sin (28.17)$$

$$= -185.6 \text{ MPa}$$

∴ $\sigma_{104.08} = 11.69$ MPa

Example 2.12 *For the element shown (Fig. 2.29)* $\sigma_x = 420$ *MPa,* $\sigma_y = 0$. *What is the maximum permissible magnitude of the shear stresses* τ *if the larger principal stress* σ_1 *is not to exceed 360 MPa.*

Solution:

$\sigma_x = 420$ MPa, $\sigma_y = 0$

$\sigma_1 = 630$ MPa

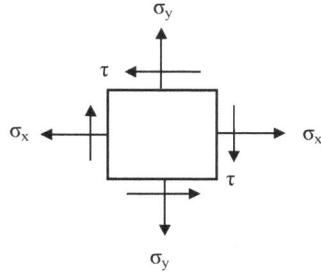

Fig. 2.29

$$\sigma_1 = \frac{\sigma_y + \sigma_x}{2} + \sqrt{\left(\frac{\sigma_y - \sigma_x}{2}\right)^2 + \tau^2}$$

$$630 = \frac{0 + 420}{2} \pm \sqrt{\left(\frac{0 - 420}{2}\right)^2 + \tau^2}$$

∴ $\tau = 363.73$ MPa

Example 2.13 *At a cross section of a beam there is a longitudinal bending stress of 120 MPa (tension) and a transverse shear stress of 50 MPa. Find the resultant stress in magnitude and direction on a plane inclined at 30° to the longitudinal axis.*

Solution:

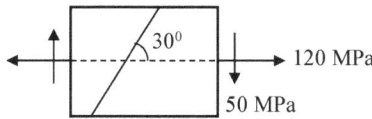

Fig. 2.30

Refer Fig. 2.30

$\sigma_x = 120$ MPa, $\sigma_y = 0$

$\tau = -50$ MPa, $\theta = 30°$

$$\sigma_\theta = \frac{\sigma_y + \sigma_x}{2} + \left(\frac{\sigma_y - \sigma_x}{2}\right) \cos 2\theta - \tau \sin 2\theta$$

$$= \frac{0 + 120}{2} + \frac{0 - 120}{2} \cos(60) - (-50)\sin(60) = 73.3 \text{ MPa}$$

$$\tau_\theta = \left(\frac{\sigma_y - \sigma_x}{2}\right) \sin 2\theta + \tau \cos 2\theta$$

$$= \frac{0 - 120}{2} \sin(60) + (-50)\cos(60) = -76.96 \text{ MPa}$$

Resultant stress:

$$= \sigma_r = \sqrt{\sigma_\theta^2 + \tau_\theta^2} = \sqrt{(73.3)^2 + (-76.96)^2} = 106.28 \text{ MPa}$$

$$\tan\varphi = \frac{\sigma_\theta}{\tau_\theta} = \frac{73.3}{76.96}$$

$$\varphi = 43.6^0$$

With longitudinal axis, the direction for the resultant stress
$= 43.6^0 - 30^0 = 13.6^0$

Example 2.14 *Two wooden pieces 100 mm × 100 mm in cross section, are glued together along line a-b as shown (Fig. 2.31). What
maximum axial force P can be applied if the allowable shear stress
along ab line is 1.2 N/mm².*

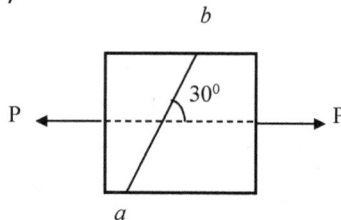

Fig. 2.31

Solution:

Cross sectional area, $A = 100 \times 100 = 10^4 \text{ mm}^2$

Allowable shear stress along *a-b*, $\tau_\theta = 1.2 \text{ N/mm}^2$

Normal stress, $\sigma_x = \dfrac{P}{A} = \dfrac{P}{10^4} \text{ N/mm}^2$

However, $\quad \tau_\theta = \dfrac{\sigma_x}{2} \sin 2\theta$

$$1.2 = \dfrac{P}{2 \times 10^4} \sin(2 \times 30)$$

$\therefore P = 27712 \text{ N} = 27.712 \text{ kN}$

Example 2.15 *At a point in a material under stress, the intensity of resultant stress on a vertical plane is 1000 Kgf/cm² inclined at 30° to the normal to that plane and the stress on horizontal plane has a normal tensile component of intensity 600 Kgf/cm² as shown (Fig. 2.32). Find the magnitude and direction of resultant stress on horizontal plane and the principal stresses.*

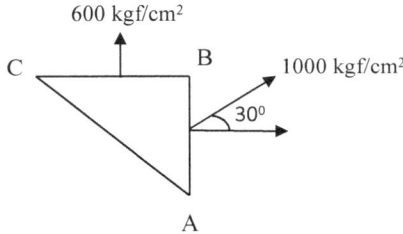

Fig. 2.32

Solution:

Refer Fig. 2.33

$\sigma_x = 1000 \cos 30$
$\qquad\quad = 866.02 \text{ kgf} / \text{cm}^2$

$\sigma_y = 600 \;\; \text{kgf} / \text{cm}^2$

$\tau = 1000 \; \sin \; 30 = 500 \text{ kgf} / \text{cm}^2$

Resultant stress on horizontal plane

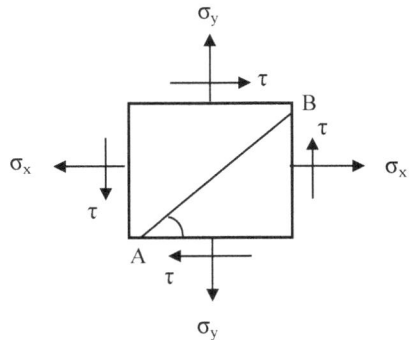

Fig. 2.33

$\sigma_r = \sqrt{600^2 + 500^2} = 781.02 \text{ kgf} / \text{cm}^2$

$\tan \varphi = \dfrac{600}{500}$

$\therefore \varphi = 50.19°$

Principal stresses:

$$\sigma_1 , \sigma_2 = \frac{\sigma_y + \sigma_x}{2} \pm \sqrt{\left(\frac{\sigma_y - \sigma_x}{2}\right)^2 + \tau^2}$$

$$= \frac{600 + 866.02}{2} \pm \sqrt{\left(\frac{600 - 866.02}{2}\right)^2 + 500^2} = 733.01 \pm 517.38$$

$\therefore \sigma_1 = 733.01 + 517.38 = 1250.39 \text{ kgf / cm}^2$
$\qquad\qquad \rightarrow$ Major principal stress

$\sigma_2 = 733.01 - 517.38 = 215.63 \text{ kgf / cm}^2$
$\qquad\qquad \rightarrow$ Minor principal stress

Example 2.16 *A 50 mm diameter circle is drawn on a mild steel plate. The plate is stressed as shown (Fig. 2.34). Find the length of major and minor axes of an ellipse formed as a result of deformation. Take E = 2×10⁵ N/mm², v = 0.25.*

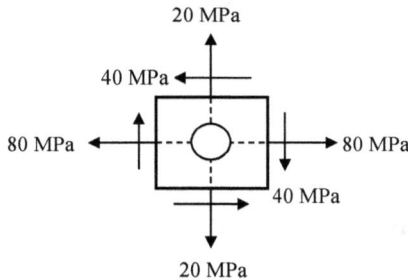

Fig. 2.34

Solution:

$\sigma_x = 80 \text{ MPa}, \sigma_y = 20 \text{ MPa}, \tau = -40 \text{ MPa}$

Principal stresses:

$$\sigma_1 , \sigma_2 = \frac{\sigma_y + \sigma_x}{2} \pm \sqrt{\left(\frac{\sigma_y - \sigma_x}{2}\right)^2 + \tau^2}$$

$$= \frac{20 + 80}{2} \pm \sqrt{\left(\frac{20 - 80}{2}\right)^2 + (-40)^2}$$

$$= 50 \pm 50$$

$$\therefore \sigma_1 = 100 \text{ MPa}, \qquad \sigma_2 = 0$$

Referring Fig. 2.35, diagonal BD will be elongated and diagonal AC will be shortened.

Therefore, circle will become an ellipse having major axis along BD and minor axis along AC.

σ_1 acts along BD and σ_2 acts along AC.

Strain along BD,

$$\varepsilon_{BD} = \frac{\sigma_1}{E} - v\frac{\sigma_2}{E} = \frac{100}{2 \times 10^5} - 0 = 5 \times 10^{-4}$$

$\therefore$ Increase in diameter (along BD)
$= \varepsilon_{BD} \times$ original diameter

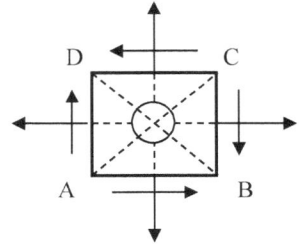

Fig. 2.35

$$= 5 \times 10^{-4} \times 50 = 25 \times 10^{-3} \text{mm}$$

Similarly strain along AC,

$$\varepsilon_{AC} = \frac{\sigma_2}{E} - v\frac{\sigma_1}{E} = 0 - 0.25 \times \frac{100}{2 \times 10^5} = -1.25 \times 10^{-4}$$

$\therefore$ Decrease in diameter (along AC) $= \varepsilon_{AC} \times$ original diameter

$$= 1.25 \times 10^{-4} \times 50 = 6.25 \times 10^{-3} \text{mm}$$

Example 2.17 *For the stressed element (Fig. 2.36)*
(i) Show that one of the principal stresses is zero if $\tau^2 = \sigma_x \sigma_y$
(ii) Find the condition for the principal stresses to have same sign.

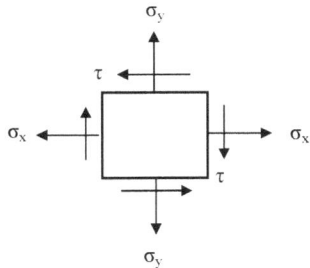

Fig. 2.36

Solution:

(i) Principal stresses:

$$\sigma_1, \sigma_2 = \frac{\sigma_y + \sigma_x}{2} \pm \sqrt{\left(\frac{\sigma_y - \sigma_x}{2}\right)^2 + \tau^2}$$

$$= \frac{\sigma_y + \sigma_x}{2} \pm \sqrt{\left(\frac{\sigma_y - \sigma_x}{2}\right)^2 + \sigma_x \sigma_y}$$

$$= \frac{\sigma_y + \sigma_x}{2} \pm \sqrt{\frac{\left(\sigma_y - \sigma_x\right)^2 + 4\,\sigma_x \sigma_y}{4}} = \frac{\sigma_y + \sigma_x}{2} \pm \sqrt{\left(\frac{\sigma_y + \sigma_x}{2}\right)^2}$$

Major principal stress, $\sigma_1 = \dfrac{\sigma_y + \sigma_x}{2} + \dfrac{\sigma_y + \sigma_x}{2} = \sigma_x + \sigma_y$

Minor principal stress $\sigma_2 = \dfrac{\sigma_y + \sigma_x}{2} - \dfrac{\sigma_y + \sigma_x}{2} = 0$

Therefore, one of the principal stresses is zero.   (Proved)

(ii) The condition for the principal stresses to have same sign is

$$\sqrt{\left(\frac{\sigma_y - \sigma_x}{2}\right)^2 + \tau^2} < \frac{\sigma_y + \sigma_x}{2}$$

$$\left\{\left(\frac{\sigma_y - \sigma_x}{2}\right)^2 + \tau^2\right\} < \left(\frac{\sigma_y + \sigma_x}{2}\right)^2$$

$$\rightarrow \left\{\left(\sigma_y - \sigma_x\right)^2 + 4\tau^2\right\} < \left(\sigma_y + \sigma_x\right)^2$$

$$(\sigma_x)^2 + \left(\sigma_y\right)^2 - 2\,\sigma_x \sigma_y + 4\tau^2 < (\sigma_x)^2 + \left(\sigma_y\right)^2 + 2\,\sigma_x \sigma_y$$

$$4\tau^2 < 4\,\sigma_x \sigma_y$$

$$\therefore \ \tau^2 < \sigma_x \sigma_y$$

This is the required condition.

2.8 MOHR'S STRESS CIRCLE

Normal stress and shear stress on any plane in an element subjected to biaxial stresses and shear stress have already been calculated analytically. These stresses can be found out graphically by Mohr's stress circle. Principal normal stresses and principal shear stresses in magnitude and direction can also be obtained.

Normal stress on a plane is expressed as

$$\sigma_\theta = \frac{\sigma_y + \sigma_x}{2} + \left(\frac{\sigma_y - \sigma_x}{2}\right)\cos 2\theta - \tau \sin 2\theta$$

$$\rightarrow \quad \left(\sigma_\theta - \frac{\sigma_y + \sigma_x}{2}\right) = \left(\frac{\sigma_y - \sigma_x}{2}\right)\cos 2\theta - \tau \sin 2\theta \tag{1}$$

Shear stress on a plane is given by

$$\tau_\theta = \left(\frac{\sigma_y - \sigma_x}{2}\right)\sin 2\theta + \tau \cos 2\theta \tag{2}$$

Squaring both sides of Eqs. (1) and (2) and then adding

$$\left(\sigma_\theta - \frac{\sigma_y + \sigma_x}{2}\right)^2 + \tau_\theta^2$$

$$= \left(\frac{\sigma_y - \sigma_x}{2}\cos 2\theta - \tau \sin 2\theta\right)^2 + \left(\frac{\sigma_y - \sigma_x}{2}\sin 2\theta + \tau \cos 2\theta\right)^2$$

$$= \left(\frac{\sigma_y - \sigma_x}{2}\right)^2 \cos^2 2\theta + \tau^2 \sin^2 2\theta - 2\left(\frac{\sigma_y - \sigma_x}{2}\right)\cos 2\theta \ \tau \sin 2\theta$$

$$+ \left(\frac{\sigma_y - \sigma_x}{2}\right)^2 \sin^2 2\theta + \tau^2 \cos^2 2\theta$$

$$+ 2\left(\frac{\sigma_y - \sigma_x}{2}\right)\sin 2\theta \ \tau \cos 2\theta$$

$$= \left(\frac{\sigma_y - \sigma_x}{2}\right)^2 (\cos^2 2\theta + \sin^2 2\theta) + \tau^2 (\cos^2 2\theta + \sin^2 2\theta)$$

$$\left(\sigma_\theta - \frac{\sigma_y + \sigma_x}{2}\right)^2 + \tau_\theta^2 = \left(\frac{\sigma_y - \sigma_x}{2}\right)^2 + \tau^2 \tag{3}$$

Equation (3) becomes the equation of circle in the form

$$(\sigma_\theta - a)^2 + \tau_\theta^2 = R^2$$

where $a = \dfrac{\sigma_y + \sigma_x}{2}$ representing the centre and

$R = \sqrt{\left(\dfrac{\sigma_y - \sigma_x}{2}\right)^2 + \tau^2}$ being the radius of circle.

This circle is known as **Mohr's stress circle**.

Consider an element subjected to biaxial stresses σ_x and σ_y and shear stress τ as shown (Fig. 2.37a). Mohr's stress circle (Fig. 2.37b) is plotted for this following the procedure as explained below:

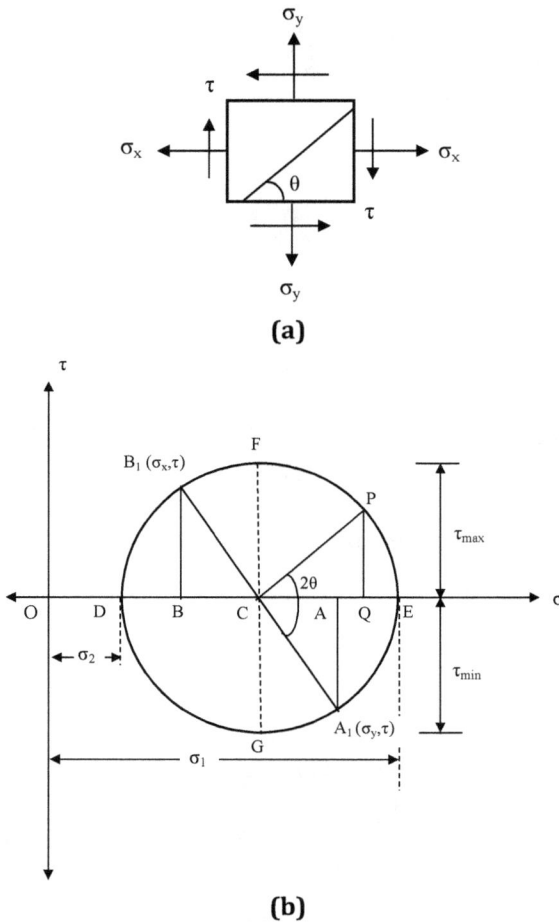

(a)

(b)

Fig. 2.37

In Mohr's stress circle, normal stress (σ) is plotted along x- axis and shear stress (τ) is plotted along y- axis (Refer Fig. 2.37b).

(i) Let O be the origin. Considering suitable scale, take OA equal to σ_y and OB equal to σ_x along x-axis. Tensile stress is plotted along positive axis and compressive stress is plotted along negative axis. (Fig. 2.37b is shown assuming $\sigma_y > \sigma_x$)

(ii) At A and B, draw perpendiculars AA_1 and BB_1 upon x-axis such that $AA_1 = BB_1 = \tau$ considering the same scale. The shear stress associated with σ_x if tends to rotate the element clockwise is taken positive and that tends to rotate the element anticlockwise is taken negative. Similar convention is also followed for the shear stress associated with σ_y.

(iii) Join A_1 and B_1 and let A_1B_1 intersects the x-axis (σ axis) at C. Now with C as centre and CA_1 (or CB_1) as radius draw a circle.

This is Mohr's stress circle (Fig. 2.37b). It may be noted that the radius of Mohr's stress circle represents particular plane on element and a point on Mohr's stress circle represents the state of stress on that particular plane in the element. For example, point A_1 represents the state of stress on plane CA_1 (horizontal plane) and point B_1 represents the state of stress on plane CB_1 (vertical plane) of the element. The Mohr's stress circle intersects the x-axis at points D and E and at D and E, the shear stress is zero (y coordinate is zero). Therefore, points D and E represent state of stress on principal planes.

CE $\Rightarrow$ Major principal plane (OE will represent major principal stress)
CD $\Rightarrow$ Minor principal plane (OD will represent minor principal stress)

At points F and G, the shear stress value is maximum and minimum respectively.

CF $\Rightarrow$ Major principal shear plane (also represents maximum shear stress)
CG $\Rightarrow$ Minor principal shear plane (also represents minimum shear stress)

It can be observed that major principal plane and minor principal planes are at $180°$ to each other in Mohr's stress circle but these are at $90°$ to each other in the element which indicates that angle θ in element is represented as 2θ in Mohr's stress circle.

Normal stress and shear stress on any plane

To find the normal stress (σ_θ) and shear stress (τ_θ) on any plane inclined at an angle θ (Fig. 2.38), draw radius CP at an angle 2θ with CA_1 (which represents horizontal plane in the present case). From P draw PQ perpendicular to OE. Now OQ represents normal stress (σ_θ) and PQ represents shear stress (τ_θ).

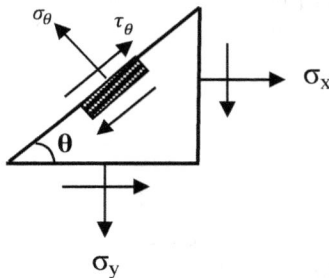

Fig. 2.38

When PQ lies above line OE, shear stress τ_θ produces a clockwise rotation on the plane of shear stress as shown (Fig. 2.38).

Example 2.18 *Solve the problem in Example 2.5 using Mohr's stress circle and verify the results found analytically.*

Solution:

Refer Fig. 2.39

Scale: 1 cm = 20 MPa

Take OA = 47 MPa = 2.35 cm
(To represent σ_y)
and OB = 110 MPa = 5.5 cm
(To represent σ_x)

Draw AA_1 and BB_1 perpendiculars

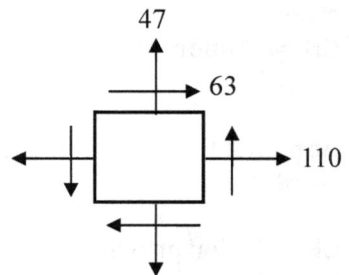

Fig. 2.39

at A and B respectively on AB each equal to 63 MPa i.e. 3.15 cm.

Join A_1 and B_1.

Let $A_1 B_1$ intersect AB at C.

Now taking C as centre and CA_1 (or CB_1) as radius draw a circle.

This is Mohr's stress circle (Fig. 2.40).

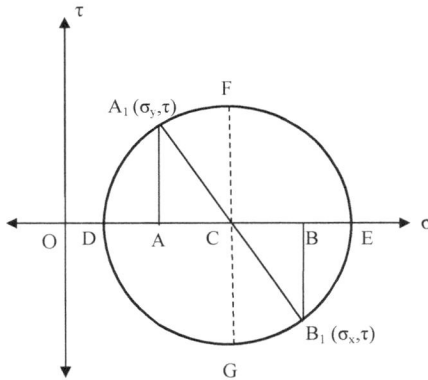

Fig. 2.40

By measurement, major principal stress, $\sigma_1 = OE \times$ scale $= 7.4 \times 20 = 148$ MPa

Minor principal stress, $\sigma_2 = OD \times$ scale $= 0.4 \times 20 = 8$ MPa

Inclination of major principal plane $= \frac{1}{2} \times \angle A_1 CE = \frac{1}{2} \times 242 = 121^0$ (measured anticlockwise)

Inclination of minor principal plane $= \frac{1}{2} \times \angle A_1 CD = \frac{1}{2} \times 62 = 31^0$ (measured anticlockwise)

Maximum shear stress $\tau_{max} = CF \times$ scale $= 3.5 \times 20 = 70$ MPa

Note: CA_1 represents horizontal plane and CB_1 represents vertical plane.

$\therefore$ All angles are measured from CA_1 in anticlockwise direction.

Example 2.19 *Draw Mohr's stress circle for the problem in Example 2.2 and verify the results obtained analytically.*

Solution:

Refer Fig. 2.41

Scale: 1 cm = 5 MPa

Take OA = - 30 MPa = - 6 cm
(to represent σ_y)
OB = 15 MPa = 3 cm (to represent σ_x)

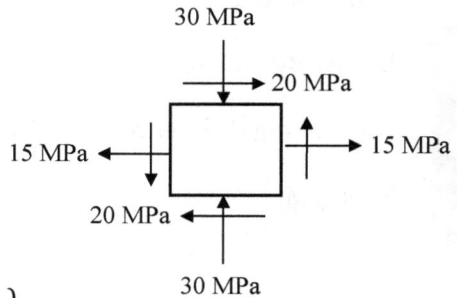

Fig. 2.41

Draw AA_1 and BB_1 perpendicular to AB each equal to 20 MPa i.e. 4 cm. Join A_1B_1.

Let A_1B_1 intersects AB at C.

Taking C as centre and CA_1 (or CB_1) as radius draw a circle.

This is Mohr's stress circle (Fig. 2.42)

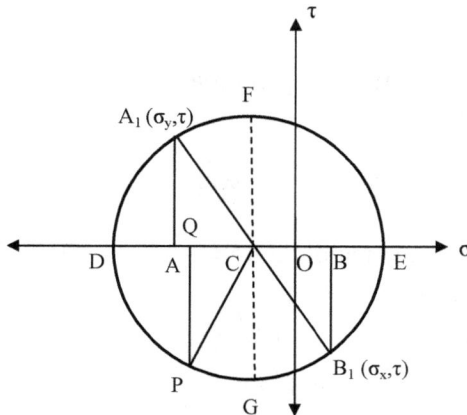

Fig. 2.42

From CA_1 taking $2\theta = 2\times55 = 110°$, mark radius CP and draw PQ perpendicular to AB.

Now $\sigma_\theta = OQ = -3.8$ cm $= -19$ MPa (compressive)   and $\tau_\theta = PQ = -5.6$ cm $= -28$ MPa

Major principal stress $\sigma_1 = OE = 4.5$ cm $= 22.5$ MPa (tensile)

Minor principal stress, $\sigma_2 = OD = -7.5$ cm $= -37.5$ MPa (compressive)

Inclination of principal planes:

$$\theta_{n_1} = \frac{1}{2} \times \angle A_1 CE = \frac{1}{2} \times 222 = 111°$$

$$\theta_{n_2} = \frac{1}{2} \times \angle A_1 CD = \frac{1}{2} \times 42 = 21°$$

Maximum shear stress, $\tau_{max} = CF = 6$ cm $= 30$ MPa

Inclination of major principal shear plane

$$\theta_{s_1} = \frac{1}{2} \times \angle A_1 CF = \frac{1}{2} \times 312 = 156°$$

Example 2.20 *For the given state of stress, find the resultant stress on plane AB inclined at 50⁰ to the horizontal as shown (Fig. 2.43a) using Mohr's stress circle. Also verify the results obtained by analytical method.*

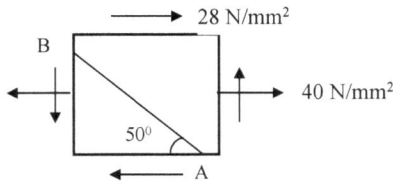

Fig. 2.43a

Solution:

$\sigma_x = 40$ N/mm², $\sigma_y = 0$, $\tau = 28$ N/mm²

Scale: 1 cm = 5 N/mm²

Refer Fig. 2.43b

OB = 40 N/mm² = 8 cm (to represent σ_x)

$\sigma_y = 0$ (here point A coincides with point O).

Draw OA_1 and BB_1 perpendicular to σ axis (x- axis) so that $OA_1 = BB_1 = 5.6$ cm (to represent 28 N/mm²).

Join A_1B_1. Let it intersect σ axis at C. Taking C as centre and CA_1 (or CB_1) as radius draw the Mohr's stress circle (Fig. 2.43b).

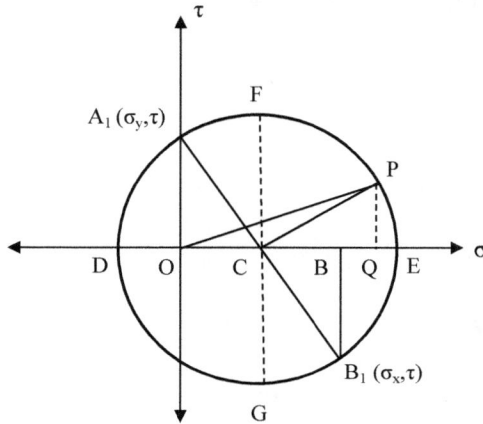

Fig. 2.43b

Mark radius CP taking $\angle A_1CP = 100°$ (clockwise) and draw PQ perpendicular upon σ axis.

Now $\sigma_\theta = OQ = 10.2$ cm $= 51$ N/mm² (tensile)

$\tau_\theta = PQ = 2.9$ cm $= 14.5$ N/mm²

Resultant stress $\sigma_r = OP = 10.6$ cm $= 53$ N/mm²

Analytical method:

$$\sigma_x = 40 \text{ N/mm}^2, \ \sigma_y = 0, \quad \tau = 28 \text{ N/mm}^2, \theta = 130°$$

$$\sigma_\theta = \sigma_{130} = \frac{\sigma_y + \sigma_x}{2} + \frac{\sigma_y - \sigma_x}{2} \cos 2\theta - \tau \sin 2\theta$$

$$= \frac{0 + 40}{2} + \frac{0 - 40}{2} \cos(2 \times 130) - (28) \sin(2 \times 130) = 51 \text{ N/mm}^2$$

$$\tau_\theta = \frac{\sigma_y - \sigma_x}{2} \sin 2\theta + \tau \cos 2\theta$$

$$= \frac{0 - 40}{2}\sin(2\times130) + (28)\cos(2\times130)$$

$$= 14.83 \text{ N/mm}^2$$

Resultant stress, $\sigma_r = \sqrt{\sigma_\theta^2 + \tau_\theta^2} = \sqrt{51^2 + 14.83^2} = 53.1 \text{ N/mm}^2$

Example 2.21 *Using Mohr's circle, obtain the expression for principal stresses in terms of σ_x, σ_y and τ where the symbols have their usual meaning.*

Solution:

The Mohr's circle is shown (Fig. 2.44).

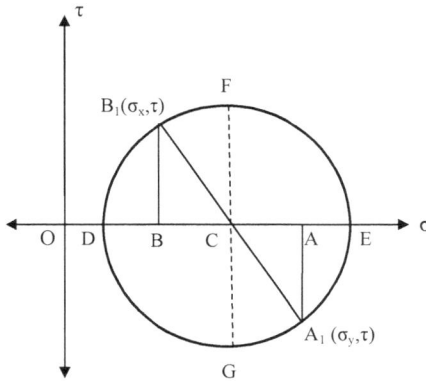

Fig. 2.44

Principal stresses are:

$\sigma_1 = OE$ and $\sigma_2 = OD$

$$CA = CB = \frac{\sigma_y - \sigma_x}{2}$$

$$CA_1 = CB_1 = \sqrt{BC^2 + BB_1{}^2}$$

$$= \sqrt{\left(\frac{\sigma_y - \sigma_x}{2}\right)^2 + \tau^2}$$

$AA_1 = BB_1 = \tau$

$$OC = OB + BC = \sigma_x + \frac{\sigma_y - \sigma_x}{2} = \frac{\sigma_y + \sigma_x}{2}$$

$$OE = OC + CE = OC + CB_1 = \frac{\sigma_y + \sigma_x}{2} + \sqrt{\left(\frac{\sigma_y - \sigma_x}{2}\right)^2 + \tau^2}$$

$$\therefore \sigma_1 = \frac{\sigma_y + \sigma_x}{2} + \sqrt{\left(\frac{\sigma_y - \sigma_x}{2}\right)^2 + \tau^2}$$

$$OD = OC - CD = OC - CB_1 = \frac{\sigma_y + \sigma_x}{2} - \sqrt{\left(\frac{\sigma_y - \sigma_x}{2}\right)^2 + \tau^2}$$

$$\therefore \sigma_2 = \frac{\sigma_y + \sigma_x}{2} - \sqrt{\left(\frac{\sigma_y - \sigma_x}{2}\right)^2 + \tau^2}$$

$$\therefore \text{Principal stresses,} \qquad \sigma_1, \sigma_2 = \frac{\sigma_y + \sigma_x}{2} \pm \sqrt{\left(\frac{\sigma_y - \sigma_x}{2}\right)^2 + \tau^2}$$

2.9 ELEMENT SUBJECTED TO NORMAL STRAINS AND SHEAR STRAIN

2.9.1 Normal strain on any plane in an element due to strains ε_x, ε_y and γ

Consider an elemental rectangular block ABCD of unit thickness subjected to two normal strains ε_x and ε_y along two mutually perpendicular directions x and y respectively along with a shear strain γ, as shown (Fig. 2.45).

We have to find out expressions for normal strain in a plane inclined at an angle θ with ε_x direction.

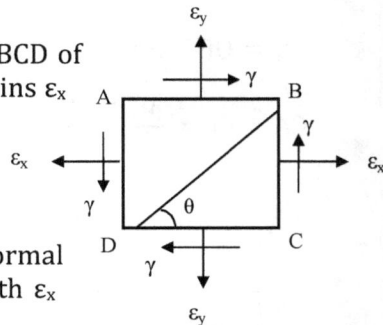

Fig. 2.45

Sign convention:

(i) Normal strains ε_x and ε_y corresponding to elongations in x and y directions respectively are positive.
(ii) Shear strain γ is positive if the 90^0 angle between the x and y axis becomes larger.
(iii) θ is positive if measured anticlockwise with respect to horizontal (similar to that for stresses considered earlier in this Chapter).

Figure 2.46 explains the effects of ε_x, ε_y and γ acting separately such that ABCD will finally be IJFD under the action of these strains.

$$CF = BE = \epsilon_x\, DC, \quad EH = \epsilon_y\, BC$$

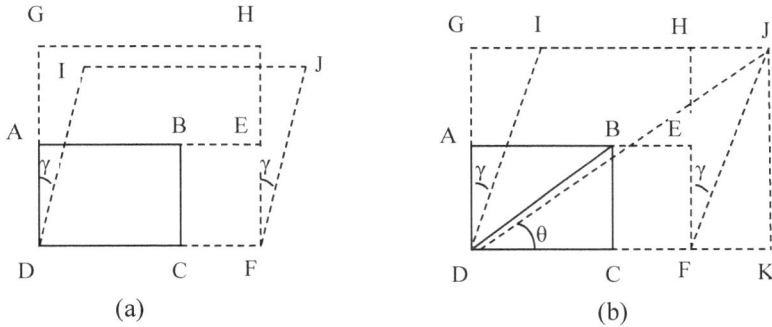

Fig. 2.46

Refer Fig. 2.46b

Let ϵ_θ = Strain along DB

$$\epsilon_\theta = \frac{DJ - DB}{DB} \qquad \therefore \quad DJ = \epsilon_\theta\, DB + DB = (1 + \epsilon_\theta)DB$$

$$DF = DC + CF = DC + \epsilon_x\, DC = (1 + \epsilon_x)DC$$

$$FH = FE + EH = FE + \epsilon_y\, BC = \left(1 + \epsilon_y\right)BC$$

Also, $KF = JK\ \tan\gamma = JK\,\gamma$ (for small γ)

From right angle triangle JKF,

$$JF = \sqrt{JK^2 + KF^2} = \sqrt{JK^2 + (JK\,\gamma)^2} = JK\sqrt{1 + \gamma^2}$$

$\therefore JF = JK$ also $JF = FH$ ($\gamma^2 \ll 1$ for small γ,)

In triangle DFJ, $DF = (1 + \epsilon_x)DC$, $JF = (1 + \epsilon_y)BC$, $DJ = (1 + \epsilon_\theta)DB$

$$DJ^2 = DF^2 + JF^2 - 2\ DF\ JF \cos(90 + \gamma)$$

$$(1 + \epsilon_\theta)^2 DB^2 = (1 + \epsilon_x)^2 DC^2 + (1 + \epsilon_y)^2 BC^2$$
$$- 2\ (1 + \epsilon_x)DC\ (1 + \epsilon_y)\ BC\ (-\sin\gamma)$$

$$(1 + \epsilon_\theta)^2 = (1 + \epsilon_x)^2 \left(\frac{DC}{DB}\right)^2 + (1 + \epsilon_y)^2 \left(\frac{BC}{DB}\right)^2$$
$$+ 2\ (1 + \epsilon_x)(1 + \epsilon_y)\left(\frac{DC}{DB}\right)\left(\frac{BC}{DB}\right)\sin\gamma$$

$$= (1 + \epsilon_x)^2 \cos^2\theta + (1 + \epsilon_y)^2 \sin^2\theta$$
$$+ 2\ (1 + \epsilon_x)(1 + \epsilon_y)\cos\theta\,\sin\theta\,\sin\gamma$$

Neglecting the squares and products of small quantities

$$1 + 2\epsilon_\theta = \cos^2\theta + 2\ \epsilon_x \cos^2\theta + \sin^2\theta + 2\ \epsilon_y \sin^2\theta$$
$$+ 2\gamma\ \sin\theta\cos\theta$$

$$= 1 + 2(\ \epsilon_x \cos^2\theta + \epsilon_y \sin^2\theta + \gamma\ \sin\theta\cos\theta)$$

$$\therefore\ \epsilon_\theta = \epsilon_x \cos^2\theta + \epsilon_y \sin^2\theta + \gamma\ \sin\theta\cos\theta$$

$$= \epsilon_x \left(\frac{1 + \cos2\theta}{2}\right) + \epsilon_y \left(\frac{1 - \cos2\theta}{2}\right) + \gamma\ \frac{\sin2\theta}{2}$$

Thus,

$$\boxed{\ \epsilon_\theta = \frac{\epsilon_y + \epsilon_x}{2} - \left(\frac{\epsilon_y - \epsilon_x}{2}\right)\cos2\theta + \frac{\gamma}{2}\sin2\theta\ }$$

2.9.2 Shear strain on any plane in an element due to strains ε_x, ε_y and γ

Consider the elemental rectangular block ABCD of unit thickness subjected to normal strains ε_x and ε_y along with shear strain γ, as explained in the previous section (Fig. 2.45). The expression for shear strain is required to be found out in a plane inclined at an angle θ. The effects of ε_x, ε_y and γ are considered separately.

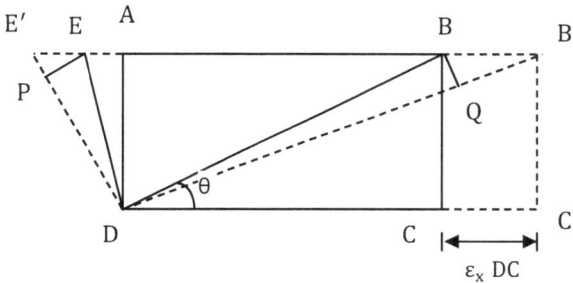

Fig. 2.47

Due to the effect of ε_x alone, the shear strain will be the change in the right angle EDB (Fig. 2.47) which is equal to the sum of angles BDB′ and EDE′

As explained in the previous section, $CC' = BB' = \varepsilon_x \, DC$

$\therefore BQ = \varepsilon_x \, DC \, \sin \theta$

Similarly, $EP = \varepsilon_x \, AE \, \cos \theta$

Due to ε_x alone, shear strain is equal to sum of angles BDB′ and EDE′ which is

$$= \frac{BQ}{DB} + \frac{EP}{DE} = \frac{\varepsilon_x \, DC \, \sin \theta}{DB} + \frac{\varepsilon_x \, AE \, \cos \theta}{DE}$$

$$= \varepsilon_x \, \cos \theta \sin \theta + \varepsilon_x \, \sin \theta \cos \theta = \varepsilon_x \, \sin 2\theta$$

Similarly, due to the action of ε_y alone, the shear strain can be found as $-\varepsilon_y \, \sin 2\theta$ and that due to the action of γ alone will be $-\gamma \, \cos 2\theta$. The required shear strain is the sum of shear strains due to ε_x, ε_y and γ.

$\gamma_\theta = \varepsilon_x \, \sin 2\theta - \varepsilon_y \, \sin 2\theta - \gamma \, \cos 2\theta = (\varepsilon_x - \varepsilon_y) \sin 2\theta - \gamma \, \cos 2\theta$

$$\frac{\gamma_\theta}{2} = -\left(\frac{\epsilon_y - \epsilon_x}{2}\right) \sin 2\theta - \frac{\gamma}{2} \cos 2\theta$$

Note:

Case 1: Element subjected to biaxial strain condition (only normal strains ϵ_x and ϵ_y and no shear strain)

$\epsilon_\theta = \dfrac{\epsilon_y + \epsilon_x}{2} - \left(\dfrac{\epsilon_y - \epsilon_x}{2}\right) \cos 2\theta = \epsilon_x \cos^2 \theta + \epsilon_y \sin^2 \theta$

$\gamma_\theta = (\varepsilon_x - \varepsilon_y) \sin 2\theta$

Case 2: Element subjected to uniaxial strain condition

$\epsilon_\theta = \epsilon_x \cos^2 \theta$ and $\gamma_\theta = \varepsilon_x \sin 2\theta$

2.10 PRINCIPAL STRAINS (Principal Normal Strains)

Principal strains are the maximum and minimum values of normal strains (like the principal stresses). The planes on which these principal strains act are called principal planes for strains.

Expression for strain: $\epsilon_\theta = \dfrac{\epsilon_y + \epsilon_x}{2} - \left(\dfrac{\epsilon_y - \epsilon_x}{2}\right) \cos 2\theta + \dfrac{\gamma}{2} \sin 2\theta$

For ϵ_θ to be maximum or minimum, $\dfrac{d\epsilon_\theta}{d\theta} = 0$

$-\left(\dfrac{\epsilon_y - \epsilon_x}{2}\right)(-\sin 2\theta)2 + \dfrac{\gamma}{2}(\cos 2\theta)2 = 0$

$\tan 2\theta_n = \dfrac{-\gamma}{\epsilon_y - \epsilon_x}$

With reference to Fig. 2.48

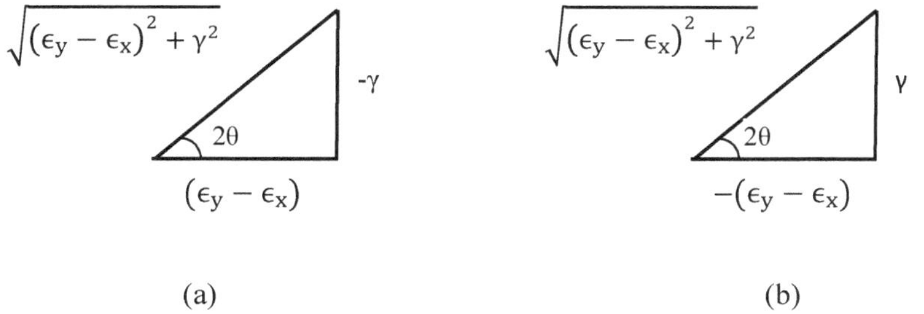

(a)   (b)

Fig. 2.48

$$\sin 2\theta = \pm \frac{\gamma}{\sqrt{\left(\epsilon_y - \epsilon_x\right)^2 + \gamma^2}} \quad \text{and} \quad \cos 2\theta = \mp \frac{\left(\epsilon_y - \epsilon_x\right)}{\sqrt{\left(\epsilon_y - \epsilon_x\right)^2 + \gamma^2}}$$

$$\therefore \quad \epsilon_\theta = \frac{\epsilon_y + \epsilon_x}{2} - \left(\frac{\epsilon_y - \epsilon_x}{2}\right) \cos 2\theta + \frac{\gamma}{2} \sin 2\theta$$

$$= \frac{\epsilon_y + \epsilon_x}{2} \pm \left(\frac{\epsilon_y - \epsilon_x}{2}\right) \left(\frac{\left(\epsilon_y - \epsilon_x\right)}{\sqrt{\left(\epsilon_y - \epsilon_x\right)^2 + \gamma^2}}\right)$$

$$\pm \frac{\gamma}{2} \left(\frac{\gamma}{\sqrt{\left(\epsilon_y - \epsilon_x\right)^2 + \gamma^2}}\right)$$

$$= \frac{\epsilon_y + \epsilon_x}{2} \pm \sqrt{\left(\frac{\epsilon_y - \epsilon_x}{2}\right)^2 + \left(\frac{\gamma}{2}\right)^2}$$

$$\epsilon_{\theta_1} = \frac{\epsilon_y + \epsilon_x}{2} + \sqrt{\left(\frac{\epsilon_y - \epsilon_x}{2}\right)^2 + \left(\frac{\gamma}{2}\right)^2}$$

$$\rightarrow \quad \text{Major principal strain } (\epsilon_1)$$

$$\epsilon_{\theta_2} = \frac{\epsilon_y + \epsilon_x}{2} - \sqrt{\left(\frac{\epsilon_y - \epsilon_x}{2}\right)^2 + \left(\frac{\gamma}{2}\right)^2}$$

$$\rightarrow \quad \text{Minor principal strain } (\epsilon_2)$$

Principal strain expression:

$$\epsilon_1, \epsilon_2 = \frac{\epsilon_y + \epsilon_x}{2} \pm \sqrt{\left(\frac{\epsilon_y - \epsilon_x}{2}\right)^2 + \left(\frac{\gamma}{2}\right)^2}$$

The expressions for ϵ_1 and ϵ_2 can be written considering $+$ and $-$ sign, respectively.

Note: Inclination of principal planes for strains:

$$\tan 2\theta_n = \frac{-\gamma}{\epsilon_y - \epsilon_x} = \frac{-\frac{\tau}{G}}{\left(\frac{\sigma_y}{E} - \nu\frac{\sigma_x}{E}\right) - \left(\frac{\sigma_x}{E} - \nu\frac{\sigma_y}{E}\right)}$$

$$= \frac{-\tau}{\frac{G}{E}\left(\sigma_y - \nu\sigma_x - \sigma_x + \nu\sigma_y\right)} = \frac{-\tau}{\frac{G}{E}\left[\left(\sigma_y - \sigma_x\right) + \nu\left(\sigma_y - \sigma_x\right)\right]}$$

$$= \frac{-\tau}{\frac{G}{E}(1 + \nu)\left(\sigma_y - \sigma_x\right)} = \frac{-\tau}{\frac{1}{2}\left(\sigma_y - \sigma_x\right)} = \frac{-2\tau}{\left(\sigma_y - \sigma_x\right)}$$

Therefore, principal planes for strains are same as principal planes for stresses and that is why they are simply called **principal planes**.

2.11 PRINCIPAL SHEAR STRAIN

Expression for shear strain, $\quad \gamma_\theta = -\left(\epsilon_y - \epsilon_x\right)\sin 2\theta - \gamma \cos 2\theta$

For γ_θ to be maximum or minimum, $\quad \dfrac{d\gamma_\theta}{d\theta} = 0$

$\left(\epsilon_y - \epsilon_x\right)(\cos 2\theta)2 - \gamma\,(-\sin 2\theta)2 = 0$

$\tan 2\theta_s = \dfrac{\epsilon_y - \epsilon_x}{\gamma}$

With reference to Fig. 2.49

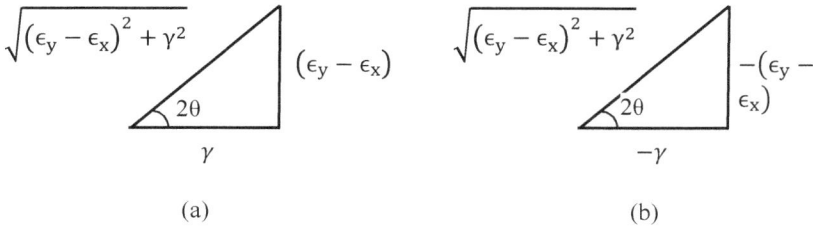

(a)

(b)

Fig. 2.49

$$\sin 2\theta = \pm \frac{\left(\epsilon_y - \epsilon_x\right)}{\sqrt{\left(\epsilon_y - \epsilon_x\right)^2 + \gamma^2}} \quad \text{and} \quad \cos 2\theta = \pm \frac{\gamma}{\sqrt{\left(\epsilon_y - \epsilon_x\right)^2 + \gamma^2}}$$

$$\therefore \quad [\gamma_\theta]_{\text{max or min}} = -\left(\epsilon_y - \epsilon_x\right)\sin 2\theta - \gamma \cos 2\theta$$

$$= \pm \frac{\left(\epsilon_y - \epsilon_x\right)\left(\epsilon_y - \epsilon_x\right)}{\sqrt{\left(\epsilon_y - \epsilon_x\right)^2 + \gamma^2}} \pm \gamma \frac{\gamma}{\sqrt{\left(\epsilon_y - \epsilon_x\right)^2 + \gamma^2}}$$

$$= \pm \sqrt{\left(\epsilon_y - \epsilon_x\right)^2 + \gamma^2}$$

$$\frac{[\gamma_\theta]_{\text{max or min}}}{2} = \pm \sqrt{\left(\frac{\epsilon_y - \epsilon_x}{2}\right)^2 + \left(\frac{\gamma}{2}\right)^2}$$

This expression is analogous to the principal stress expression by substituting σ_x by ϵ_x, σ_y by ϵ_y and τ by $\frac{\gamma}{2}$

Therefore, Maximum shear strain, $\gamma_{\text{max}} = \sqrt{\left(\epsilon_y - \epsilon_x\right)^2 + \gamma^2}$

Minimum shear strain, $\gamma_{\text{min}} = -\sqrt{\left(\epsilon_y - \epsilon_x\right)^2 + \gamma^2}$

Also, $\gamma_{\text{max}} = \epsilon_1 - \epsilon_2$

Maximum shear strain is the algebraic difference between the maximum and minimum principal strains.

Note: The expressions for strains can be written from those for the stresses by suitable substitution as

$$\sigma_x \leftrightarrow \epsilon_x, \qquad \sigma_y \leftrightarrow \epsilon_y, \qquad \tau \leftrightarrow \frac{\gamma}{2} \quad \text{and} \quad \theta \leftrightarrow (90 + \theta)$$

This comparison is tabulated below.

Stress

(1) $\sigma_\theta = \dfrac{\sigma_y + \sigma_x}{2}$
$+ \left(\dfrac{\sigma_y - \sigma_x}{2}\right) \cos 2\theta - \tau \sin 2\theta$

$\tau_\theta = \left(\dfrac{\sigma_y - \sigma_x}{2}\right) \sin 2\theta + \tau \cos 2\theta$

(2) Principal stresses:
σ_1, σ_2
$= \dfrac{\sigma_y + \sigma_x}{2} \pm \sqrt{\left(\dfrac{\sigma_y - \sigma_x}{2}\right)^2 + \tau^2}$

(3) Inclination of principal planes:
$\tan 2\theta_n = \dfrac{-2\tau}{\sigma_y - \sigma_x}$

(4) Maximum shear stress:
$\tau_{max} = \left(\dfrac{\sigma_1 - \sigma_2}{2}\right)$

Minimum shear stress:
$\tau_{min} = -\left(\dfrac{\sigma_1 - \sigma_2}{2}\right)$

(5) Inclination of principal shear planes:
$\tan 2\theta_s = \dfrac{\sigma_y - \sigma_x}{2\tau}$

Strain

(1) $\epsilon_\theta = \dfrac{\epsilon_y + \epsilon_x}{2}$
$- \left(\dfrac{\epsilon_y - \epsilon_x}{2}\right) \cos 2\theta + \dfrac{\gamma}{2} \sin 2\theta$

$\dfrac{\gamma_\theta}{2} = -\left(\dfrac{\epsilon_y - \epsilon_x}{2}\right) \sin 2\theta$
$- \dfrac{\gamma}{2} \cos 2\theta$

(2) Principal strains:
ϵ_1, ϵ_2
$= \dfrac{\epsilon_y + \epsilon_x}{2} \pm \sqrt{\left(\dfrac{\epsilon_y - \epsilon_x}{2}\right)^2 + \left(\dfrac{\gamma}{2}\right)^2}$

(3) Inclination of principal planes:
$\tan 2\theta_n = \dfrac{-\gamma}{\epsilon_y - \epsilon_x}$

(4) Maximum shear strain:
$\dfrac{\gamma_{max}}{2} = \left(\dfrac{\epsilon_1 - \epsilon_2}{2}\right)$

Minimum shear strain:
$\dfrac{\gamma_{min}}{2} = -\left(\dfrac{\epsilon_1 - \epsilon_2}{2}\right)$

(5) Inclination of principal shear planes:
$\tan 2\theta_s = \dfrac{\epsilon_y - \epsilon_x}{\gamma}$

2.12 MOHR'S STRAIN CIRCLE

From the discussion carried out earlier in this Chapter,

$$\epsilon_\theta = \frac{\epsilon_y + \epsilon_x}{2} - \left(\frac{\epsilon_y - \epsilon_x}{2}\right)\cos 2\theta + \frac{\gamma}{2}\sin 2\theta \qquad (1)$$

$$\frac{\gamma_\theta}{2} = -\left(\frac{\epsilon_y - \epsilon_x}{2}\right)\sin 2\theta - \frac{\gamma}{2}\cos 2\theta \qquad (2)$$

Rearranging Eqs. (1) and (2), taking squares and then adding result the following:

$$\left\{\epsilon_\theta - \frac{\epsilon_y + \epsilon_x}{2}\right\}^2 + \left\{\frac{\gamma_\theta}{2}\right\}^2$$
$$= \left\{-\left(\frac{\epsilon_y - \epsilon_x}{2}\right)\cos 2\theta + \frac{\gamma}{2}\sin 2\theta\right\}^2$$
$$+ \left\{-\left(\frac{\epsilon_y - \epsilon_x}{2}\right)\sin 2\theta - \frac{\gamma}{2}\cos 2\theta\right\}^2$$

$$\therefore \quad \left\{\epsilon_\theta - \frac{\epsilon_y + \epsilon_x}{2}\right\}^2 + \left\{\frac{\gamma_\theta}{2}\right\}^2 = \left(\frac{\epsilon_y - \epsilon_x}{2}\right)^2 + \left(\frac{\gamma}{2}\right)^2 \qquad (3)$$

Eq. (3) becomes the equation of a circle: $(\epsilon_\theta - a)^2 + \left(\frac{\gamma_\theta}{2}\right)^2 = R^2$

in which $a = \dfrac{\epsilon_y + \epsilon_x}{2}$ representing the centre and

$R = \sqrt{\left(\dfrac{\epsilon_y - \epsilon_x}{2}\right)^2 + \left(\dfrac{\gamma}{2}\right)^2}$ being the radius of the circle.

This circle is known as the **Mohr's strain circle**. The procedure to draw this Mohr's strain circle is very much similar to that for the Mohr's stress circle discussed earlier in this Chapter.

2.13 PRINCIPAL STRESSES IN TERMS OF PRINCIPAL STRAINS

In terms of principal stresses σ_1 and σ_2 the principal strains ϵ_1 and ϵ_2 can be expressed as

$$\epsilon_1 = \frac{\sigma_1}{E} - v\frac{\sigma_2}{E} \quad \text{and} \quad \epsilon_2 = \frac{\sigma_2}{E} - v\frac{\sigma_1}{E}$$

$$\epsilon_1 + v\epsilon_2 = \frac{\sigma_1}{E} - v\frac{\sigma_2}{E} + v\frac{\sigma_2}{E} - v^2\frac{\sigma_1}{E} = (1 - v^2)\frac{\sigma_1}{E}$$

$$\therefore \sigma_1 = \frac{E}{(1 - v^2)}(\epsilon_1 + v\epsilon_2)$$

Similarly, $\quad \sigma_2 = \dfrac{E}{(1 - v^2)}(\epsilon_2 + v\epsilon_1)$

$$\sigma_1 = \frac{E}{(1 - v^2)}(\epsilon_1 + v\epsilon_2)$$

$$\sigma_2 = \frac{E}{(1 - v^2)}(\epsilon_2 + v\epsilon_1)$$

Note: For a plane stress condition (refer Fig. 2.2b), the strain condition is in general three dimensional as

$$\epsilon_x = \frac{\sigma_x}{E} - v\frac{\sigma_y}{E}, \quad \epsilon_y = \frac{\sigma_y}{E} - v\frac{\sigma_x}{E}, \quad \epsilon_z = -v\frac{\sigma_x}{E} - v\frac{\sigma_y}{E}, \quad \gamma_{xy} = \frac{\tau_{xy}}{G}$$

Similarly, for a plane strain condition (Fig. 2.45), the stress condition is also three dimensional. Therefore, two-dimensional strain condition gives three-dimensional stress condition and two dimensional stress condition gives three dimensional strain condition.

Example 2.22 *The strains determined by rosette as shown (Fig. 2.50) are $\varepsilon_1 = +800\mu$, $\varepsilon_2 = +600\mu$, $\varepsilon_3 = +100\mu$. Determine the (i) in plane principal strains and (ii) in plane maximum shear strains.*

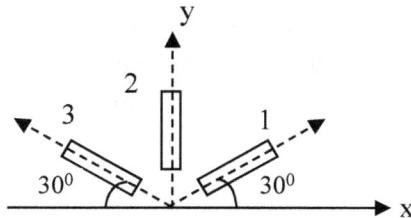

Fig. 2.50

Solution:

(Rosette instrument used for measuring strains)

$\varepsilon_{120} = 800 \times 10^{-6}$, $\qquad \varepsilon_{60} = 600 \times 10^{-6}$, $\qquad \varepsilon_0 = 100 \times 10^{-6}$

$\therefore \varepsilon_x = \text{unknown}$, $\quad \varepsilon_y = 100 \times 10^{-6}$, $\quad \gamma = \text{unknown}$

Expression for strain:

$$\varepsilon_\theta = \frac{\varepsilon_y + \varepsilon_x}{2} - \frac{\varepsilon_y - \varepsilon_x}{2}\cos2\theta + \frac{\gamma}{2}\sin2\theta$$

Corresponding to $\theta = 60°$,

$\varepsilon_{60} = 600 \times 10^{-6}$

$$= \frac{100 \times 10^{-6} + \varepsilon_x}{2} - \frac{100 \times 10^{-6} - \varepsilon_x}{2}\cos 120$$
$$+ \frac{\gamma}{2}\sin 120$$

$$= \frac{100 \times 10^{-6}}{2}(1 - \cos 120) + \left(\frac{1}{2} + \frac{1}{2}\cos 120\right)\varepsilon_x + \frac{\gamma}{2}\sin 120$$

$$= 7.5 \times 10^{-5} + 0.25\,\varepsilon_x + 0.433\,\gamma$$

Corresponding to $\theta = 120°$,

$\varepsilon_{120} = 800 \times 10^{-6}$

$$= \frac{100 \times 10^{-6} + \varepsilon_x}{2} - \frac{100 \times 10^{-6} - \varepsilon_x}{2}\cos 240$$
$$+ \frac{\gamma}{2}\sin 240$$

$$= 7.5 \times 10^{-5} + 0.25\,\varepsilon_x + 0.433\,\gamma$$

$\therefore \varepsilon_x = 2.5 \times 10^{-3}$ and $\gamma = -2.309 \times 10^{-4}$

Principal strains:

$$\varepsilon_{1,2} = \frac{\varepsilon_y + \varepsilon_x}{2} \pm \sqrt{\left(\frac{\varepsilon_y - \varepsilon_x}{2}\right)^2 + \left(\frac{\gamma}{2}\right)^2}$$

$$= \frac{100 \times 10^{-6} + 2.5 \times 10^{-3}}{2}$$

$$\pm \sqrt{\left(\frac{100 \times 10^{-6} - 2.5 \times 10^{-3}}{2}\right)^2 + \left(\frac{-2.309 \times 10^{-4}}{2}\right)^2}$$

$$= 0.0013 \pm 1.205 \times 10^{-3}$$

$$\therefore \ \varepsilon_1 = 2.505 \times 10^{-3} \text{ and } \varepsilon_2 = 9.5 \times 10^{-5}$$

Maximum shear strain:

$$\gamma_{max.} = \varepsilon_1 - \varepsilon_2 = 2.41 \times 10^{-3}$$

Example 2.23 *It is observed that an element of a body contracts 0.0005 cm per cm along x axis, elongates 0.0003 cm per cm along y axis and distorts through an angle of 0.0006 radians. Find the principal strains and determine the directions in which those strains act. Verify your answer using Mohr's circle for strain.*

Solution:

$$\varepsilon_x = -0.0005, \quad \varepsilon_y = 0.0003, \qquad \gamma = 0.0006$$

Principal strains:

$$\varepsilon_{1,2} = \frac{\varepsilon_y + \varepsilon_x}{2} \pm \sqrt{\left(\frac{\varepsilon_y - \varepsilon_x}{2}\right)^2 + \left(\frac{\gamma}{2}\right)^2}$$

$$= \left\{\frac{0.0003 - 0.0005}{2}\right\} \pm \sqrt{\left(\frac{0.0003 + 0.0005}{2}\right)^2 + \left(\frac{0.0006}{2}\right)^2}$$

$$= (-1 \pm 5) \times 10^{-4}$$

$$\therefore \ \varepsilon_1 = 4 \times 10^{-4}, \ \varepsilon_2 = -6 \times 10^{-4}$$

Inclination of principal planes:

$$\tan 2\theta = \frac{-\gamma}{\varepsilon_y - \varepsilon_x} = \frac{-0.0006}{0.0003 - (-0.0005)} = -0.75$$

$\therefore 2\theta = -36.86^0$

$\therefore 2\theta = -36.86^0 + 180^0, -36.86^0 + 180^0 + 180^0$

$\theta = 71.56°, 161.56°$

Mohr's strain circle:

Refer Fig. 2.51

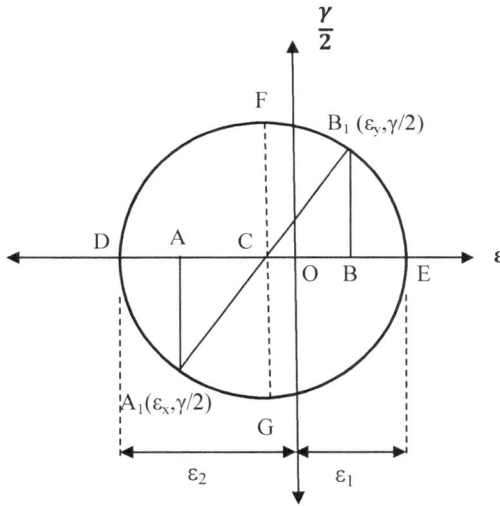

Fig. 2.51

Considering a scale of 1 cm = 0.0001

$OA = 5$ cm (to represent ε_x), $OB = 3$ cm $\left(\text{to represent } \varepsilon_y\right)$,

$AA_1 = BB_1 = 3$ cm (to represent $\gamma/2$)

Join $A_1 B_1$ and let it intersect axis at C. Now taking C as centre and CA_1 (or CB_1) as radius, Mohr's circle for strain is drawn (Fig. 2.51).

By measurement

$\varepsilon_1 = OE = 4.1$ cm $= 0.00041 \quad \rightarrow \quad$ Major principal strain

$\varepsilon_2 = OD = -6$ cm $= -6 \times 10^{-4} \rightarrow$ Minor principal strain

Measuring angles B_1CD and B_1CE, the values of $2\theta_1$ and $2\theta_2$ can be obtained.

$$\angle B_1CD = 143° , \qquad \angle B_1CE = 323°$$

$$\therefore \theta_1 = 71.5° \therefore \theta_2 = 161.5°$$

It can be observed that θ_1 and θ_2 are corresponding to the minor and major principal planes, respectively.

Example 2.24 *A thick aluminium plate is subjected to plane state of stress as shown (Fig. 2.52). Knowing that $E = 70 \times 10^9$ N/m^2 and $v = 0.3$, determine the normal strain along diagonal AB.*

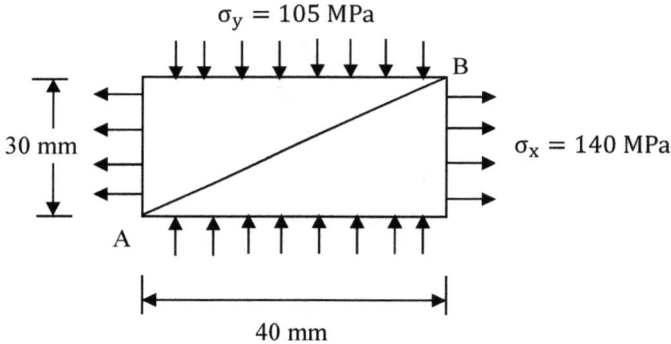

$\sigma_y = 105$ MPa

$\sigma_x = 140$ MPa

30 mm

40 mm

Fig. 2.52

Solution:

$\sigma_x = 140$ MPa (tensile), $\sigma_y = -105$ MPa (compressive), $\tau = 0$

$$\varepsilon_x = \frac{\sigma_x}{E} - v\frac{\sigma_y}{E}, \qquad \varepsilon_y = \frac{\sigma_y}{E} - v\frac{\sigma_x}{E}$$

$$\varepsilon_x = \frac{140 \times 10^6}{70 \times 10^9} - \frac{0.3(-105 \times 10^6)}{70 \times 10^9} = 2.45 \times 10^{-3}$$

$$\varepsilon_y = \frac{-105 \times 10^6}{70 \times 10^9} - \frac{0.3(140 \times 10^6)}{70 \times 10^9} = -2.1 \times 10^{-3}$$

$$\gamma = 0 \text{ (as } \tau = 0)$$

$$\therefore \ \varepsilon_\theta = \frac{\varepsilon_y + \varepsilon_x}{2} - \frac{\varepsilon_y - \varepsilon_x}{2}\cos 2\theta + \frac{\gamma}{2}\sin 2\theta$$

Reference to Fig. 2.52, $\quad \theta = \tan^{-1}\frac{30}{40} + 90 = 126.87°$

(θ is the inclination of plane on which strain acts.)

$$\therefore \ \varepsilon_{126.87} = \frac{-2.1 \times 10^{-3} + 2.45 \times 10^{-3}}{2}$$
$$-\left(\frac{-2.1 \times 10^{-3} - 2.45 \times 10^{-3}}{2}\right)\cos\,(2 \times 126.87)$$
$$+\,0$$

$$= -4.61 \times 10^{-4}$$

Example 2.25 *At a specific point on a steel machine part, measurement with an electric rectangular rosette indicates that, normal strains in the directions 0^0, $45°$ and $90°$ measured in anticlockwise direction are 0.0005 (contraction), 0.0002 (extension) and 0.0003 (extension) respectively. Assuming $E = 2.1 \times 10^5 \ N/mm^2$ and $v = 0.3$, find the principal stresses at the investigated point.*

Solution:

Refer Fig. 2.53

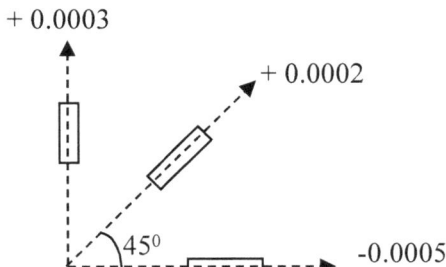

Fig. 2.53

$\varepsilon_x = -0.0005 = -5 \times 10^{-4}$

$\varepsilon_y = +0.0003 = 3 \times 10^{-4}$

$\varepsilon_\theta = \varepsilon_{135} = 0.0002 = 2 \times 10^{-4}$

$$\varepsilon_\theta = \frac{\varepsilon_y + \varepsilon_x}{2} - \frac{\varepsilon_y - \varepsilon_x}{2} \cos 2\theta + \frac{\gamma}{2} \sin 2\theta$$

$$2 \times 10^{-4} = \frac{3 \times 10^{-4} - 5 \times 10^{-4}}{2} - \frac{3 \times 10^{-4} + 5 \times 10^{-4}}{2} \cos 270 + \frac{\gamma}{2} \sin 270$$

$$\therefore \gamma = -6 \times 10^{-4}$$

Principal strains:

$$\varepsilon_{1,2} = \frac{\varepsilon_y + \varepsilon_x}{2} \pm \sqrt{\left(\frac{\varepsilon_y - \varepsilon_x}{2}\right)^2 + \left(\frac{\gamma}{2}\right)^2}$$

$$= \left[\frac{3 - 5}{2} \pm \sqrt{\left(\frac{3 + 5}{2}\right)^2 + \left(\frac{-6}{2}\right)^2}\right] \times 10^{-4} = (-1 \pm 5) \times 10^{-4}$$

$$\therefore \varepsilon_1 = 4 \times 10^{-4}, \quad \varepsilon_2 = -6 \times 10^{-4}$$

$$\sigma_1 = \frac{E}{1 - v^2}(\varepsilon_1 + v\varepsilon_2) = \frac{2.1 \times 10^5}{1 - (0.3)^2}\{4 - (0.3)6\}10^{-4}$$
$$= 50.76 \text{ N/mm}^2 \text{ (tensile)}$$

$$\sigma_2 = \frac{E}{1 - v^2}(\varepsilon_2 + v\varepsilon_1) = \frac{2.1 \times 10^5}{1 - (0.3)^2}\{-6 + (0.3)4\}10^{-4}$$
$$= -110.77 \text{ N/mm}^2 \text{ (compressive)}$$

Example 2.26 *A material is subjected to two mutually perpendicular linear strains together with a shear strain. One of the linear strains is 0.00025 (tensile). Determine the magnitudes of the other linear strain and the shear strain, if the principal strains are 0.0001 (compressive) and 0.0003 (tensile).*

Solution:

$$\varepsilon_x = 0.00025 \quad \text{(tensile)}, \quad \varepsilon_y = \text{unknown}, \quad \gamma = \text{unknown},$$

$$\varepsilon_1 = 0.0003 \text{ (tensile)}, \quad \varepsilon_2 = -0.0001 \text{ (compressive)}$$

$$\varepsilon_1 = \frac{\varepsilon_y + \varepsilon_x}{2} + \sqrt{\left(\frac{\varepsilon_y - \varepsilon_x}{2}\right)^2 + \left(\frac{\gamma}{2}\right)^2}$$

$$\therefore 0.0003 = \frac{\varepsilon_y + 0.00025}{2} + \sqrt{\left(\frac{\varepsilon_y - 0.00025}{2}\right)^2 + \left(\frac{\gamma}{2}\right)^2} \qquad (1)$$

Again,

$$\varepsilon_2 = \frac{\varepsilon_y + \varepsilon_x}{2} - \sqrt{\left(\frac{\varepsilon_y - \varepsilon_x}{2}\right)^2 + \left(\frac{\gamma}{2}\right)^2}$$

$$-0.0001 = \frac{\varepsilon_y + 0.00025}{2} - \sqrt{\left(\frac{\varepsilon_y - 0.00025}{2}\right)^2 + \left(\frac{\gamma}{2}\right)^2} \qquad (2)$$

Solving the above equations (1) and (2), the values of ε_y and γ are obtained as

$$\varepsilon_y = -5 \times 10^{-5}, \qquad \gamma = 2.65 \times 10^{-4}$$

Example 2.27 *A state of plane strain in a steel plate is defined by the following data.*
$\varepsilon_x = 0.0005, \varepsilon_y = 0.00014, \gamma_{xy} = 0.00036.$
Construct a Mohr's circle and find the magnitude and direction of principal strains.

Solution:

Scale: 1 cm = 0.0001 for strain

Refer Fig. 2.54

Take OA = 1.4 cm and OB = 5 cm (to represent ε_y and ε_x respectively)

Draw AA_1 and BB_1 perpendicular to AB such that $AA_1 = BB_1 = $ 1.8 cm (to represent $\gamma/2$).

Join A_1B_1 and let it intersect AB at C. With C as centre and CA_1 as radius describe the Mohr's strain circle (Fig. 2.54).

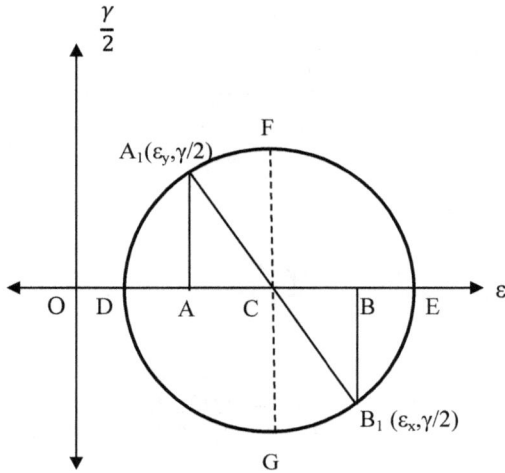

Fig. 2.54

By measurement,

$\varepsilon_1 = OE = 5.8$ cm $= 5.8 \times 0.0001$
$$= 5.8 \times 10^{-4} \rightarrow \text{Major principal strain}$$

$\varepsilon_2 = OD = 0.6$ cm $= 0.6 \times 10^{-4} \rightarrow$ Minor principal strain

Inclination of principal planes:

$2\theta_2 = \angle A_1 CD = 46°$

$\therefore \theta_2 = 23°$

$2\theta_1 = \angle A_1 CE = 226°$

$\therefore \theta_1 = 113°$

Example 2.28 *In a certain material, the maximum strain must not exceed that produced by a simple tensile stress of 90 N/mm². Show that the maximum permissible pure shear stress is 90/(1+v) where v→ Poisson's ratio.*

Solution:

For simple tensile stress of 90 N/mm²,

Major principal strain, $\qquad \varepsilon_1 = \dfrac{\sigma_1}{E} = \dfrac{90}{E}$

For pure shear stress of τ, principal stresses, $\sigma_1 = \tau$ and $\sigma_2 = -\tau$

Corresponding major pricipal strain,

$$\varepsilon_1 = \frac{\sigma_1}{E} - v\frac{\sigma_2}{E} = \frac{\tau}{E} - v\frac{(-\tau)}{E} = (1+v)\frac{\tau}{E}$$

From the above relations,

$$\frac{90}{E} = (1+v)\frac{\tau}{E}$$

$$\therefore \tau = \frac{90}{1+v} \qquad \rightarrow \quad \text{Maximum Permissible shear stress}$$

Example 2.29 *A flat brass plate was stretched by tensile forces acting in directions x and y at right angles. Strain gauges showed that the strains in the x- and y-directions respectively were 0.00072 and 0.00016. Find (a) the stresses acting in the x- and y- directions (b) the normal and shear stresses on a plane inclined at 30° to the x- direction. E = 80000 N/mm², Poisson's ratio = 0.3.*

Solution:

$$\varepsilon_x = 0.00072, \;\; \varepsilon_y = 0.00016, \;\;\; E = 80000\frac{N}{mm^2}, \;\; v = 0.3$$

(a)
$$\sigma_1 = \frac{E}{1-v^2}(\varepsilon_1 + v\varepsilon_2) = \frac{80000}{1-(0.3)^2}[0.00072 + (0.3\times0.00016)]$$
$$= 67.51 \text{ N/mm}^2$$

$$\sigma_2 = \frac{E}{1-v^2}(\varepsilon_2 + v\varepsilon_1) = \frac{80000}{1-(0.3)^2}[0.00016 + (0.3\times0.00072)]$$
$$= 33.05 \text{ N/mm}^2$$

(b)

$$\sigma_\theta = \frac{\sigma_y + \sigma_x}{2} + \frac{\sigma_y - \sigma_x}{2}\cos2\theta$$

$$= \frac{33.05 + 67.51}{2} + \frac{33.05 - 67.51}{2}\cos(2 \times 30)$$

$$= 41.66\frac{N}{mm^2} \quad \rightarrow \text{ Normal stress}$$

$$\tau_\theta = \frac{\sigma_y - \sigma_x}{2}\sin2\theta = \frac{33.05 - 67.51}{2}\sin(2 \times 30)$$

$$= -14.92\frac{N}{mm^2} \quad \rightarrow \text{ Shear stress}$$

HIGHLIGHTS

Definitions

1. *Plane stress condition*: If in a cuboidal element, stresses are acting only on two pairs of parallel planes and no stress is present on the third pair of parallel planes, then the element is said to be under plane stress condition. This is also known as two-dimensional stress condition.

2. *Principal stresses* (principal normal stresses): These are the maximum and minimum value of normal stresses. The maximum normal stress is called the major principal stress and the minimum normal stress is called the minor principal stress.

3. *Principal planes* (principal normal planes): These are the planes on which shear stress is zero and only normal stress is acting. These are the planes on which principal stresses act. The plane on which major principal stress acts is called major principal plane and the plane on which minor principal stress acts is called minor principal plane. These are perpendicular to each other.

4. *Principal shear stresses*: These are the maximum and minimum values of shear stresses. These are numerically equal but opposite in sign. The maximum shear stress is called major principal shear

stress and the minimum shear stress is called minor principal shear stress.

5. *Principal shear planes*: These are the planes on which the shear stress is maximum or minimum. These are perpendicular to each other. The plane on which maximum shear stress acts is called major principal shear plane and the plane on which minimum shear stress acts is called minor principal shear plane.

6. *Mohr's stress circle*: A graphical method used to find normal and shear stresses on any plane in an element. Principal normal stresses and principal shear stresses can also be found out. Mohr's stress circle is a circle with center at a $= \frac{\sigma_y + \sigma_x}{2}$ and radius R $=$ $\sqrt{\left(\frac{\sigma_y + \sigma_x}{2}\right)^2 + \tau^2}$ drawn to a scale in a two dimensional Cartesian system with normal stress σ along x axis and shear stress τ along y axis.

7. *Principal strains* (principal normal strains): These are the maximum and minimum values of normal strains. The maximum normal strain is called major principal strain and the minimum normal strain is called minor principal strain.

8. *Principal shear strains*: There are the maximum and minimum values of shear strains. The maximum value is called major principal shear strain and the minimum value is called minor principal shear strain.

9. *Principle of complementary shear stresses*: According to this, a shear stress is automatically accompanied by a shear stress of equal intensity but of opposite turning moment.

Concepts and Formulae

1. When an element is subjected to biaxial stresses σ_x and σ_y along with a shear stress τ, then on any plane inclined at θ with the direction of σ_x.

Normal stress, $\quad \sigma_\theta = \dfrac{\sigma_y + \sigma_x}{2} + \dfrac{\sigma_y - \sigma_x}{2} \cos 2\theta - \tau \sin 2\theta$

Shear stress, $\tau_\theta = \dfrac{\sigma_y - \sigma_x}{2} \sin 2\theta + \tau \cos 2\theta$

Resultant stress, $\sigma_r = \sqrt{\sigma_\theta^2 + \tau_\theta^2}$

2. Principal stresses:

$$\sigma_1, \sigma_2 = \frac{\sigma_y + \sigma_x}{2} \pm \sqrt{\left(\frac{\sigma_y - \sigma_x}{2}\right)^2 + \tau^2}$$

3. Inclinations of principal planes (principal normal planes):

$\tan 2\theta_n = \dfrac{-2\tau}{\sigma_y - \sigma_x}$ (This will give θ_{n_1} and θ_{n_2})

4. Inclination of principal shear planes:

$\tan 2\theta_s = \dfrac{\sigma_y - \sigma_x}{2\tau}$ (This will give θ_{s_1} and θ_{s_2})

5. Maximum shear stress, $\tau_{max} = \dfrac{\sigma_1 - \sigma_2}{2}$

Minimum shear stress, $\tau_{min} = -\left(\dfrac{\sigma_1 - \sigma_2}{2}\right)$

6. Normal stress on principal shear planes $= \dfrac{\sigma_y + \sigma_x}{2}$

7. Principal shear planes are inclined at 45^0 to the principal normal planes.

8. On principal normal planes, normal stress (σ) is maximum or minimum and shear stress (τ) is zero. On principal shear planes, shear stress (τ) is maximum or minimum but normal stress (σ) need not be zero $\left(\sigma = \frac{\sigma_1 + \sigma_2}{2}\right)$.

9. Sum of normal stresses on two mutually perpendicular planes at a point is invariant.

10. Radius of Mohr's stress circle represents a particular plane on element and a point on it represents the state of stress corresponding to the particular plane. Also, θ in element is represented as 2θ in Mohr's stress circle.

11. Due to strains ε_x, ε_y and γ in an element,

Normal strain on any plane,

$$\varepsilon_\theta = \frac{\varepsilon_y + \varepsilon_x}{2} - \left(\frac{\varepsilon_y - \varepsilon_x}{2}\right)\cos 2\theta + \frac{\gamma}{2}\sin 2\theta$$

Shear strain on any plane, $\gamma_\theta = (\varepsilon_x - \varepsilon_y)\sin 2\theta - \gamma\cos 2\theta$

12. Principal strains:

$$\varepsilon_{1,2} = \frac{\varepsilon_y + \varepsilon_x}{2} \pm \sqrt{\left(\frac{\varepsilon_y - \varepsilon_x}{2}\right)^2 + \left(\frac{\gamma}{2}\right)^2}$$

13. Principal planes for strains are same as principal planes for stresses.

14. Principal shear strains:

$$\gamma_{max} = \varepsilon_1 - \varepsilon_2$$

$$\gamma_{min} = -(\varepsilon_1 - \varepsilon_2)$$

15. Principal stresses in terms of principal strains:

$$\sigma_1 = \frac{E}{1 - v^2}(\varepsilon_1 + v\varepsilon_2), \quad \sigma_2 = \frac{E}{1 - v^2}(\varepsilon_2 + v\varepsilon_1)$$

SHORT TYPE QUESTIONS

1. In case of biaxial state of normal stresses, the maximum shear stress is equal to
(a) the sum of normal stresses
(b) the difference of normal stresses

(c) half the sum of normal stresses

(d) half the difference of normal stresses.

[Ans. (d)]

2. If the tensile stress in a specimen under uniaxial tension is σ, what will be the value of maximum shear stress in the specimen?

[Ans. $\sigma/2$]

3. The normal stress assumes the maximum value on the plane on which _____ is zero.

[Ans. shear stress]

4. If the normal cross section A of a member is subjected to a tensile force P, the resulting normal stress on an oblique plane inclined at an angle θ to transverse plane will be

(a) $\dfrac{P}{A} \sin^2 \theta$ (b) $\dfrac{P}{A} \cos^2 \theta$ (c) $\dfrac{P}{2A} \sin 2\theta$ (d) $\dfrac{P}{2A} \cos 2\theta$

[Ans. (b)]

5. Consider the following statements:

State of stress at a point when completely specified enables one to determine the

(i) principal stresses at the point.

(ii) maximum shear stress at the point.

(iii) stress components on any plane containing the point

Of these statements,

(a) (i), (ii) and (iii) are correct

(b) (i) and (iii) are correct

(c) (ii) and (iii) are correct

(d) (i) and (ii) are correct

[Ans. (a)]

6. State of stress at a point in a strained body is shown (Fig. 2.55). Which one of the following figures given below represents correctly the Mohr's circle for the state of stress?

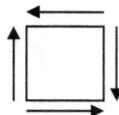

Fig. 2.55

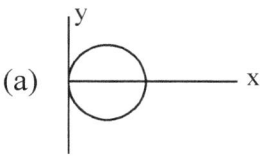

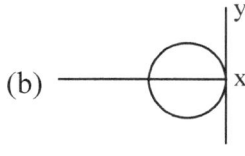

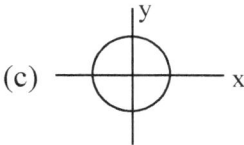

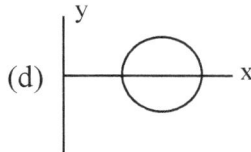

(a) (b) (c) (d)

[Ans. (c)]

7. Radius of Mohr's stress circle gives
(a) major principal stress
(b) minor principal stress
(c) maximum shear stress
(d) none of these

[Ans. (c)]

8. When two like equal principal stresses act then the radius of Mohr's stress circle will be _____.

[Ans. zero]

9. If ε_1 and ε_2 are the maximum and minimum strains in the neighborhood of a point in a stressed material of Young's modulus, E and Poisson's ratio, v then the maximum principal stress is given by

(a) $E\,\varepsilon_1$ (b) $E\,(\varepsilon_1 + \varepsilon_2)$ (c) $\dfrac{E(\varepsilon_1 + v\varepsilon_2)}{1 - v^2}$ (d) $\dfrac{E(\varepsilon_2 + v\varepsilon_1)}{1 - v^2}$

[Ans. (c)]

10. A Mohr's circle reduces to a point when the body is subjected to the following:
(a) pure stress
(b) uniaxial stress only
(c) equal & opposite stresses on two mutually perpendicular planes, the planes being free of shear.
(d) equal stresses on two mutually perpendicular planes, the planes being free of shear.

[Ans. (d)]

11. The state of plane stress at a point is described by $\sigma_x = \sigma_y = \sigma$ and $\tau_{xy} = 0$. The normal stress on the plane inclined at 45^0 to the x plane will be

(a) σ (b)$\sqrt{2}\,\sigma$ (c) $\sqrt{3}\,\sigma$ (d) 2σ

[Ans. (a)]

12. If normal stress on a normal plane is σ, then normal stress on a plane inclined at θ to the normal plane will be

(a) $\sigma \cos^2 \theta$ (b) $\sigma \sin^2 \theta$ (c) $\sigma \sin \theta \cos \theta$ (d) $\sigma \sin 2\theta$

[Ans. (a)]

13. If normal stress on a normal plane is σ, then shear stress on a plane inclined at θ to the normal plane will be

(a) $\sigma \sin^2 \theta$ (b) $\sigma \cos^2 \theta$ (c) $\dfrac{\sigma}{2} \sin 2\theta$ (d) $\sigma \sin 2\theta$

[Ans. (c)]

14. In case of biaxial state of normal stresses, the normal stress of 45^0 plane is equal to

(a) sum of normal stresses
(b) difference of normal stresses
(c) half the sum of normal stresses
(d) half the difference of normal stresses

[Ans. (c)]

15. In case of pure shear, the principal stresses are _____.
[Ans. equal in magnitude to the shear stress but opposite in nature]

EXERCISE PROBLEMS

1. At a point in a strained material the principal stresses are 100 N/mm² (tensile) and 60 N/mm² (compressive). Determine the normal stress, shear stress and resultant stress on a plane inclined at 50^0 to the axis of major principal stress. Also, determine the maximum shear stress at the point.

[Ans. $\sigma_\theta = 33.89$ N/mm², $\tau_\theta = 78.78$ N/mm², $\sigma_r = 85.76$ N/mm², $\tau_{max} = 80$ N/mm²]

2. In an elastic material at a certain point, on planes at right angles to one another, direct stresses of 120 N/mm² (tensile) and 100

N/mm^2 (compressive) are acting. The major principal stress in the material is to be limited to 160 N/mm^2. To what shearing stress the material may be subjected on the given planes. Also find the minimum principal stress and the maximum shearing stress at that point?

[Ans. τ = 101.98 N/mm^2, σ_2 = -140 N/mm^2, τ_{max} = 150 N/mm^2]

3. At a certain point in a strained material the stresses on two mutually perpendicular planes are 20 N/mm^2 and 10 N/mm^2 both tensile. They are accompanied by shear stress of magnitude of 10 N/mm^2. Find the location of principal planes and evaluate the principal stresses.

[Ans. θ = 31.71^0, 121.71^0, σ_1= 26.18 N/mm^2, σ_2= 3.82 N/mm^2]

4. For the state of stress shown (Fig. 2.56), find the principal stresses and directions. Also find the maximum shearing stress and directions along with the normal stresses on the planes of maximum shear stress.

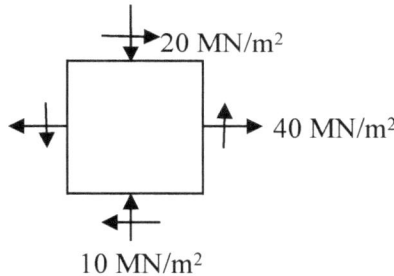

20 MN/m^2

40 MN/m^2

10 MN/m^2

Fig. 2.56

[Ans. $\sigma_1 = 50 \dfrac{MN}{m^2}$, $\sigma_2 = 0$, $\theta_{n_1} = 116.5^0$, $\theta_{n_2} = 26.5^0$,

$\tau_{max} = 25\dfrac{MN}{m^2}$, $\theta_s = 71.5^0, 161.5^0$, $\sigma_n = 25 \dfrac{MN}{m^2}$]

5. Solve problem 4, by using Mohr's stress circle.

6. Draw the Mohr's stress circle for direct stress of 65 MN/m^2 (tensile) and 35 MN/m^2 (compressive) and estimate the magnitude and direction of the resultant stresses on planes making angles of 20^0 and 65^0 with the plane of the first principal stress. Find also the normal and tangential stresses on these planes.

[Ans. σ_r = 62.2 MN/m^2, ϕ = 31^0 for 20^0 plane. σ_r = 41.9 MN/m^2, ϕ = 114^0 for 65^0 plane, σ = 53.3 MN/m^2, τ = 32.1 MN/m^2 for 20^0 plane, σ = 17.1 MN/m^2, τ = 38.3 MN/m^2 for 65^0 plane]

7. Mutually perpendicular faces of a square element of a thin plate are subjected to normal and shear stresses of 63 MN/m^2 (tensile), 47.2 MN/m^2 (compressive) and 39.4 MN/m^2 (shear). Determine graphically or otherwise the magnitudes and directions of principal stresses and the greatest shear stress.

[Ans. σ_1 = 75.6 MN/m^2 (tensile), σ_2 = 59.8 MN/m^2 (comp.), θ_1 = 17.76^0, θ_2 = 107.76^0, τ_{max} = 7.9 MN/m^2]

8. At a point in a strained material, the major and minor principal stresses are 100 MPa (tensile) and 50 MPa (compressive) respectively. Find the planes on which the stress is wholly shear.

[Ans. θ = 35.26^0 and 125.26^0]

9. A block of material is subjected to a tensile strain of 12×10^{-6} and a compressive strain of 15×10^{-6} on planes at right angles to each other. There is also a shear strain of 12×10^{-6} and there is no strain on planes at right angles to the above planes. Calculate the principal strains in magnitude and direction.

[Ans. 13.27×10^{-6} (tensile) and 16.27×10^{-6} (compressive), 12^0 and 102^0]

10. At a point in an elastic material under strain, there are normal stresses of 50 N/mm^2 and 30 N/mm^2 respectively at right angles to each other with a shear stress of 25 N/mm^2. Find the principal stresses and position of principal planes if (i) 50 N/mm^2 is tensile and 30 N/mm^2 is also tensile (ii) 50 N/mm^2 is tensile and 30 N/mm^2 is compressive. Find also maximum shear stress and its plane in both the cases.

[Ans. (i) σ_1 = 66.9 N/mm^2 (tensile), σ_2 = 13 N/mm^2 (tensile), θ = 34^0 and 124^0, τ_{max} = 26.9 N/mm^2 at 76^0 (ii) σ_1 = 57.1 N/mm^2 (tensile), σ_2 = 37 N/mm^2 (comp.), θ = 16^0 and 106^0, τ_{max} = 47.1 N/mm^2 at 61^0]

Chapter 3

Bending Moments and Shear Forces

Learning Objectives

After going through this chapter, the reader will be able to
- understand different types of beams along with various types of supports and loadings for the beams.
- determine the shear force, bending moment and axial force at any beam cross section due to external loading.
- draw shear force, bending moment and axial force diagrams for the beams.
- derive and apply the relationship between the intensity of loading, shear force and bending moment.

3.1 INTRODUCTION

Beams are very useful components employed in many structural applications. Regardless of the type of loading, the resulting action leads to effects that are of major significance in the analysis and design of beams. In this Chapter, we will analyze the beams in detail. First, different types of beams and various loadings will be discussed followed by the concept of shear force and bending moment. Next, the procedure to draw the shear force, bending moment and axial force diagrams for the beams will be explained. The relationship between the intensity of loading, shear force and bending moment will also be presented. In this Chapter, we will limit our discussion to statically determinate beams only.

3.2 BEAM

Beam is a structural component or machine member which is mainly used to carry transverse loads (Fig. 3.1). Its one dimension is very large as compared to the other two. The lateral forces acting on the beams cause bending of the beams.

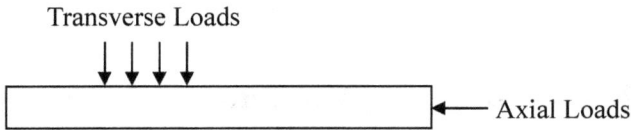

Transverse Loads

Fig. 3.1

3.2.1 Types of beams

(i) *Statically determinate beams*:

Statically determinate beam is a type of beam which can be analyzed by using the equilibrium equations alone. In such a type of beam, the number of reactions that can be offered by the supports are three. Examples of such beams are cantilever, simply supported beam, simply supported beam with overhang.

(a) Cantilever

A beam fixed at one end and free (unsupported) at the other end is called a cantilever (Fig. 3.2a).

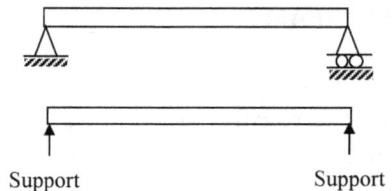

(b) Simply supported beam

A beam having its ends freely resting on supports is known as a simply supported beam (Fig. 3.2b).

(c) Overhanging beam

A beam having its end portion extended beyond the support(s) is known as overhanging beam (Fig. 3.2c).

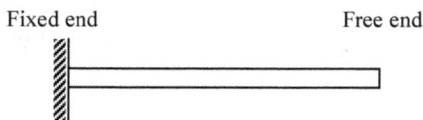

(a) Cantilever   (b) Simply supported beam

Support   Support   Support   Support

Overhanging from one support   Overhanging from both supports

(c) Overhanging beam
Fig. 3.2

(ii) *Statically indeterminate beams*:

Statically indeterminate beam is a beam which cannot be analyzed using equilibrium equations alone. In addition to equilibrium equations, compatibility equations are also required. Examples of such beams are propped cantilever, fixed beam and continuous beam.

(a) Propped cantilever

When a support is provided at some suitable point of a cantilever beam to resist its deflection then it is known as propped cantilever (Fig. 3.3a)

(b) Fixed beam

A beam whose both ends are fixed or built in walls is known as fixed beam (Fig. 3.3b). This beam is also known as built-in beam or encastred beam.

(c) Continuous beam

When more than two supports are provided for a beam, it is said to be continuous beam (Fig. 3.3c).

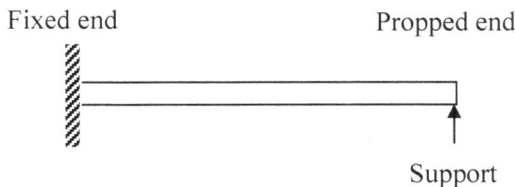

Fixed end   Propped end

Support

(a) Propped cantilever

(b) Fixed beam

Support   Support   Support

(c) Continuous beam
Fig. 3.3

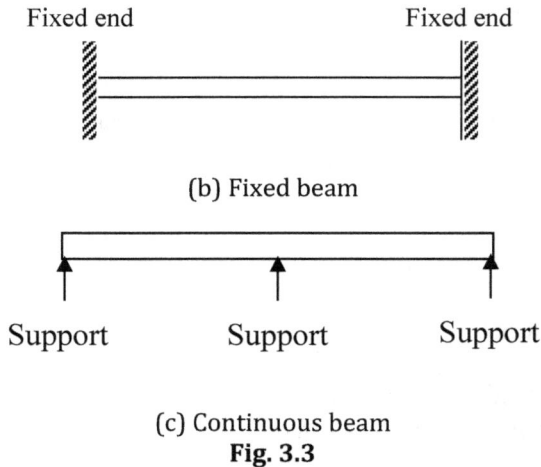

3.3 TYPES OF SUPPORTS AND LOADS

3.3.1 Types of supports for the beam

The following are the types of supports for the beams:

(i) Simple support – (a) Roller support   (b) Hinged support
(ii) Fixed support or built-in support or encastred support

Roller support:

This support offers one reaction which acts in a direction perpendicular to the plane on which the roller rests (Fig. 3.4a).

Hinged support or pinned support:

This support offers reaction whose direction is dependent upon the direction of loads. This can have two independent reaction components resolved along any two mutually perpendicular directions (Fig. 3.4b).
(a) Vertical reaction (V)
(b) Horizontal reaction (H)
Beam can rotate at hinge support and can rotate as well as slide at roller support.

Fixed support

This is also called built-in or encastred support or clamped support. This support offers three reaction components (Fig. 3.4c).
(a) Vertical reaction (V)
(b) Horizontal reaction (H)
(c) Moment reaction (M)
This support does not allow the beam end to rotate or slide.

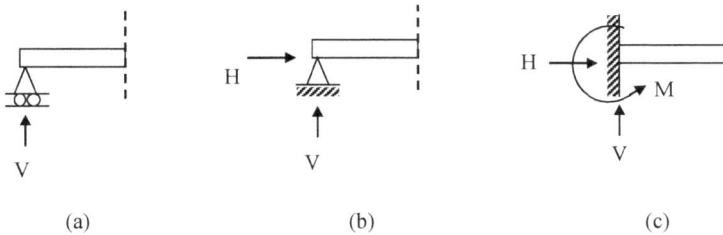

(a)   (b)   (c)

Fig. 3.4

Table 3.1 summarizes the reactions offered by the various supports.

Table 3.1

Type of support	Rotation	Sliding
(i) Roller	Allowed	Allowed
(ii) Hinge	Allowed	Not allowed
(iii) Fixed	Not allowed	Not allowed

3.3.2 Types of Loads

(i) *Point Load (or concentrated load)*

It is assumed to act at a point. Practically this type of load is applied over a very small area.

(ii) *Uniformly distributed load (u.d.l)*

It is distributed or spread uniformly over some length. The intensity of load remains constant.

(iii) *Uniformly varying load (u.v.l)*

It is distributed uniformly over some length of beam but the intensity of load varies linearly. The load varies from some value at a position to some other value at another position on the beam in such a way that the change in load per unit length is same over the loaded portion of the beam.

Figure 3.5 explains the different types of loads applied to a beam.

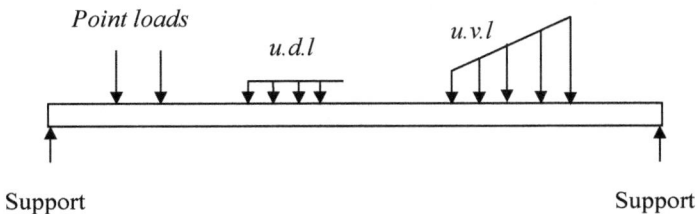

Fig. 3.5

3.4 SHEAR FORCE AND BENDING MOMENT

Due to the applied loads, shear forces and bending moments are developed internally in the beams. They vary from section to section along the length of beam. For proper design of the beams, it becomes essential to determine these developed shear forces and bending moments. Shear force at a section is defined as the algebraic sum of transverse forces acting to any one side (left or right) of the section while bending moment at a section is equal to the algebraic sum of moments of transverse forces acting to any one side (left or right) of the section about that section.

For example, a beam supported at A and B is carrying loads of 10 kN and 30 kN at C and D respectively, as shown (Fig. 3.6a). The reactions R_A and R_B at supports A and B, respectively can be found out from the moment equilibrium equations:

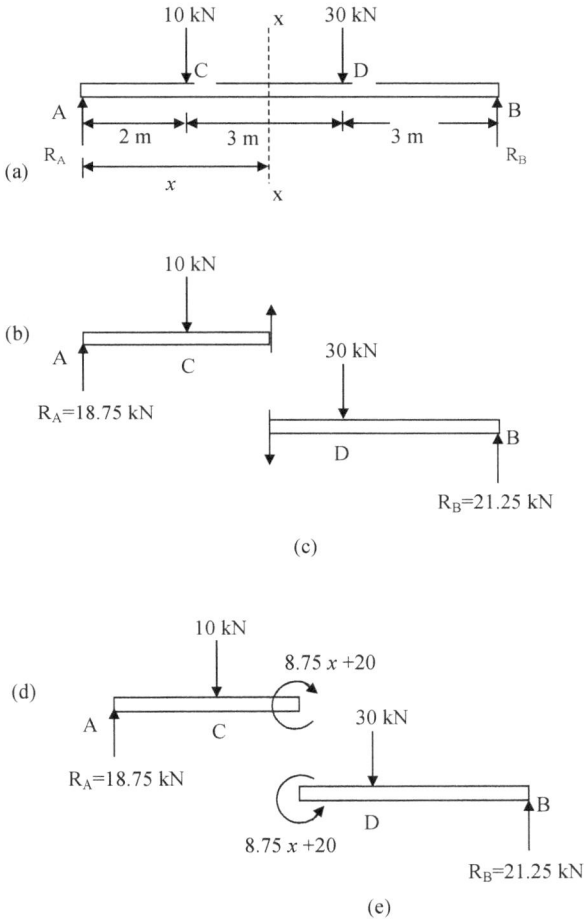

Fig. 3.6

Taking any section x-x at a distance x from A and considering the portion of beam to the left of the section (Fig. 3.6b),
Sum of vertical forces $= 18.75 - 10 = 8.75$ kN (↑)

Considering the portion of beam to the right of the section (Fig. 3.6c)
Sum of vertical forces $= 30 - 21.25 = 8.75$ kN (↓)

Therefore a force of 8.75 kN is considered to act tangentially at the section x-x which tries to shear off the section. This force is known as **shear force** and is obtained as the algebraic sum of the vertical forces acting to any one side (left or right) of the section.

Now considering again the left portion of section x-x (see Fig. 3.6d),
Sum of moments of vertical forces about the section x-x
$= 18.75\,x - 10\,(x - 2) = (8.75\,x + 20)\text{ kNm }(\curvearrowright)$

Considering the right portion of section x-x (see Fig. 3.6e)
Sum of moments of vertical forces about section x-x
$= 21.25\,(8 - x) - 30\,(5 - x) = (8.75\,x + 20)\text{ kNm }(\curvearrowleft)$

Therefore a moment of $(8.75\,x + 20)$ kNm is considered to act at
section x-x which tries to bend the section. This moment is known as
bending moment and is obtained as the algebraic sum of the moments of vertical forces acting to any one side (left or right) of the
section about that section.

Sign convention:

(i) Shear Force:

When the shear force is calculated using the loads acting to the left
side of the section, upward forces are positive and downward forces
are considered negative. When the shear force is calculated using
the loads acting to the right side of the section, downward forces are
positive and upward forces are considered negative (Fig. 3.7a).

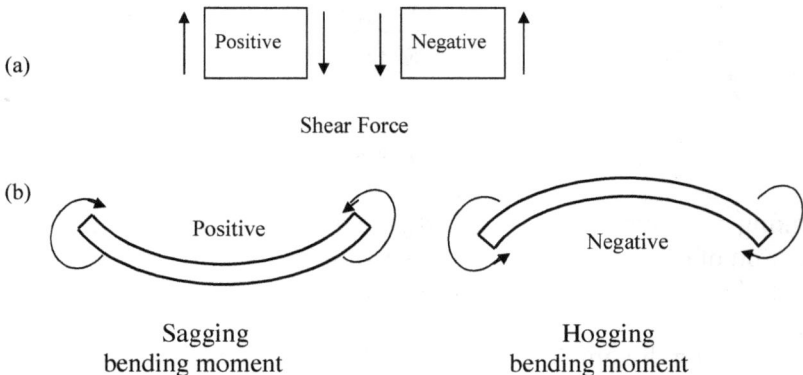

(a)

Shear Force

(b)

Sagging
bending moment

Hogging
bending moment

Fig. 3.7

(ii) Bending moment:

When the bending moment is calculated using loads acting to the
left portion of the section, clockwise moments are positive and anti-

clockwise moments are considered negative. When the bending moment is calculated using the loads acting to the right portion of the section, anticlockwise moments are positive and clockwise moments are considered negative.

In other words, sagging bending moment is positive while hogging bending moment is considered negative (Fig. 3.7b)

Note: Although the choice of a sign convention for shear force and bending moment may be arbitrary, the one presented here is often used in engineering practice.

3.5 SHEAR FORCE DIAGRAM AND BENDING MOMENT DIAGRAM

As discussed in the previous section, for proper design of the beams, it becomes essential to determine the developed shear forces and bending moments so as to find out the maximum value of shear force and bending moment. One way to do this is by plotting the shear force and bending moment diagrams.

A shear force diagram (SFD) is one which shows the variation of shear force along the length of beam while a bending moment diagram (BMD) is one which shows the variation of bending moment along the length of the beam.

As these diagrams provide detailed information regarding the variation of shear forces and bending moments along the beam axis, these are often used by the designers to proportion the size of beam along its length.

3.5.1 SFD and BMD for a cantilever subjected to a point load at free end

A cantilever beam of span length l is carrying a point load (or concentrated load) W at its free end as shown (Fig. 3.8a).

Consider a section at a distance x from the fixed end A (Fig. 3.8a). Shear force at any section between A and B is V=W. Shear force diagram is shown in Fig. 3.8b.

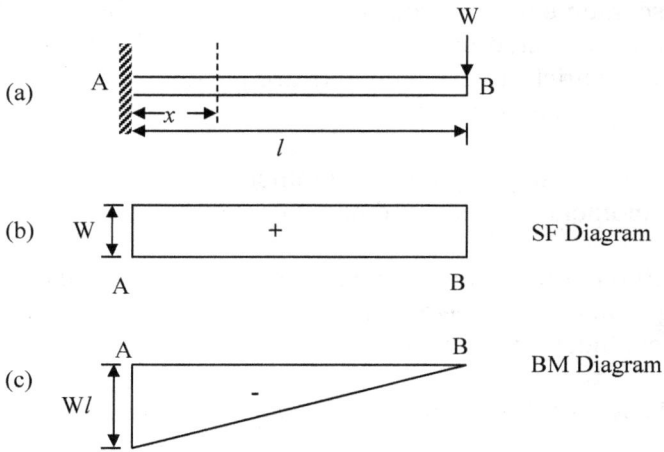

Fig. 3.8

Bending moment at any section at a distance x from A is $M = -W(l - x)$

At B, $x = l$ and $M = 0$

At A, $x = 0$ and $M = -Wl$

Bending moment varies linearly from A to B and the bending moment diagram is shown in Fig. 3.8c.

3.5.2 SFD and BMD for a cantilever subjected to uniformly distributed load

A cantilever beam of span length l is carrying uniformly distributed load of intensity w per unit length over its entire span as shown (Fig. 3.9a).

Shear force at a section at a distance x from fixed end A is $V = w(l - x)$

At A, $x = 0$ and $V = wl$,   At B, $x = l$ and $V = 0$

Shear force varies linearly between A and B and the shear force diagram is shown in Fig. 3.9b.

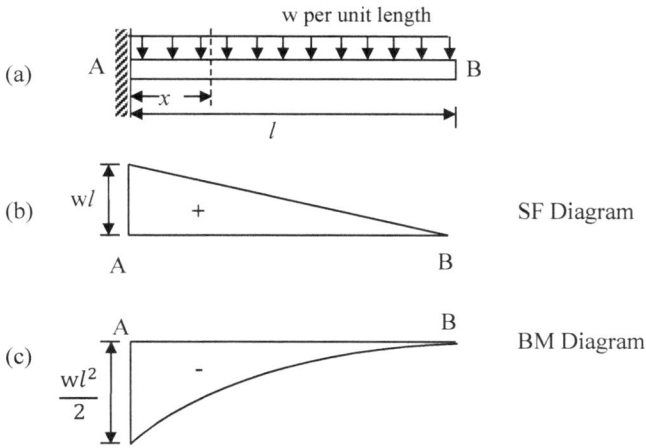

Fig. 3.9

Bending moment at a section at a distance x from A is

$$M = -\frac{w(l-x)(l-x)}{2} = -\frac{w(l-x)^2}{2}$$

At B, $x = l$ and $M = 0$

At A, $x = 0$ and $M = -wl^2/2$

Bending moment varies parabolically from A to B and the bending moment diagram is shown in Fig. 3.9c.

3.5.3 SFD and BMD for a simply supported beam subjected to point load at mid span

A simply supported beam of span length l is carrying a point load W at its mid span as shown (Fig. 3.10a).

Let R_A and R_B be the reactions at supports A and B respectively.

Taking moments about A,

$$\sum M_A = 0$$

$$R_B \left(\frac{l}{2} + \frac{l}{2} \right) - W \frac{l}{2} = 0$$

$$\therefore R_B = \frac{W}{2} \quad \text{and} \quad R_A = \frac{W}{2}$$

Shear force at any section between A and C is W/2 and that between C and B is −W/2 .

The shear force diagram is shown in Fig. 3.10b.

Bending moment at A ,
$$M_A = 0 \quad \text{and Bending moment at B}, \quad M_B = 0$$

Bending moment at C , $\quad M_c = \dfrac{W}{2} \times \dfrac{l}{2} = \dfrac{Wl}{4}$

Bending moment varies linearly having zero value at A and B and maximum value at midpoint C. The bending moment diagram is shown in Fig. 3.10c.

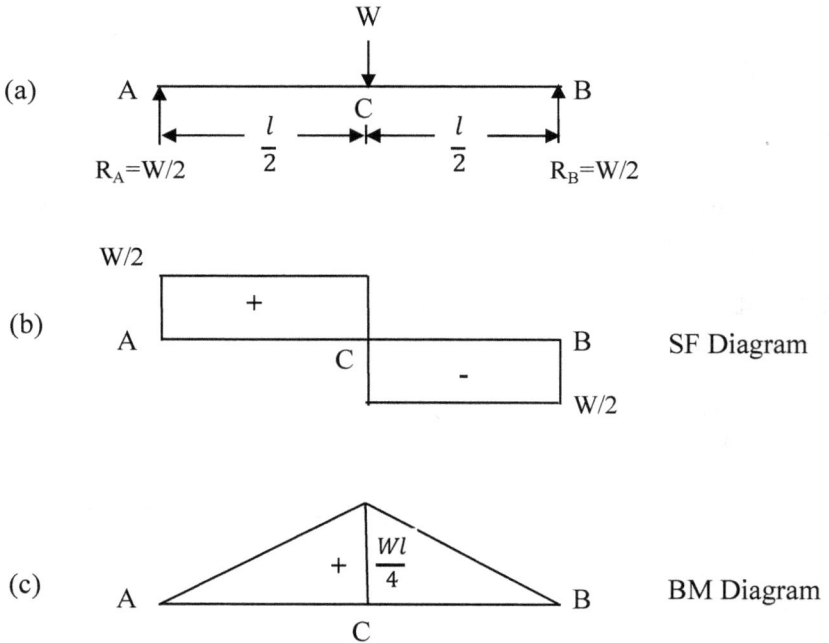

Fig. 3.10

3.5.4 SFD and BMD for a simply supported beam subjected to point load off mid span

A simply supported beam of span length l is carrying a point load W as shown (Fig. 3.11a).

To find out the reactions at supports A and B

$$\Sigma\, M_A = 0$$

$$R_B \times l - W \times a = 0$$

$$\therefore\ R_B = \frac{Wa}{l} \quad \text{and} \quad R_A = \frac{Wb}{l}$$

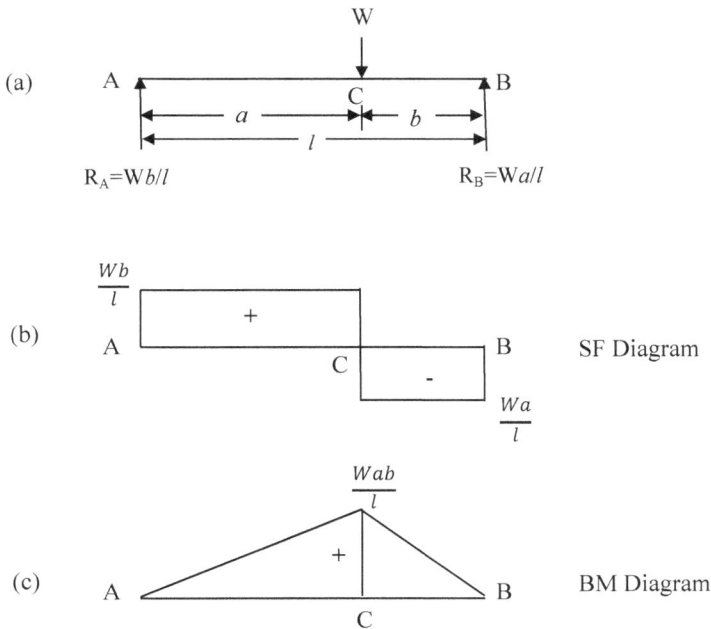

Fig. 3.11

Shear force at any section between A and C is $\dfrac{Wb}{l}$ and that between C and B is $-\dfrac{Wa}{l}$

$$V_A = \frac{Wb}{l} \quad \text{and} \quad V_B = -\frac{Wa}{l}$$

Shear force at a section just to the left of C, $V_C^L = \frac{Wb}{l}$

Shear force at a section just to the right of C,
$$V_C^R = \frac{Wb}{l} - W = -\frac{Wa}{l}$$

The shear force diagram is shown in Fig. 3.11b.

Bending moment at A,
$$M_A = 0 \quad \text{and Bending moment at B,} \quad M_B = 0$$

Bending moment at C, $M_c = \frac{Wb}{l} \times a = \frac{Wab}{l}$

Bending moment varies linearly having zero value at A and B and maximum value at C. The bending moment diagram is shown in Fig. 3.11c.

3.5.5 SFD and BMD for a simply supported beam carrying uniformly distributed load

A simply supported beam of span length l is subjected to uniformly distributed load of intensity w per unit length over its entire span as shown (Fig. 3.12a).

First the reactions R_A and R_B respectively at the supports A and B are required to be found out.

$$\sum M_A = 0$$

$$R_B \times l - (w \times l) \times \frac{l}{2} = 0$$

$$\therefore \ R_B = \frac{wl}{2} \quad \text{and} \quad R_A = \frac{wl}{2}$$

Shear force at a section at a distance x from fixed end A is

$$V = \frac{wl}{2} - wx = w\left(\frac{l}{2} - x\right)$$

At A, $\quad x = 0 \quad$ and $\quad V_A = \dfrac{wl}{2}$,

At B, $\quad x = l$ and $\quad V_B = -\dfrac{wl}{2}$

Shear force varies linearly between A and B and the shear force diagram is shown in Fig. 3.12b.

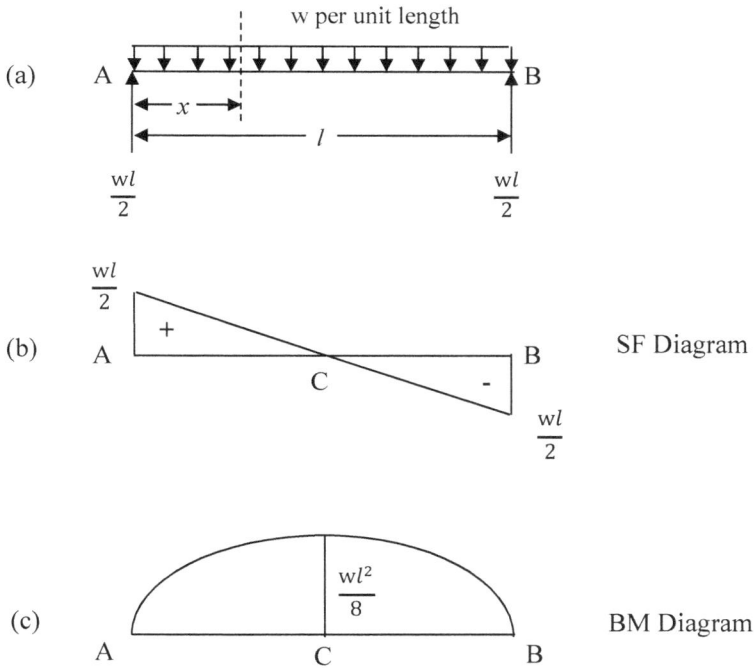

Fig. 3.12

Bending moment at a section at a distance x from A is

$$M = \frac{wl}{2}x - (wx)\frac{x}{2} = \frac{w(lx - x^2)}{2}$$

At A, $x = 0$ and $M_A = 0$

At B, $x = l$ and $M_B = 0$

At C, $x = \dfrac{l}{2}$ and $M_C = \dfrac{wl^2}{8}$

Bending moment varies parabolically and the bending moment diagram is shown in Fig. 3.12c.

Example 3.1 *Draw the shear force and bending moment diagrams for the beam loaded as shown in Fig. 3.13a. Also find the position and magnitude of the maximum bending moment.*

Solution:

Refer Fig. 3.13a

Taking moments about A

$$R_B \times 5 = 4 \times 1.5 \times \left(1.5 + \dfrac{1.5}{2}\right)$$

$\therefore$ $R_B = 2.7$ kN   and $R_A = (4 \times 1.5) - 2.7 = 3.3$ kN

Shear force calculation:

Shear force between A and C is R_A i.e. 3.3 kN (+ve)

Shear force between D and B is -2.7 kN

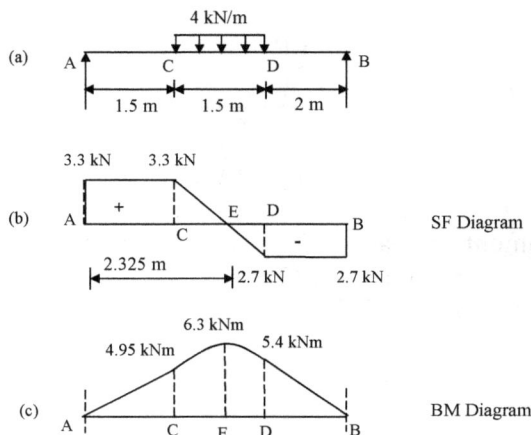

Fig. 3.13

Let shear force is zero at a section at a distance x from A (say point E).

$$V_E = 3.3 - 4(x - 1.5) = 0 \quad \rightarrow \quad x = 2.325 \text{ m}$$

Shear force variation between C and D is linear.

Shear force diagram is shown in Fig. 3.13b.

Bending moment calculation:

Between A and C, bending moment
$M = R_A x = 3.3 x$ (linear variation)

At C, $x = 1.5$, $M_C = 3.3 \times 1.5 = 4.95 \text{ kNm}$

Between C and D, $M = 3.3 x - 2(x - 1.5)^2$

At D, $x = 3$, $M_D = 3.3 \times 3 - 2(3 - 1.5)^2 = 5.4 \text{ kNm}$

Bending moment varies parabolically between C and D.

Maximum bending moment

$$M_E = 3.3 \times 2.325 - 2(2.325 - 1.5)^2 = 6.31 \text{ kNm}$$

Bending moment diagram is shown in Fig. 3.13c.

Example 3.2 *Draw the shear force and bending moment diagrams for the simply supported beam loaded as shown in Fig. 3.14a.*

Solution:

Refer Fig. 3.14a

Taking moments about A

$$R_B \times 8 = (4 \times 1.5) + (10 \times 4) + (7 \times 6)$$

$$\therefore \ R_B = 11 \text{ kN} \quad \text{and} \quad R_A = 10 \text{ kN}$$

Shear force calculation:

Shear force At A, $V_A = 10$ kN

Shear force at section just to left of C, $V_C^L = 10$ kN

Shear force at section just to right of C, $V_C^R = 10 - 4 = 6$ kN

Similarly, $V_D^L = 6$ kN, $\qquad V_D^R = 6 - 10 = -4$ kN

$\qquad\qquad V_E^L = -4$ kN, $\qquad V_E^R = -4 - 7 = -11$ kN,

Shear force At B, $V_B = -11$ kN

Shear force diagram is shown in Fig. 3.14b.

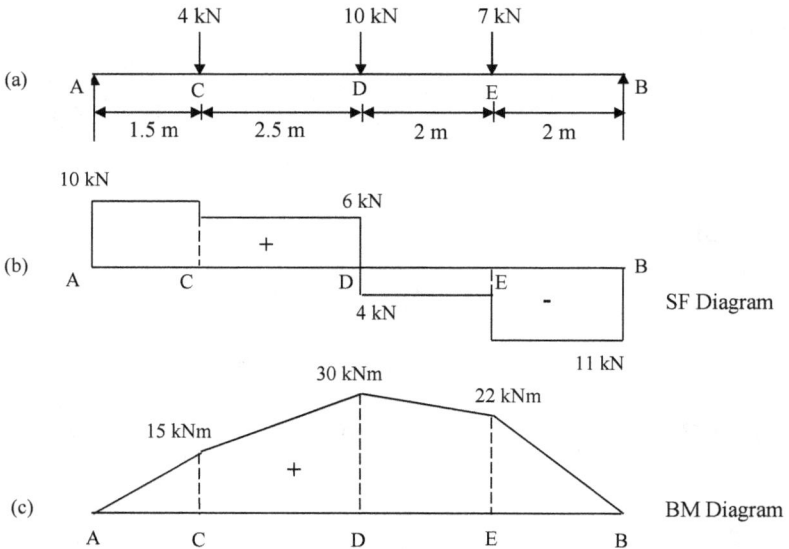

Fig. 3.14

Bending moment calculation:

At A, $M_A = 0$

At C, $M_C = 10 \times 1.5$
$\qquad\qquad = 15$ kNm   (Between A and C linear variation)

At D, $M_D = 10 \times 4 - 4 \times 2.5$
$\qquad = 30$ kNm (Between C and D linear variation)

At E, $M_D = 11 \times 2$
$\qquad = 22$ kNm (Between D and E linear variation)

At B, $M_B = 0$ (Between E and B linear variation)

Bending moment diagram is shown in Fig. 3.13c.

Example 3.3 *Draw the shear force and bending moment diagrams for the beam loaded as shown in Fig. 3.15a.*

Solution:

Shear force calculation:

Shear force between D and E = 2.5 kN

Between C and D, shear force at a section at distance x from free end

$$V_x = 2.5 + 1(x - 0.5) = 2 + x$$

∴ Shear force varies linearly between C and D.

At C, $V_C = 2 + 2.5 = 4.5$ kNm

At B, $V_B^R = 4.5$ kNm, $V_B^L = 4.5 + 3 = 7.5$ kNm

At A, $V_A = 7.5$ kN

Shear force diagram is shown in Fig. 3.15b.

Bending moment calculation:

At E, $M_E = 0$

At D, $M_D = -2.5 \times 0.5 = -1.25$ kNm

Between C and D, bending moment at a section at distance *x* from free end

$$M_x = -2.5\,x - 1(x - 0.5)\frac{(x - 0.5)}{2} = -2.5\,x - \frac{(x - 0.5)^2}{2}$$

∴ Bending moment varies parabolically between C and D.

$$\text{At C, } M_C = -2.5 \times 2.5 - \frac{(2.5 - 0.5)^2}{2} = -8.25 \text{ kNm}$$

$$\text{At B, } M_B = -2.5 \times 4 - 1 \times 2(1.5 + 1) = -15 \text{ kNm}$$

$$\text{At A, } M_A = -2.5 \times 5 - 1 \times 2(2.5 + 1) - 3 \times 1 = -22.5 \text{ kNm}$$

Bending moment varies linearly between A and B, B and C and D and E.

Bending moment diagram is shown in Fig. 3.15c.

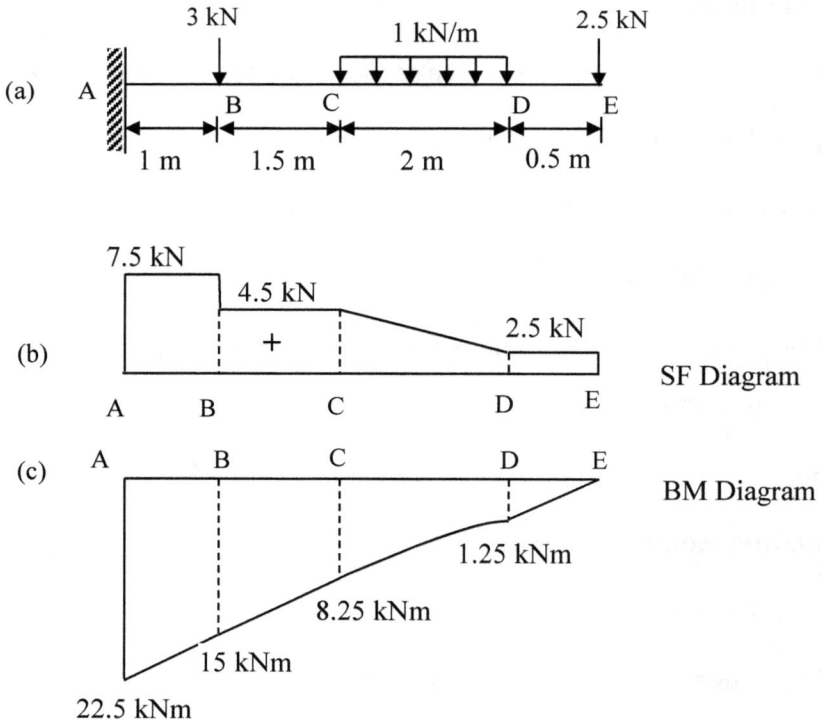

Fig. 3.15

Example 3.4 *Draw the shear force and bending moment diagrams for the simply supported beam loaded as shown in Fig. 3.16a.*

Solution:

Refer Fig. 3.16a

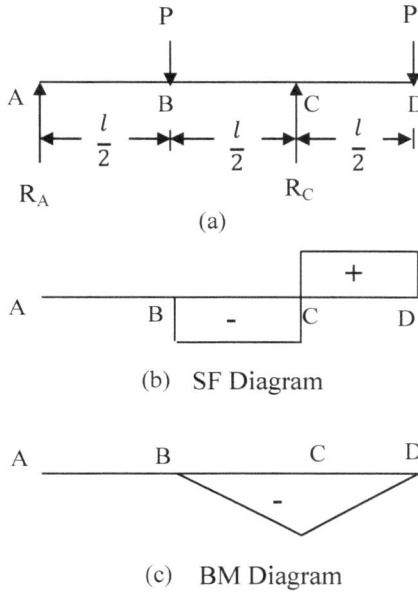

(a)

(b) SF Diagram

(c) BM Diagram

Fig. 3.16

Taking moments about A

$$R_C \times l = P \times \frac{l}{2} + P \times \frac{3l}{2}$$

$$\therefore \ R_C = 2\,P \quad \text{and} \quad R_A = 0$$

Shear force calculation:

$$V_A = 0, \quad V_B^L = 0$$

Shear force is zero between A and B.

$$V_B^R = 0 - P = -P$$

$$V_C^L = -P \ , \ V_C^R = -P + 2P = P \ , \ V_D = P$$

Shear force diagram is shown in Fig. 3.16b.

Bending moment calculation:

$$M_A = 0, \quad M_B = 0, \quad M_C = -P\frac{l}{2}, \quad M_D = 0$$

Bending moment diagram is shown in Fig. 3.16c.

3.6 POINT OF CONTRAFLEXURE

Point of contraflexure (or point of inflexion) is the point at which the bending moment changes sign i.e. from positive to negative or vice versa. This point is of zero bending moment. At this point the beam flexes in opposite directions. This point is also called a virtual hinge. To locate the point of contraflexure, the expression for bending moment at a distance x from one end support is made equal to zero and solving for the value of x, the point(s) of contraflexure can be located.

Example 3.5 *Draw the shear force and bending moment diagrams for the beam loaded as shown in Fig. 3.17a.*

Solution:

Refer Fig. 3.17a

Taking moments about A

$$R_B \times 8 + (2 \times 2 \times 1) = (10 \times 4) + (2 \times 9)$$

$$\therefore \ R_B = 6.75 \text{ kN} \quad \text{and} \quad R_A = 9.25 \text{ kN}$$

Shear force calculation:

$$V_C = 0 \ ,$$

At just left of A, $\quad V_A^L = -4 \text{ kN}$

At just right of A, $V_A^R = -4 +$ 9.25
$= 5.25$ kN

$V_E^L = 5.25$ kN

$V_E^R = 5.25 - 10 = -4.75$ kN

$V_B^L = -4.75$ kN

$V_B^R = -4.75 + 6.75 = 2$ kN

$V_D^L = 2$ kN, $\qquad V_D^R = 0$

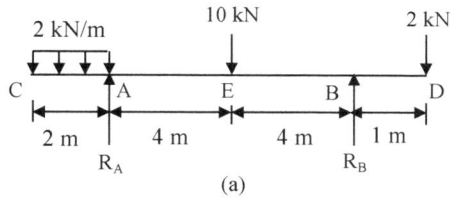

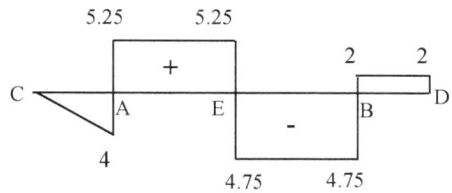

Shear force diagram is shown in Fig. 3.17b.

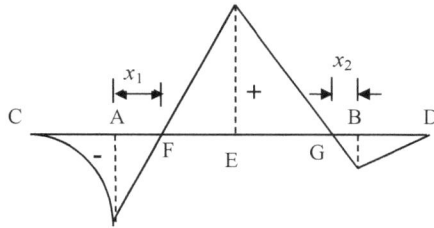

(a)

(b) SF Diagram

(c) BM Diagram

Bending moment calculation:

Fig. 3.17

At C, $M_C = 0$

$M_A = -2 \times 2 \times 1 = -4$ kNm

Between C and A, bending moment at a section at distance x from C is

$M_x = -2\,(x)\,\left(\dfrac{x}{2}\right) = -x^2$

$\therefore$ Bending moment varies parabolically between C and A.

At E, $M_E = -(2 \times 2)(1 + 4) + (9.25 \times 4) = 17$ kNm

At B, $M_B = -2 \times 1 = -2$ kNm

$M_D = 0$

Bending moment varies linearly between A and E, between E and B and between B and D.

Bending moment diagram is shown in Fig. 3.17c.

Point of contraflexure: This is the point where bending moment changes sign.

Here F and G are points of contraflexure (Fig. 3.17c).

Bending moment at F,
$$M_F = 9.25\, x_1 - (2 \times 2)(1 + x_1) = 0 \quad \rightarrow \quad x_1$$
$$= 0.761 \text{ m}$$

Bending moment at G,
$$M_G = -2(1 + x_2) + 6.75\, x_2 = 0 \quad \rightarrow \quad x_2$$
$$= 0.421 \text{ m}$$

Example 3.6 *Draw the shear force and bending moment diagrams for the beam loaded as shown in Fig. 3.18. Indicate all the important features.*

Solution:

Refer Fig. 3.19a

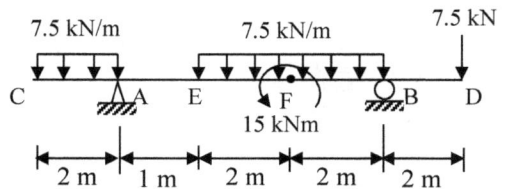

Fig. 3.18

Taking moments about A

$$(R_B \times 5) + 15 + (7.5 \times 2 \times 1)$$
$$= (7.5 \times 4 \times 3) + (7.5 \times 7)$$

$$\therefore R_B = 22.5 \text{ kN} \quad \text{and} \quad R_A = 30 \text{ kN}$$

Shear force calculation:

$$V_C = 0 \,,$$

$$V_A^L = -7.5 \times 2 = -15 \text{ kN}$$

$$V_A^R = -15 + 30 = 15 \text{ kN}$$

$V_E = 15$ kN

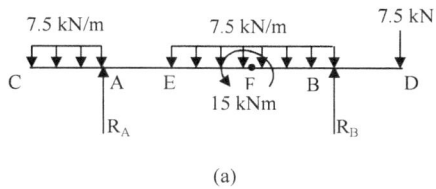

(a)

$V_F = 15 - (7.5 \times 2) = 0$

$V_B^L = 0 - (7.5 \times 2) = -15$ kN

$V_B^R = -15 + 22.5 = 7.5$ kN

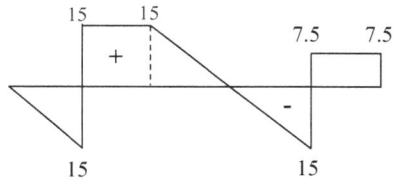

$V_D^L = 7.5$ kN, $\qquad V_D^R = 0$

Shear force diagram is shown in Fig. 3.19b.

(b) SF Diagram

Bending moment calculation:

Between C and A, bending moment at a section at distance x from C is

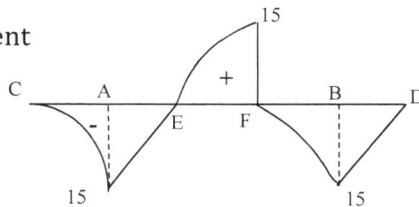

(c) BM Diagram

$$M_x = -7.5\,(x)\,\left(\frac{x}{2}\right)$$

At C, $M_C = 0$

At A, $x = 2$,

Fig. 3.19

$M_A = -7.5 \times 2 \times 1 = -15$ kNm

Bending moment varies parabolically between C and A.

Between A and E, bending moment at a section at distance x from A is

$$M_x = -7.5 \times 2\,(1 + x) + (30 \times x)$$

At E, $x = 1$, $M_E = 0$

Bending moment varies linearly between A and E.

At just left of F, $M_F^L = (30 \times 3) - (7.5 \times 2 \times 4) - (7.5 \times 2 \times 1)$
$\qquad\qquad = 15$ kNm

Bending moment varies parabolically between E and F.

At just right of F, $M_F^R = 15 - 15 = 0$

$M_B = -7.5 \times 2 = -15$ kNm

Bending moment varies parabolically between F and B.

$M_D = 0$

Bending moment diagram is shown in Fig. 3.19c.

Example 3.7 *Draw the shear force and bending moment diagrams for the beam loaded as shown (Fig. 3.20).*

Solution:

Refer Fig. 3.21a

Taking moments about A

$R_B \times l = M$

$\therefore R_B = \dfrac{M}{l}$ and $R_A = -\dfrac{M}{l}$

Shear force from A to B $= -\dfrac{M}{l}$

Shear force diagram is shown in Fig. 3.21b.

Bending moment at A, $M_A = 0$

At just left of C, $M_C^L = -\dfrac{M}{l} \times \dfrac{l}{2} = -\dfrac{M}{2}$

At just right of C, $M_C^R = -\dfrac{M}{2} + M = \dfrac{M}{2}$

Bending moment at B, $M_B = 0$

Bending moment diagram is shown in Fig. 3.21c.

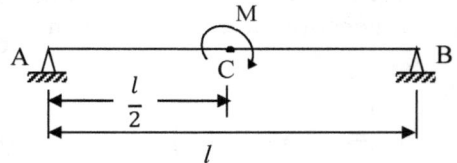

Fig. 3.20

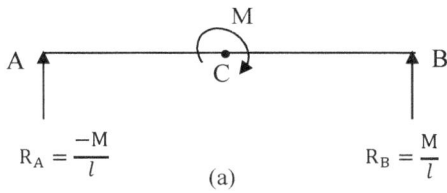

(a)

(b) SF Diagram

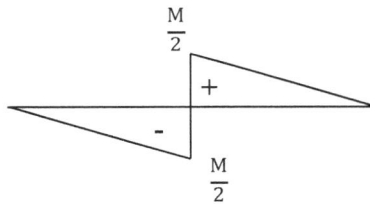

(c) BM Diagram

Fig. 3.21

Example 3.8 *Draw the shear force and bending moment diagrams for the beam loaded as shown (Fig. 3.22).*

Solution:

Refer Fig. 3.23a

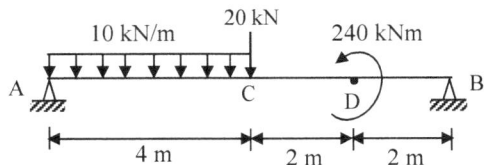

Fig. 3.22

Taking moments about A

$(R_B \times 8) + 240 = (10 \times 4 \times 2) + (20 \times 4)$

$\therefore$ $R_B = -10$ kN and $R_A = 70$ kN

Shear force calculation:

$V_A = 70$ kN

$V_C^L = 70 - 40 = 30$ kN

$V_C^R = 30 - 20 = 10$ kN

$V_D = 10$ kN

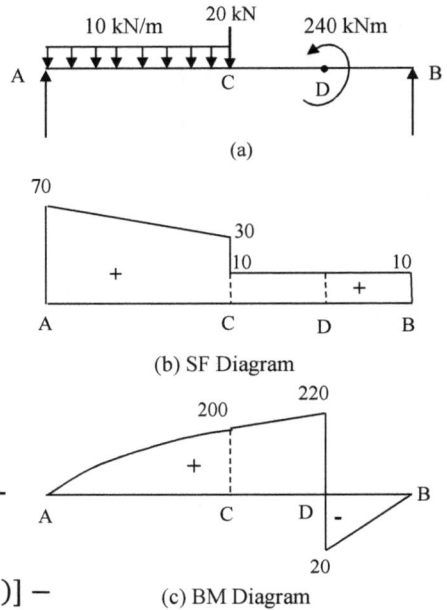

(a)

$V_B^L = 10$ kN, $V_B^R = 0$

Shear force diagram is shown
in Fig. 3.23b.

Bending moment calculation:

(b) SF Diagram

$M_A = 0$

$M_C = (70 \times 4) - (10 \times 4 \times 2)$
$\qquad = 200$ kNm

Bending moment varies parabolically between A and C.

(c) BM Diagram

$M_D^L = (70 \times 6) - [10 \times 4 \times (2 + 2)] -$
$(20 \times 2) = 220$ kNm

Fig. 3.23

$M_D^R = 220 - 240 = -20$ kNm

$M_B = 0$

Bending moment diagram is shown in Fig. 3.23c.

Example 3.9 *A wooden log (Fig. 3.24) of
length 4 m carries two persons each
weighing 750 N at 1 m from ends and
float on water. Draw SF and BM
diagrams for the wooden log.*

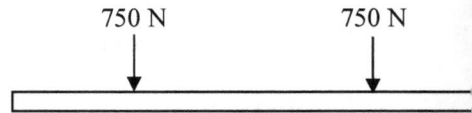

Solution: **Fig. 3.24**

Weight of the wooden log balanced by water reaction is the uniformly distributed load (Fig. 3.25a).

$$\text{Water reaction} = \frac{750 + 750}{4} = 375 \text{ N/m}$$

Shear force calculation:

$V_C = 0$

$V_A^L = 375 \times 1 = 375$ N

$V_A^R = 375 - 750 = -375$ N

$V_B^L = -375 + (375 \times 2)$
$\qquad = 375$ N

$V_B^R = 375 - 750 = -375$ N

$V_D = 0$

Shear force diagram is shown in Fig. 3.25b.

Bending moment calculation:

$M_C = 0$

$M_A = \left(375 \times 1 \times \dfrac{1}{2}\right) = 187.5$ Nm

Bending moment varies parabolically between C and A.

$M_E = \left(375 \times 2 \times \dfrac{2}{2}\right) - 750 \times 1 = 0$

$M_B = \left(375 \times 1 \times \dfrac{1}{2}\right) = 187.5$ Nm

$M_D = 0$

Bending moment varies parabolically throughout the length.

Bending moment diagram is shown in Fig. 3.25c.

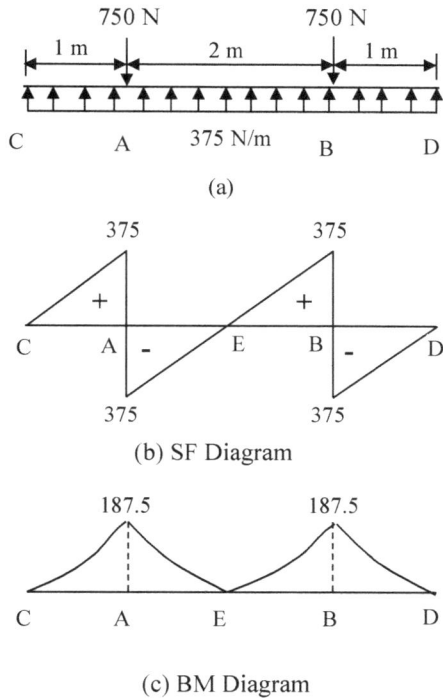

(a)

(b) SF Diagram

(c) BM Diagram

Fig. 3.25

Example 3.10 *An electric pole of length L is to be lifted in horizontal position. For this purpose, two slings separated by distance l are attached to the pole. The pole weight is w kN/m. Two wires are used to connect slings to the hook (Fig. 3.26). Find the distance l between the slings so that maximum bending moment in the pole is minimized. Also draw the SF and BM diagrams.*

Solution:

Refer Fig. 3.27a

$M_C = 0$

$$M_A = -w\left(\frac{L-l}{2}\right)\left(\frac{L-l}{4}\right)$$

$$= \frac{-w(L-l)^2}{8}$$

$$M_E = \left(\frac{wL}{2}\right)\frac{l}{2} - \left(\frac{wL}{2}\right)\frac{L}{4}$$

$$= \frac{wLl}{4} - \frac{wL^2}{8}$$

$M_B = M_A$ (symmetry)

$M_D = 0$

For the bending moment in the beam to be as small as possible, the maximum positive and negative bending moment shall be numerically equal.

$$\frac{-w(L-l)^2}{8} = \frac{wLl}{4} - \frac{wL^2}{8}$$

$$(L-l)^2 = 2Ll - L^2$$

$$\rightarrow l^2 - 4Ll + 2L^2 = 0$$

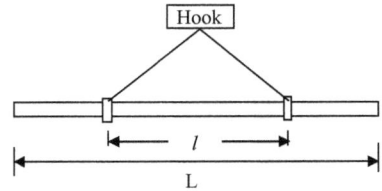

Fig. 3.26

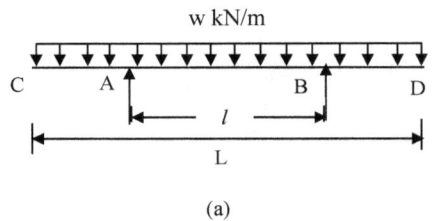

(a)

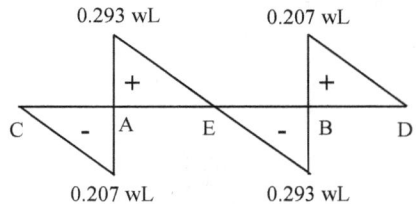

(b) SF Diagram

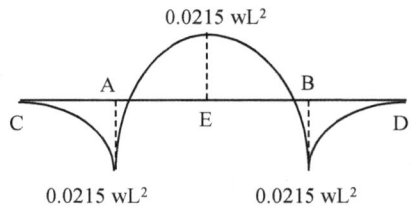

(c) BM Diagram

Fig. 3.27

$$l = \frac{4L \pm \sqrt{(-4L)^2 - 4 \times 1 \times 2L^2}}{2}$$

$$= \frac{4L \pm \sqrt{8L^2}}{2}$$

$$l = (0.586) \, L$$

Shear force calculation:

$$V_C = 0$$

Between C and A, shear force at a section at distance x from C is

$$V = -w \, x$$

$$V_A^L = -w \left(\frac{L - l}{2} \right) = (-0.207) \, wL \qquad (l = 0.586 \, L)$$

Linear variation of shear force observed from C to A.

$$V_A^R = -0.207 \, wL + \frac{wL}{2} = (0.293) \, wL$$

Between A and B, shear force at a section at distance x from A is

$$V = 0.293 \, wL - wx$$

$$V_B^L = 0.293 \, wL - wl = 0.293 \, wL - 0.586 \, wL = (-0.293) \, wL$$

$$V_B^R = -0.293 \, wL + \frac{wL}{2} = (0.207) \, wL$$

Shear force also varies linearly between B and D.

$$V_D = 0$$

Shear force diagram is shown in Fig. 3.27b.

Bending moment calculation:

$$M_C = 0$$

$$M_A = -w(0.207\ L)\ \frac{(0.207\ L)}{2} = (-0.0215)\ wL^2$$

$$M_E = \frac{wL}{2}\frac{(0.586\ L)}{2} - w\left(\frac{L}{2}\right)\left(\frac{L}{4}\right) = (-0.0215)\ wL^2$$

$$M_B = -w(0.207\ L)\ \frac{(0.207\ L)}{2} = (-0.0215)\ wL^2$$

$$M_D = 0$$

Bending moment varies parabolically throughout the length.

Bending moment diagram is shown in Fig. 3.27c.

Example 3.11 *Draw shear force and bending moment diagrams for the beam loaded as shown (Fig. 3.28).*

Solution:

Refer Fig. 3.29a

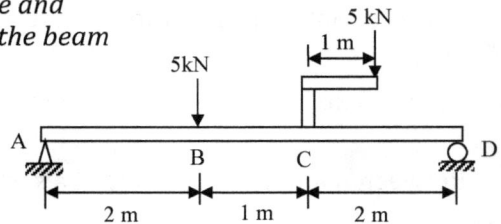

Fig. 3.28

Taking moments about A

$$(5 \times 2) + (5 \times 3) + 5 = (R_D \times 5)$$

$$\therefore\ R_D = 6\ kN \quad and \quad R_A = 4\ kN$$

Shear force calculation:

$$V_A = 4\ kN$$

$$V_B^L = 4\ kN$$

$$V_B^R = 4 - 5 = -1\ kN$$

$V_C^L = -1$ kN,

$V_C^R = -1 - 5 = -6$ kN

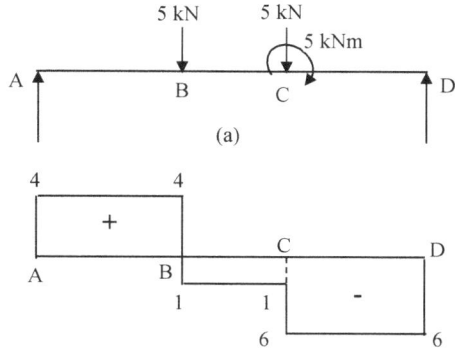

(a)

$V_D = -6$ kN
Shear force diagram is
shown in Fig. 3.29b.

Bending moment calcula-tion:

$M_A = 0$

$M_B = 4 \times 2 = 8$ kNm

$M_C^L = (4 \times 3) - (5 \times 1)$
$\qquad = 7$ kNm

$M_C^R = 7 + 5 = 12$ kNm

$M_D = 0$

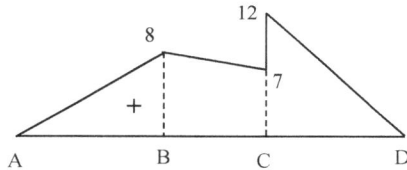

(b) SF Diagram

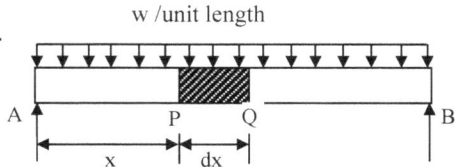

(c) BM Diagram

Fig. 3.29

Bending moment diagram is shown in Fig. 3.29c.

3.7 RELATIONSHIP BETWEEN INTENSITY OF LOADING, SHEAR FORCE AND BENDING MOMENT

Consider an elemental length dx of
a beam simply supported at its
ends as shown (Fig. 3.30).

Let V and (V+dV) be the shear
force at distance x and (x+dx)
respectively from A.

M and (M+dM) be the bending moment
at the respective sections.

For equilibrium of beam element PQ

$V - (V + dV) - w\, dx = 0$

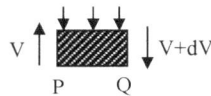

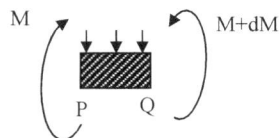

Fig. 3.30

$$\therefore \ w = -\frac{dV}{dx}$$

i.e. Intensity of loading = – rate of change of shear force

Also $V = -\int w \, dx$

Shear force is obtained by integrating the intensity of loading.

w dx = area of load diagram

dV = change of shear force

Area of loading diagram between two sections
        = change of shear force between those two sections

For equilibrium of beam element PQ

$$V \, dx + M - w \, dx \, \frac{dx}{2} - (M + dM) = 0$$

$$\therefore \ V = \frac{dM}{dx}$$

i.e. shear force = Rate of change of bending moment

Also $M = \int V \, dx$

Bending moment is obtained by integrating the shear force.

V dx = area of shear force diagram between P and Q

dM = change of bending moment

Area of shear force diagram between two sections
        = change of bending moment between those two sections

(1) $w = -\dfrac{dV}{dx}$ or $V = -\int w \, dx$

(2) $\quad V = \dfrac{dM}{dx} \quad$ or $\quad M = \displaystyle\int V\, dx$

(3) $\quad w = -\dfrac{d^2 M}{dx^2} \quad$ or $\quad M = -\displaystyle\int\int w\, dx$

Example 3.12 *The bending moment diagram for a beam ABCD supported at B and C is shown (Fig. 3.31a). Draw the shear force diagram and the loading diagram.*

Solution:

Refer Fig. 3.31a

Segment AB

$$M_A = 0, \qquad M_B = -40,$$
$$V = \frac{M_B - M_A}{AB} = \frac{-40 - 0}{2} = -20 \text{ kN}$$

Segment BE

$$M_B = -40, M_E = 110, \quad V = \frac{M_E - M_B}{BE} = \frac{110 + 40}{3} = 50 \text{ kN}$$

Segment EF

$$M_E = 110, M_F = 130, \quad V = \frac{M_F - M_E}{EF} = \frac{130 - 110}{2} = 10 \text{ kN}$$

Segment FC

$$M_F = 130, M_C = -20, \quad V = \frac{M_C - M_F}{FC} = \frac{-20 - 130}{5}$$
$$= -30 \text{ kN}$$

Segment CD

$$M_C = -20, M_D = 0, \quad V = \frac{M_D - M_C}{CD} = \frac{0 + 20}{2} = 10 \text{ kN}$$

In shear force diagram, a vertical line represents a point load, whereas an inclined line represents *udl* and a horizontal line represents no load.

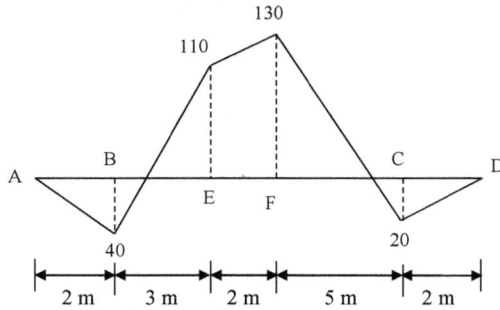

(a)

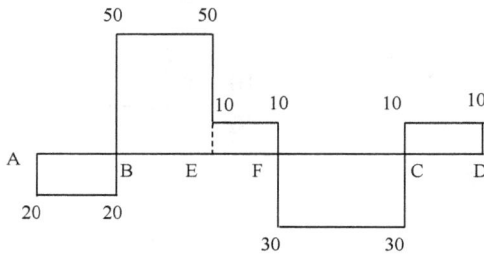

(b) SF Diagram

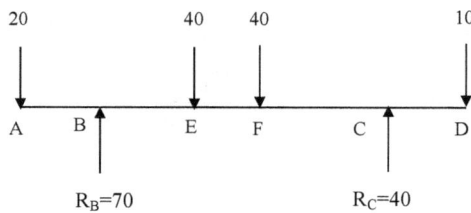

(c) Loading Diagram

Fig. 3.31

The shear force diagram and the loading diagram are shown in Figs. 3.31b and 3.31c, respectively.

Example 3.13 *The bending moment diagram (BMD) for a beam is shown (Fig. 3.32a). Draw the shear force diagram and the loading diagram.*

Solution:

Refer Fig. 3.32a

Segment AC

BMD is straight inclined line.

So shear force is constant.

$$V = \frac{M_C - M_A}{AC} = \frac{3 - (-6)}{3}$$
$$= 3 \text{ kN}$$

Segment CB

$$y = k\,x^2$$

$$x = 3, y = 3 \quad \rightarrow \quad k = 1/3$$

$$y = \frac{1}{3}x^2$$

$$M = 3 - y = 3 - \frac{x^2}{3}$$

$$V = \frac{dM}{dx} = \frac{-2x}{3}$$

$$w = \frac{-dV}{dx} = \frac{2}{3}$$

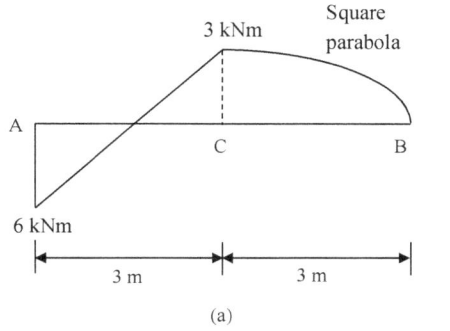

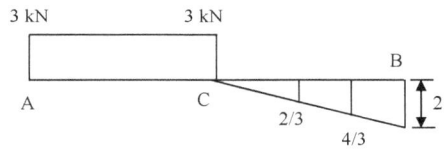

Point	x	$V = \dfrac{-2x}{3}$
C	0	0
	1	-2/3
	2	-4/3
B	3	-2

The shear force diagram is shown in Fig. 3.32b and the loading diagram in Fig. 3.32c.

3 kNm Square parabola

A   C   B

6 kNm

3 m   3 m

(a)

3 kN   3 kN

A   C   B

2/3   4/3   2

(b) SF Diagram

3 kN

6 kNm   (2/3 kNm

A   C   B

$R_A = 3$ kN $R_A = 2$ kN

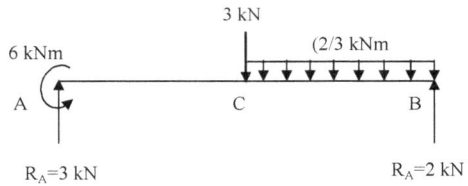

(c) Loading Diagram

Fig. 3.32

Example 3.14 *A simply supported beam of span l is subjected to hydrostatic pressure of intensity zero at one end and w at the other end (Fig. 3.33).*
Draw shear force and bending moment diagrams for the beam.

Solution:

Refer Fig. 3.34a

Fig. 3.33

Taking moments about A

$$R_B \times l = \left(\frac{1}{2} w l\right)\frac{2 l}{3}$$

$$\therefore \ R_B = \frac{wl}{3} \ , \qquad R_A = \frac{wl}{6}$$

Intensity of loading at C at a distance x from A is $\dfrac{wx}{l}$

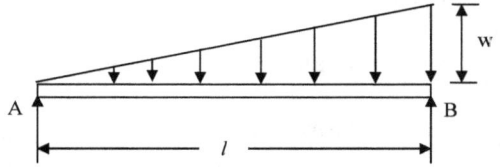

(a)

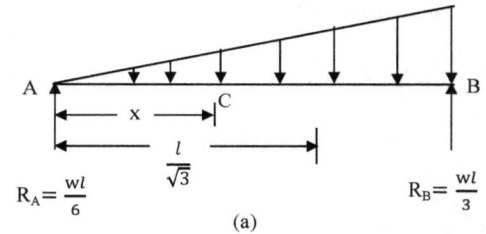

Shear force calculation:

$$V_A = \frac{wl}{6} , V_B = \frac{-wl}{3}$$

$$V_C = \frac{wl}{6} - \frac{1}{2}\left(\frac{wx}{l}\right)x$$

$$= \frac{wl}{6} - \frac{wx^2}{2 l}$$

The SF diagram is shown in Fig. 3.34b.

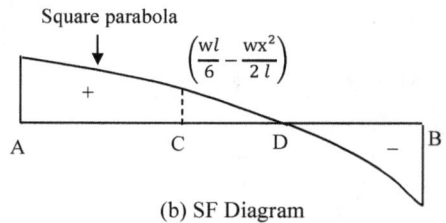

Shear force is zero at $\ x = \dfrac{l}{\sqrt{3}}$

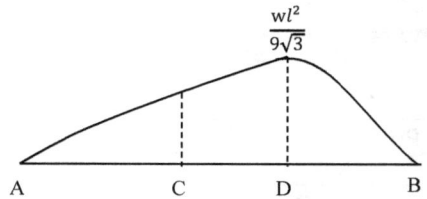

Bending moment calculation:

$$M_A = 0 , M_B = 0$$

Fig. 3.34

$$M_C = \left(\frac{wl}{6}\right)x - \frac{1}{2}\left(\frac{wx}{l}\right)x\left(\frac{x}{3}\right)$$

$$= \frac{wlx}{6} - \frac{wx^3}{6l}$$

$$M_D = \frac{wl}{6}\left(\frac{l}{\sqrt{3}}\right) - \frac{w}{6l}\left(\frac{l}{\sqrt{3}}\right)^3 = \frac{wl^2}{9\sqrt{3}}$$

The BM diagram is shown in Fig. 3.34 c.

Example 3.15 *The diagram given (Fig. 3.35a) is the shear force diagram for a beam which rests on two supports one being on the left end. Deduce directly from this diagram (i) bending moment at 2 m interval (ii) the loading on the beam. Also draw the loading and the bending moment diagrams.*

Solution:

Refer Fig. 3.35a

Segment AB:

$$V = 10 - \frac{10 - 5.5}{6}x$$

$$= 10 - 0.75\,x$$

$$w = -\frac{dV}{dx} = 0.75\ \text{kN/m}$$

$$M = \int V\,dx = 10\,x - 0.75\,\frac{x^2}{2} + C_1$$

At left support $x = 0, M = 0$

$$\therefore\ C_1 = 0$$

$$M = 10\,x - \frac{0.75}{2}x^2$$

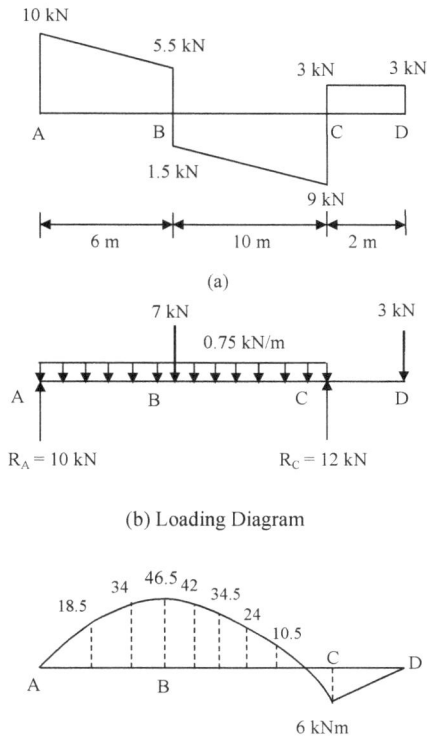

(a)

(b) Loading Diagram

(c) BM Diagram

Fig. 3.35

Point	x	M
A	0	0
	2	18.5
	4	34
B	6	46.5

Segment BC:

$$V = -\left(1.5 + \frac{9 - 1.5}{10}\, x\right) = -1.5 - 0.75\, x$$

$$w = -\frac{dV}{dx} = 0.75 \text{ kN/m}$$

$$M = \int V\, dx = -1.5\, x - \frac{0.75}{2}\, x^2 + C_2$$

At B,   x = 0,   M = 46.5

$\therefore\ C_2 = 46.5$

$$M = -1.5\, x - \frac{0.75}{2}\, x^2 + 46.5$$

Point	x	M
B	0	46.5
	2	42
	4	34.5
	6	24
	8	10.5
C	10	-6

Segment CD:

$V = 3\ kN$

$$w = -\frac{dV}{dx} = 0 \quad \therefore \quad \text{No udl between C and D}$$

$$M = \int V\ dx = \int 3\ dx = 3\ x + C_3$$

At C, $x = 0,$ $M = -6$ $\therefore\ C_3 = -6$

$M = 3\ x - 6$

Point	x	M
C	0	-6
D	2	0

The Loading diagram and BM diagram are shown (Figs. 3.35 b and 3.35 c).

Example 3.16 *The shear force diagram for an overhung beam supported at A and C is shown (Fig. 3.36a). Draw the loading diagram and the bending moment diagram.*

Solution:

Refer Fig. 3.36a

A and C are simple supports
(i) At A there is a reaction

$R_A = 7\ kN$

(ii) There is udl from A to B

$$w = \frac{V_A - V_B}{AB}$$

$$= \frac{7-1}{3} = 2 \text{ kN/m}$$

(iii) At B there is point load

$$V_B^R = V_B^L - W$$

$$-3 = 1 - W$$

$\rightarrow$ $W = 4$ kN

(iv) There is udl from B to C

$$w = \frac{V_B - V_C}{BC} = \frac{-3 - (-9)}{3}$$

$$= 2 \text{ kN/m}$$

(v) At C there is a reaction

$$R_C = 12 \text{ kN}$$

(vi) There is no udl from C to D

(vii) There is a point load of 3 kN at D

(a)

(b) Loading Diagram

(c) BM Diagram

Fig. 3.36

Check :

$$\sum F_y = 0$$

upward forces: $R_A + R_C = 7 + 12 = 19$ kN

downward forces: $(2 \times 6) + 4 + 3 = 19$ kN

$$\sum M_A = 0$$

Clockwise moments: $\left(2 \times 6 \times \frac{6}{2}\right) + (4 \times 3) + (3 \times 8) = 72$ kNm

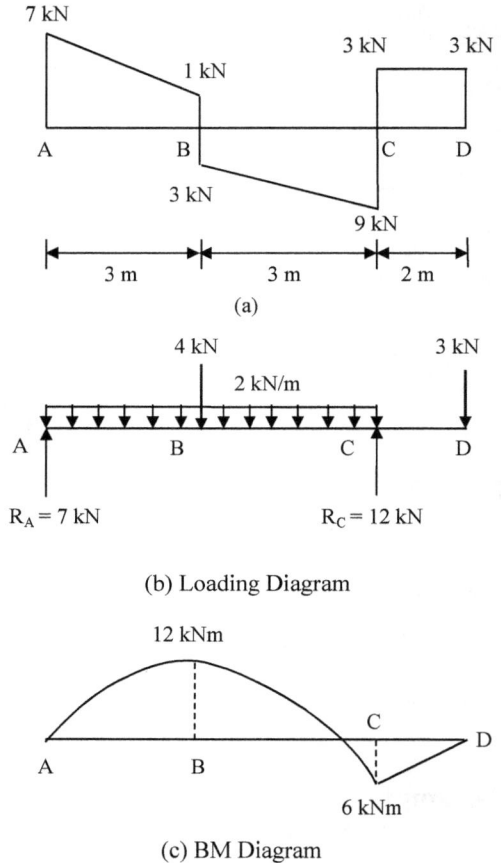

Anticlockwise moments: $12 \times 6 = 72$ kNm

The loading diagram is shown in Fig. 3.36b.

Bending moment calculation:

$M_A = 0$

$M_B = (7 \times 3) - \left(2 \times 3 \times \dfrac{3}{2}\right) = 12$ kNm

$M_C = -3 \times 2 = -6$ kNm

$M_D = 0$

Parabolic variation of bending moment between A to C

The BM diagram is shown in Fig. 3.36c.

Example 3.17 *A beam ABCD (AB = BC = CD = 3m) is simply supported at B and C. It carries a point load of 1 kN at free end A , a uniformly distributed load of 2 kN/m between B and C and anticlockwise moment of 24 kNm in the plane of beam at D. Draw the shear force and bending moment diagrams.*

Solution:

Refer Fig. 3.37a

Taking moments about B

$R_C \times 3 + 24 + (6 \times 3) = \left(2 \times 3 \times \dfrac{3}{2}\right)$

$\therefore R_C = -11$ kN and $R_B = 23$ kN

Shear force calculation:

$V_A = -6$ kN

$V_B^L = -6$ kN

$V_B^R = -6 + 23 = 17$ kN

Between B to C, shear force varies linearly.

$V_C^L = 17 - (2 \times 3)$

$\quad = 11$ kN

$V_C^R = 11 - 11 = 0$

$V_D = 0$

Shear force diagram is shown in Fig. 3.37b.

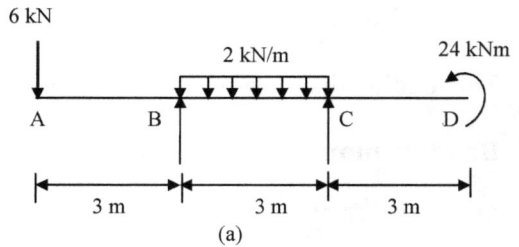

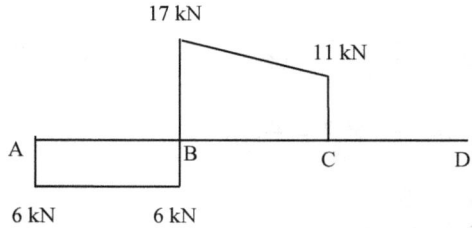

Bending moment calculation:

$M_A = 0$

$M_B = -6 \times 3 = -18$ kNm

Bending moment varies parabolically between B and C.

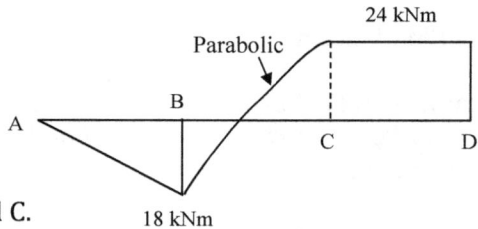

(a)

(b) SF Diagram

(c) BM Diagram

Fig. 3.37

$M_C = -(6 \times 6) + (23 \times 3) - \left(2 \times 3 \times \dfrac{3}{2}\right) = 24$ kNm

$M_D^L = -(6 \times 9) + (23 \times 6) - 2 \times 3 \times \left(\dfrac{3}{2} + 3\right) - (11 \times 3) = 24$ kNm

$M_D^R = 24 - 24 = 0$

Bending moment diagram is shown in Fig. 3.37c.

Example 3.18 *Calculate the reactions at A and D for the beam loaded as shown (Fig. 3.38). Also draw the shear force and bending moment diagrams.*

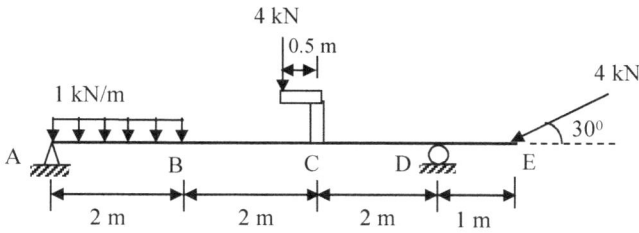

Fig. 3.38

Solution:

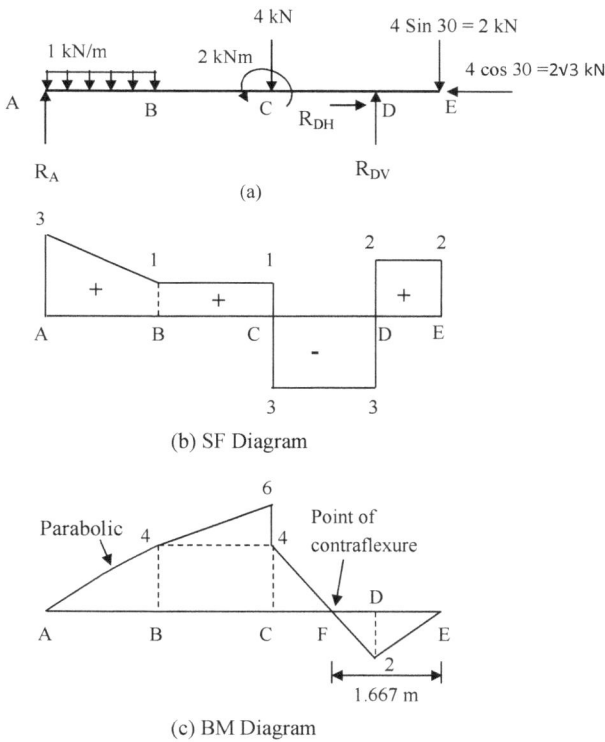

(a)

(b) SF Diagram

(c) BM Diagram

Fig. 3.39

The loaded beam is shown in Fig. 3.39a.

Let R_{DV}, R_{DH} = Vertical and horizontal components of reaction at D

Taking moments about A

$(R_{DV} \times 6) + 2 = (1 \times 2 \times 1) + (4 \times 4) + (2 \times 7)$

$\therefore R_{DV} = 5 \text{ kN}$

$R_A = 3 \text{ kN}$ and $R_{DH} = 2\sqrt{3} \text{ kN}$

Resultant reaction at D, $R_D = \sqrt{R_{DV}^2 + R_{DH}^2} = 6.08 \text{ kN}$

Shear force calculation:

$V_A = 3 \text{ kN}, \ V_B = 3 - (1 \times 2) = 1 \text{ kN}$

Between A to B, shear force varies linearly.

$V_C^L = 1 \text{ kN} \ , \quad V_C^R = 1 - 4 = -3 \text{ kN}$

$V_D^L = -3 \text{ kN} \ , \quad V_D^R = -3 + 5 = 2 \text{ kN}$

$V_E^L = 2 \text{ kN} \ , \quad V_E^R = 2 - 2 = 0$

Shear force diagram is shown in Fig. 3.39b.

Bending moment calculation:

$M_A = 0$

$M_B = (3 \times 2) - (1 \times 2 \times 1) = 4 \text{ kNm}$

Bending moment varies parabolically between A and B.

$M_C^L = (3 \times 4) + (1 \times 2 \times 3) = 6 \text{ kNm}$

$M_C^R = 6 - 2 = 4 \text{ kNm}$

$M_D = -(2 \times 1) = -2 \text{ kNm}$

$M_E = 0$

Bending moment diagram is shown in Fig. 3.39c.

Point of contraflexure (where BM changes sign):

$$M_F = 5(x - 1) - 2x = 0 \qquad \rightarrow \qquad x = 1.667 \text{ m}$$

Therefore, Point of contraflexure is at a distance of 1.667 m from right end E.

Example 3.19 *A beam is loaded and supported as shown (Fig. 3.40a). Find the magnitude of the clockwise moment M to be applied at C so that the reactions at B will be 30 kN upward. Then draw the shear force and bending moment diagrams.*

Solution:

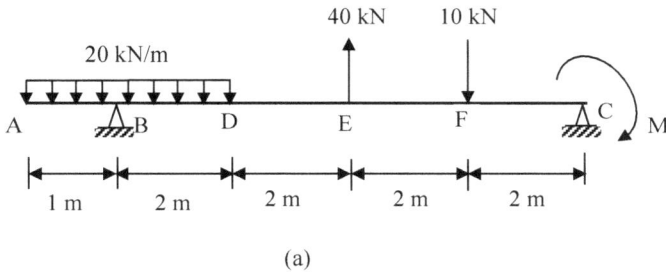

(a)

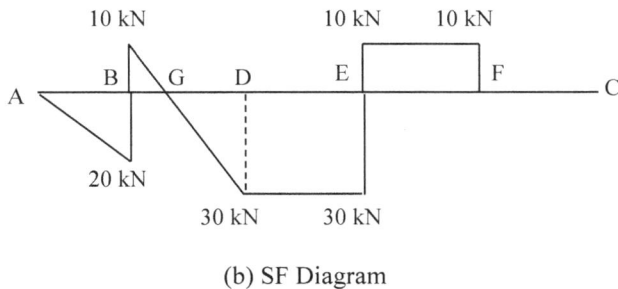

(b) SF Diagram

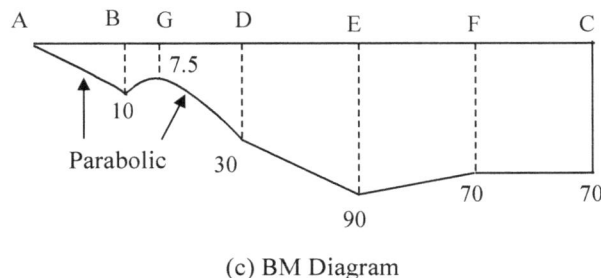

(c) BM Diagram

Fig. 3.40

Refer Fig. 3.40a

Given $R_B = 30$ kN

Taking moments about C

$$(R_B \times 8) + (40 \times 4) + M = (10 \times 2) + (20 \times 3)\left(6 + \frac{3}{2}\right)$$

$\therefore$ $M = 70$ kNm

Again $R_B + R_C + 40 = (20 \times 3) + 10$ $\therefore$ $R_C = 0$

Shear force calculation:

$V_A = 0$, $V_B^L = -(20 \times 1) = -20$ kN

$V_B^R = -20 + 30 = 10$ kN

Between A to B, shear force varies linearly.

$V_D = 10 - (20 \times 2) = -30$ kN

$V_E^L = -30$ kN   , $V_E^R = -30 + 40 = 10$ kN

$V_F^L = 10$ kN , $V_F^R = 10 - 10 = 0$

Shear force diagram is shown in Fig. 3.40b.

Bending moment calculation:

$M_A = 0$

$$M_B = \left(-20 \times 1 \times \frac{1}{2}\right) = -10 \text{ kNm}$$

$$M_D = \left(-20 \times 3 \times \frac{3}{2}\right) + (30 \times 2) = -30 \text{ kNm}$$

$$M_E = -20 \times 3 \times \left(\frac{3}{2} + 2\right) + (30 \times 4) = -90 \text{ kNm}$$

$$M_F = -20 \times 3 \times \left(\frac{3}{2} + 4\right) + (30 \times 6) + (40 \times 2) = -70 \text{ kNm}$$

$$M_C = -70 \text{ kNm}$$

Shear force changes sign at G and BG = 0.5

$$M_G = \left(-20 \times 1.5 \times \frac{1.5}{2}\right) + (30 \times 0.5) = -7.5 \text{ kNm}$$

Bending moment diagram is shown in Fig. 3.40c.

Example 3.20 *Two beams AB and BC are connected by a frictionless joint to form beam ABC (Fig. 3.41). Draw shear force and bending moment diagrams for the beam.*

Solution:

Free body diagrams for portions AB and BC of beam are shown (Fig. 3.42a).

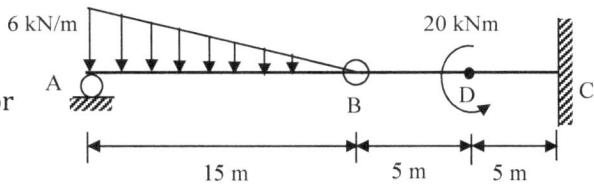

Fig. 3.41

Consider beam AB:

Taking moments about B

$$R_A \times 15 = \left(\frac{1}{2} \times 15 \times 6\right) \times \frac{2}{3} \times 15$$

$$\therefore \ R_A = 30 \text{ kN}$$

$$F = \left(\frac{1}{2} \times 15 \times 6\right) - 30 = 15 \text{ kN}$$

Shear force calculation:

$$V_A = 30 \text{ kN}$$

$$V_B = -15 \text{ kN}$$

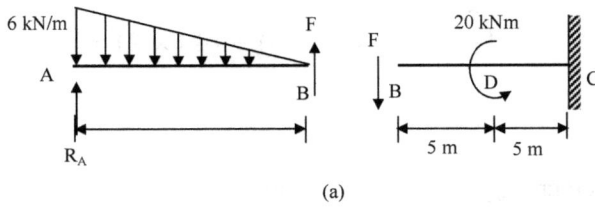

(a)

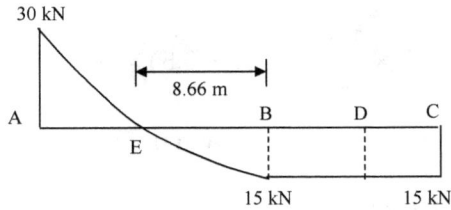

(b) SF Diagram

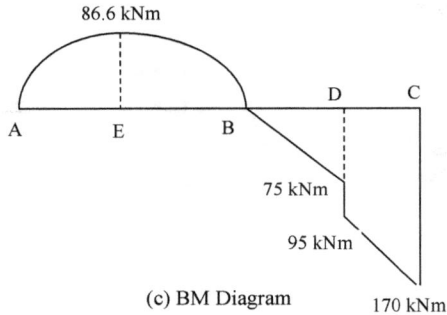

(c) BM Diagram

Fig. 3.42

Shear force at distance x from B is

$$V_x = -15 + \left(\frac{1}{2} \times x \times 0.4\,x\right) = -15 + 0.2\,x^2$$

$$V_{x=5} = -10 \text{ kN}$$

$$V_{x=10} = 5 \text{ kN}$$

It may be noted that $V_{x=15} = 30$ kN corresponds to V_A

Shear force is zero when x = 8.66 m (at E)

Considering BC portion of the beam

$V_D = -15 \text{ kN}$

$V_C = -15 \text{ kN}$

Shear force diagram is shown in Fig. 3.42b.

Bending moment calculation:

Bending moment at a section at distance x from B is

$$M_x = 15\,x - \left(\frac{1}{2} \times x \times x \times 0.4\,x\right)\frac{x}{3} = 15\,x - \frac{x^3}{15}$$

At B , x = 0 $\quad \therefore M_B = 0$

$$M_{x=5} = 15 \times 5 - \frac{5^3}{15} = 66.67 \text{ kNm}$$

$$M_{x=8.66} = 15 \times 8.66 - \frac{8.66^3}{15} = 86.6 \text{ kNm}$$

$$M_{x=10} = 15 \times 10 - \frac{10^3}{15} = 83.3 \text{ kNm}$$

$$M_{x=15} = M_A = 15 \times 15 - \frac{15^3}{15} = 0$$

Considering BC portion of the beam

$$M_D^L = -(15 \times 5) = -75 \text{ kNm}$$

$$M_D^R = -75 - 20 = -95 \text{ kNm}$$

$$M_C = -(15 \times 10) - 20 = -170 \text{ kNm}$$

Bending moment diagram is shown in Fig. 3.42c.

$$\boxed{\textbf{HIGHLIGHTS}}$$

Definitions

1. **Shear force:** Shear force at a section is equal to the algebraic sum of the transverse forces acting to any one side of the section (left or right of the section).

2. **Bending Moment:** Bending Moment at a section is equal to the algebraic sum of moments of transverse forces acting to any one side of the section (left or right of the section) about that section.

3. **Shear Force Diagram:** It shows the variation of shear force along the length of the beam.

4. **Bending Moment Diagram:** It shows the variation of bending moment along the length of the beam.

5. **Point of contraflexure:** It is the point on the bending moment diagram at which bending moment changes its sign (from positive to negative and vice versa).

Concepts and Formulae

1. For cantilever subjected to point load (W) at free end

Maximum shear force $V_{max} = W$ at any section

Maximum bending moment $M_{max} = -Wl$ at fixed end

For cantilever subjected to u.d.l. of w/unit length over entire span

$V_{max} = wl$ at fixed end

$M_{max} = -wl^2/2$ at fixed end

2. For simply supported beam subjected to point load W at mid span

$V_{max} = W/2$

$M_{max} = Wl/4$ at mid span

For simply supported beam subjected to point load W off mid span

$V_{max} = Wa/l$ or Wb/l whichever is higher

$M_{max} = Wab/l$ under the load

For simply supported beam subjected to udl over entire span

$V_{max} = wl/2$ at the supports

$M_{max} = wl^2/8$ at midspan

3. For point of contraflexure, the expression for bending moment at a distance x from one end support is made zero and value(s) of x is (are) found out and point(s) of contraflexure is (are) located.

4. Relationship between intensity of loading, shear force and bending moment:

$$w = -\frac{dV}{dx} \quad , \quad V = \frac{dM}{dx} \quad , \quad w = -\frac{d^2M}{dx^2}$$

5. The bending moment is maximum or minimum wherever shear force is zero. Such points are salient points and are to be indicated in BM diagram. The point of contraflexure is also an important point and if exists should be indicated in BM diagram.

6. To find loading from SF Diagram,
A vertical line of SF diagram represents a point load, an inclined line represents uniformly distributed load (u.d.l.) and a horizontal line represents no load in a portion.

SHORT TYPE QUESTIONS

1. Point of contraflexure is where
(a) bending moment is maximum
(b) shear force is maximum
(c) shear force is zero
(d) bending moment is zero

[Ans: (d)]

2. The maximum bending moment in a simply supported beam of length *l* subjected to point load W at mid-point is ____
(a) W*l*/2   (b) W*l*   (c) W*l*/8   (d) W*l*/4

[Ans: (d)]

3. Bending moment at end supports in case of simply supported beam is always
(a) zero   (b) less than unity (c) more than unity   (d) maximum

[Ans: (a)]

4. The moment diagram for a cantilever beam subjected to bending moment at the end of the beam will be ____
(a) rectangle (b) triangle   (c) parabola   (d) ellipse

[Ans : (a)]

5. The value of the slope of shear force diagram is equal to the value of ____ and slope of ____ diagram is equal to the value of shear force.

[Ans: intensity of loading, bending moment]

6. A cantilever beam is loaded uniformly throughout its length. The shape of the shear force diagram will be
(a) A rectangle (b) A right angle triangle (c) An isosceles triangle

[Ans: (b)]

7. If the shear force acting at every section of a beam is of the same magnitude and of the same direction then it represents a
(a) simply supported beam with concentrated loaded at the centre
(b) overhanging beam having equal overhang at both supports and carrying equal concentrated loads acting in the same direction at the free ends
(c) cantilever subjected to concentrated load at free end
(d) simply supported beam having concentrated loads of equal magnitude and in the same direction acting at equal distance from supports.

[Ans : (c)]

8. Match list I with list II and select the correct answer from the codes given below

List – I
(Condition of beam)

List – II
(bending moment dia-gram)

A – Cantilever subjected to mo-   1. Triangle
ment at the free end
B – Cantilever carrying *udl* over   2. Cubic parabola
whole length
C – Cantilever carrying *uvl* from   3. Parabola
zero at free end to maximum at
support
D – A beam having load at the   4. Rectangle
centre and supported at the ends

Codes :

	A	B	C	D
(a)	4	1	2	3
(b)	4	3	2	1
(c)	3	4	2	1
(d)	3	4	1	2

[Ans : (b)]

9. If a beam is subjected to a constant bending moment along its length then the shear force will
(a) also have a constant value everywhere along its length
(b) be zero at all section along the beam
(c) be maximum at the centre and zero at ends
(d) be zero at the centre and maximum at the ends

[Ans : (b)]

10. Area of the shear force diagram between two points is equal to the ___

[Ans : Change of BM between the two points]

11. For the shear force to be uniform throughout the span of a simply supported beam, it should carry
(a) u.d.l. over its entire span   (b) a point load at mid span
(c) a couple anywhere in the span   (d) two concentrated loads-
equally spaced

[Ans : (c)]

EXERCISE PROBLEMS

1. A simply supported beam of 10 m long carries a uniformly distrib-

tuted load of intensity 2 kN/m over entire length and point loads 1 kN and 2 kN at distances 2 m and 5 m from the left support. Draw the shear force and bending moment diagrams for the loaded beam. Also find the value of maximum bending moment.

[Ans : M_{max} = 31 kNm]

2. A cantilever 1.5 m long is loaded with a uniformly distributed load of 2 kN/m run over a length of 1.25 m from the free end. It also carries a point load of 3 kN at a distance of 0.25 m from the free end. Draw the shear force and bending moment diagrams for the cantilever. What is the value of maximum bending moment.

[Ans : M_{max} = -5.94 kNm]

3. A simply supported beam of span 4 m carries a uniformly distributed load of 2 kN/m over a length of 1.5 m and a point load of 2 kN at a distance of 3 m from the left support. The beam is also subjected to a clockwise couple of 3 kNm at a distance 2 m from left support. Draw the SF and BM diagrams.

[Ans : R_a = 2.188 kN, R_b = 2.812 kN, M_{max} = 3.615 kNm]

4. A beam of length 6 m is simply supported at its ends. It is loaded with gradually varying load of 750 N/m from left hand support to 1500 N/m to the right hand support. Construct the SF and BM diagrams and also find the amount and position of maximum bending moment over the beam.

[Ans : M_{max} = 5077.5 Nm at 3.16 m from the left hand support]

5. A horizontal beam AB of length 8 m is simply supported at A and B. It carries udl of 3 kN/m over the entire span and a clockwise moment of 12 kNm applied in the plane of the beam at a point C, 5 m from A. Draw the shear force and bending moment diagrams and determine the position and magnitude of maximum bending moment.

[Ans : 27 kNm at C]

6. A horizontal beam AB of 8 m length is supported at A. The beam supports a u.d.l. of 1.5 KN/m over its entire length and also concentrated loads of 3 KN and 1.5 KN at D and B respectively, D being 2 m from A. Draw SF and BM diagrams for the beam. What is the value of maximum bending moment and where does it occur?

[Ans : 8 kNm at 2 m from A]

7. Draw the SF and BM diagrams for the beam loaded as shown (Fig. 3. 43).

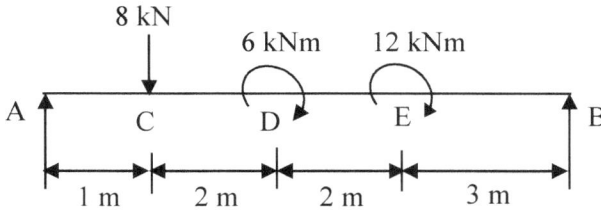

Fig. 3.43

[Ans : Reactions are 4.75 kN and 3.25 kN , V_{max} = 4.75 kN , M_{max} = 9.75 kNm]

8. A simply supported beam of length 5 m, carries a uniformly distributed load of 1000 N/m extending from the left end to a point 2 m away. There is also a clockwise couple of 1500 Nm applied at the center of the beam. Draw the SF and BM diagrams for the beam and find the maximum bending moment.

[Ans : 845 Nm at a distance of 1.3 m from the left end]

9. Given figure (Fig. 3.44) shows the SF diagram for a beam which rests on two supports, one being at the left hand end. Deduce from this diagram the loading on the beam. Also draw the bending moment diagram showing the important points.

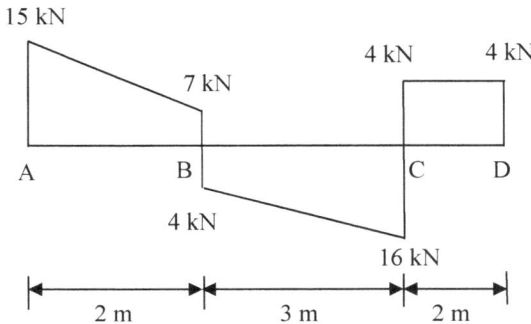

Fig. 3.44

Chapter 4

Bending Stresses in Beams

Learning Objectives

After going through this chapter, the reader will be able to
- describe the theory of simple bending.
- derive the bending stress equation and discuss the significance of the assumptions made in deriving this equation.
- draw the bending stress distribution diagram and calculate the maximum normal stresses in critical sections.
- explain the behavior of composite beams and determine the stresses in such beams.
- derive the equation for shear stress distribution and calculate the maximum shear stress in beam sections.
- design beam sections for various loading conditions.

4.1 INTRODUCTION

Whenever some lateral loads act on a beam, shear force (SF) and bending moment (BM) are developed in each cross section of the beam (excepting a few sections where the SF and/or BM may be zero). In the previous chapter, we discussed the procedure to calculate the shear force and bending moment at any section of the beam and also to find their variation along the length of beam. The beam resists the shear force and bending moment by developing internal stress resultants. The resistance offered to the bending moment is known as bending stress and that offered to the shear force is known as shear stress. This chapter deals with the bending and shear stresses developed in beams. The analysis of the composite beams and beams of uniform strength are also discussed.

4.2 PURE BENDING (or Simple Bending)

In bending of beams when both bending moments as well as shear forces are present, the bending is known as ordinary bending. When there is only constant bending moment acting on a beam, it is known as pure bending. In pure bending, the beam is subjected to constant bending moment and no shear force and due to this beam bends as an arc of a circle.

Let's consider a simply supported beam loaded as shown in Fig. 4.1a for which the shear force diagram and bending moment diagram are explained in Figs. 4.1b and 4.1c, respectively.

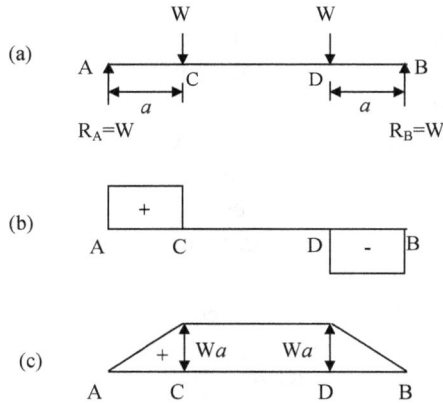

Fig. 4.1

It may be observed from these figures that in the region from C to D of the beam, pure bending occurs (zero shear force and constant bending moment). Similar observations can be made for the beam loaded as shown in Fig. 4.2a for which the shear force and bending moment diagram are explained in Figs. 4.2b and 4.2c respectively. In this case, pure bending occurs throughout the length of the beam.

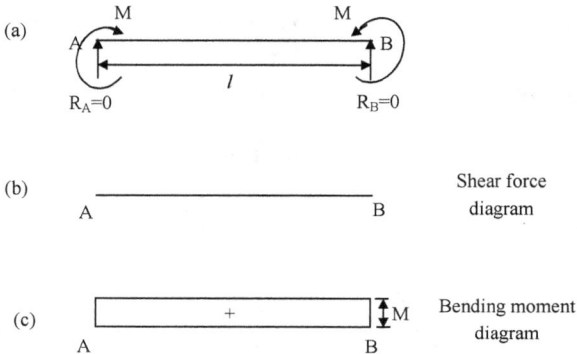

Fig. 4.2

4.3 THEORY OF SIMPLE BENDING

Whenever a beam is subjected to simple bending (pure bending) the longitudinal fibers on one side of beam are subjected to compression while the fibers on the other side are subjected to tension. In between the top and bottom fibers, there exists a surface where bending stress is zero. This is known as Neutral surface (or neutral layer).

Neutral Layer (or **Neutral Plane**):

It is a layer in which the longitudinal fibers do not change in length. At neutral layer, stress and strain are zero. On one side of the neutral layer, longitudinal fibers will elongate and on its other side, the longitudinal fibers will contract.

Neutral Axis:

It is the line of intersection of neutral layer with the cross section of plane. At neutral axis, stress and strain are zero.

Figure 4.3 explains the neutral layer and neutral axis which are denoted by plane EFF'E' and line FF' respectively.

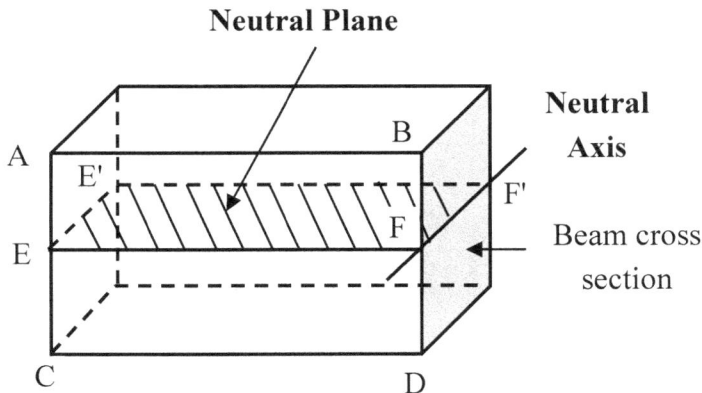

Fig. 4.3

4.4   DERIVATION OF BENDING STRESS EQUATION   (FLEXURE FORMULA)

Assumptions:

Following assumptions are taken for deriving the bending stress equation:
(i) The plane cross section of beam before bending remains plane after bending i.e. no warping of cross section.
(ii) The material of the beam is homogeneous and isotropic and follows Hooke's law.
(iii) Longitudinal fibers of the beam are free to elongate and contract.
(iv) Modulus of elasticity of beam material is same in compression and tension.
(v) The beam is subjected to pure bending (i.e. only constant bending moment and no shear force).
(vi) The radius of curvature of beam is very large as compared to the dimensions of beam.

Derivation:

Consider a beam subjected to pure bending as shown (Fig. 4.4).

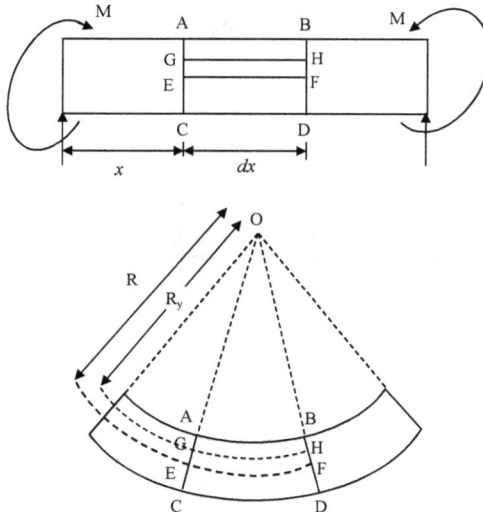

Fig. 4.4

O   = Centre of curvature

R   = Radius of curvature

EF   = Longitudinal fiber at neutral axis

GH   = Longitudinal finer at a distance y from neutral axis

I   = Moment of inertia

M   = Bending moment

σ = Bending stress

E   = Young's modulus

EF fiber doesn't change in length (Fiber at neutral axis).

Initial length of GH fiber = R dθ

Final length of GH fiber = (R-y) dθ

$\therefore$ Strain in GH fiber,

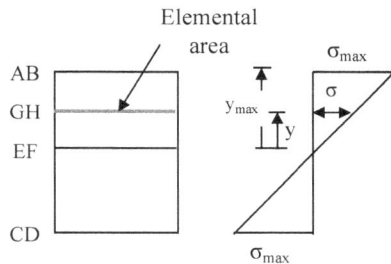

Fig. 4.5

$$\varepsilon = \frac{R\,d\theta - (R-y)\,d\theta}{R\,d\theta} = \frac{y}{R}$$

Stress , $\sigma = E\,\varepsilon = E\dfrac{y}{R}$

$\therefore \dfrac{\sigma}{y} = \dfrac{E}{R}$

Now considering the beam cross section (refer Fig. 4.5)

Elemental force,

$$dF = \sigma\,dA = E\frac{y}{R}\,dA$$

(Stress on elemental area, $\sigma = E\dfrac{y}{R}$)

Elemental moment,

$$dM = dF\, y = \left(E\frac{y}{R}\, dA\right) y = \frac{E}{R}\, y^2\, dA$$

$$\therefore \;\; \text{Total moment,} \qquad M = \int dM = \int \frac{E}{R}\, y^2\, dA = \frac{E}{R} \int y^2\, dA = \frac{E}{R}\, I$$

where $\;\; I = \int y^2\, dA = $ Moment of inertia

$$\rightarrow \quad \frac{M}{I} = \frac{E}{R}$$

$$\boxed{\frac{\sigma}{y} = \frac{M}{I} = \frac{E}{R}} \quad \longrightarrow \quad \textbf{Bending Stress Equation}$$

4.5 SECTION MODULUS (Z)

Section modulus is defined as the ratio of moment of inertia of beam cross section about the centroidal axis to the distance of extreme fiber from the centroidal axis. It is denoted by Z and is written as

$$Z = \frac{I}{y_{max}}$$

Section moduli for some standard cross sections are computed as:

(i) **Rectangular cross section** (Fig. 4.6a)

b and d are width and depth of rectangular cross section.

$$I = \frac{bd^3}{12} \;, \qquad y_{max} = \frac{d}{2}$$

$$\therefore \;\; Z = \frac{I}{y_{max}} = \frac{bd^2}{6}$$

(ii) **Hollow Rectangular cross section** (Fig. 4.6b)

$$I = \frac{BD^3}{12} - \frac{bd^3}{12} = \frac{1}{12}\left(BD^3 - bd^3\right), \qquad y_{max} = \frac{D}{2}$$

$$\therefore \ Z = \frac{I}{y_{max}} = \frac{(BD^3 - bd^3)}{6D}$$

(iii) **Circular cross section** (Fig. 4.6c)

$$I = \frac{\pi d^4}{64}, \qquad y_{max} = \frac{d}{2}$$

$$\therefore \ Z = \frac{I}{y_{max}} = \frac{\pi d^3}{32}$$

(iv) **Hollow circular cross section** (Fig. 4.6d)

$$I = \frac{\pi(D^4 - d^4)}{64}, \qquad y_{max} = \frac{D}{2}$$

$$\therefore Z = \frac{I}{y_{max}} = \frac{\pi}{32}\left(\frac{D^4 - d^4}{D}\right)$$

(v) **Triangular cross section** (Fig. 4.6e)

$$I = \frac{bh^3}{36}, \qquad y_{max} = \frac{2h}{3}$$

$$\therefore \ Z = \frac{I}{y_{max}} = \frac{bh^2}{24}$$

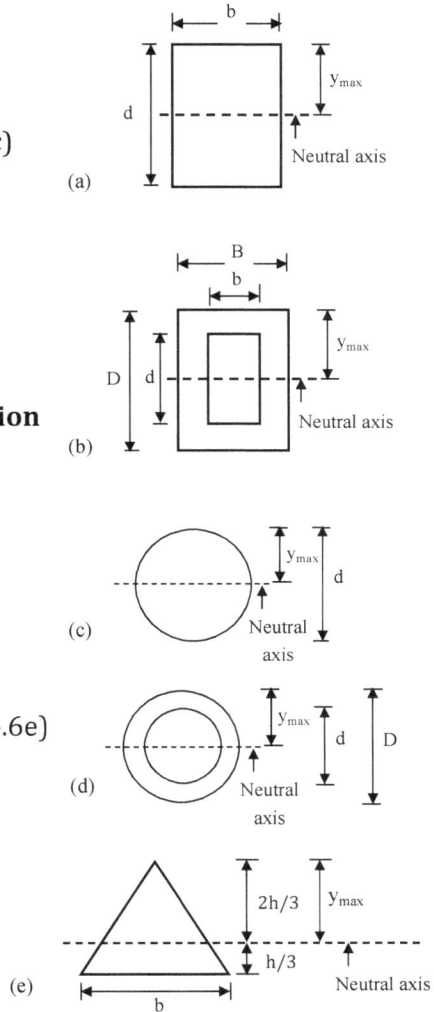

Fig. 4.6

4.6 MAXIMUM BENDING STRESS

Maximum bending stress developed in the beam for the symmetrical as well as unsymmetrical sections can be found out as follows:

For symmetrical sections (sections having both axes of symmetry):

$$\frac{\sigma_{max}}{y_{max}} = \frac{M}{I}$$

$$\sigma_{max} = \frac{M}{I} \, y_{max} = \frac{M}{(I/y_{max})} = \frac{M}{Z}$$

For unsymmetrical sections (sections having one axis of symmetry):

$$(\sigma_{max})_U = \frac{M}{I} \, (y_{max})_U = \frac{M}{Z_U}$$

$$(\sigma_{max})_L = \frac{M}{I} \, (y_{max})_L = \frac{M}{Z_L}$$

For an initially straight beam following Hooke's law, neutral axis passes through the centroid of cross section and neutral axis and centroidal axes are coinciding. For initially curved beams, neutral axis and centroidal axis are not coinciding with each other. In this chapter we shall limit our discussion to initially straight beams only.

4.7 MOMEMNT OF RESISTANCE (or **Bending strength** or **Safe moment**)

Bending strength or moment of resistance is the product of allowable bending stress and the section modulus.

Bending strength, $M = \sigma_{all} \times Z$

where σ_{all} = Allowable stress or maximum permissible stress and Z = Section modulus.

This bending strength is also called *moment carrying capacity* of the beam.

Note:

1. The bending stress equation or flexure formula derived in this chapter is based on the assumption of pure bending (i.e. no shear force and only constant bending moment) and normal stresses (bending stresses) induced can be computed. If the beam is subjected to ordinary bending i.e. non-uniform bending, shear force will also be present thereby causing shear stress in addition to the bending stress. But the presence of this shear stress does not affect much

to the normal stress that found from pure bending. Therefore, the theory of pure bending is also used for calculating normal stresses in case of ordinary bending.

2. Bending moment is developed due to external loading and moment of resistance is resisting its action. Moment of resistance of beam depends upon the material's allowable tensile and compressive stress. When the bending moment developed in a member is greater than the moment of resistance, the member fails.

Example 4.1 *Find the maximum flexural stress at a section 250 mm from the support for the cantilever beam loaded as shown in Fig. 4.7. The beam weighs 350 N per meter of length and P = 450 N.*

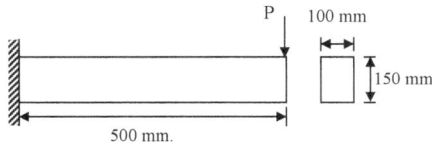

Fig. 4.7

Solution:

Refer Fig. 4.7a

Bending moment at a section C at 250 mm from A (i.e. mid-span)

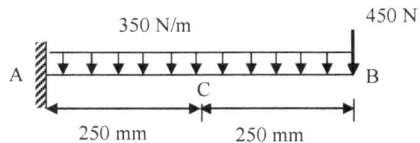

Fig. 4.7a

$$M_C = -450 \times 250 - 350 \times 250 \times \frac{250}{2}$$

$$= -123.43 \times 10^3 \text{ Nmm}$$

Moment of Inertia

$$I = \frac{bd^3}{12} = \frac{100 \times 150^3}{12} = 28.12 \times 10^6 \text{ mm}^4$$

$$\therefore \frac{\sigma}{y} = \frac{M}{I} \text{ or } \frac{\sigma}{\left(\frac{150}{2}\right)} = \frac{123.43 \times 10^3}{28.12 \times 10^6}$$

$$\therefore \ \sigma = 0.329 \frac{N}{mm^2} \rightarrow \text{Maximum stress}$$

Note:

When it is required to find σ_{max} instead of σ_{max} at 250 mm from support, first the maximum bending moment (in this case at A) may be found and then accordingly σ_{max} can be calculated.

Example 4.2 *A steel wire of 10 mm diameter is wound around a circular drum of radius 1000 mm. if* $E = 2 \times 10^5 N/mm^2$, *find the maximum bending stress.*

Solution:

Moment of Inertia of wire cross section

$$I = \frac{\pi d^4}{64} = \frac{\pi \times 10^4}{64} = 490.87 \text{ mm}^4$$

Here $y_{max} = \dfrac{10}{2} = 5$ mm

Radius of curvature, R = Drum radius + Radius of wire = 1000+(10/2)=1005 mm

$$\frac{M}{I} = \frac{\sigma_{max}}{y_{max}} = \frac{E}{R}$$

$$\therefore \ \frac{M}{490.87} = \frac{\sigma_{max}}{5} = \frac{2 \times 10^5}{1005}$$

$\therefore \ \sigma_{max} = 995 \text{ N/mm}^2 \rightarrow$ Maximum Bending stress
(Maximum bending moment, M = 97.68×10^3 Nmm)

Example 4.3 *A steel scale is 320 mm long, 25 mm wide and 1 mm thick. It is bent in its flexible plane (in a circular plane) by the application of end couple of magnitude 250 Nmm. Given E=2.1 $\times 10^{11}$ N/m^2, determine the radius of bent arc and also find the maximum bending stress set up in the scale.*

Solution:

$$I = \frac{bd^3}{12} = \frac{25 \times (1)^3}{12} = 2.083 \text{ mm}^4$$

$$y_{max} = \frac{1}{2} = 0.5 \text{ mm}$$

$$E = 2.1 \times 10^{11} \text{ N/m}^2$$

$$M = 250 \text{ N mm}$$

$$\frac{M}{I} = \frac{\sigma_{max}}{y_{max}} = \frac{E}{R}$$

$$\frac{250}{2.083} = \frac{\sigma_{max}}{0.5} = \frac{2.1 \times 10^5}{R}$$

$$\therefore \ \sigma_{max} = 60 \ \frac{N}{cm^2} \quad \rightarrow \text{Maximum stress}$$

$$R = 1750 \text{ mm} \rightarrow \text{Radius of bent arc}$$

Example 4.4 *A steel beam simply supported at its ends (Fig. 4.8) is of I section having the following dimensions:*
Flange = 250 mm wide and 24 mm thick. Web = 12 mm thick, overall depth = 600 mm.
If this beam carries a uniformly distributed load of 50 kN/m on a span of 8 m, calculate the maximum stress produced due to bending.

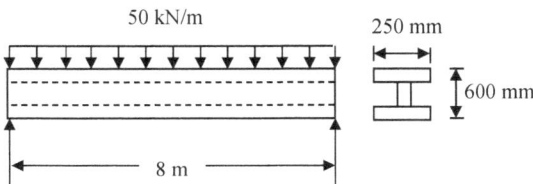

Fig. 4.8

Solution:

Refer Fig. 4.8a

Moment of inertia of the cross section

$$I = 2 \times \left[\frac{250 \times 24^3}{12} + 250 \times 24 \times (300 - 12)^2 + \frac{12 \times 276^3}{12} \right.$$
$$\left. + 12 \times 276 \times \left(\frac{276}{2} \right)^2 \right]$$

$= 1162 \times 10^6 \text{ mm}^4$

$y_{max} = 300 \text{ mm}$

$$M = \frac{wl}{2} \times \frac{l}{2} - w \times \frac{l}{2} \times \frac{l}{4} = \frac{wl^2}{8}$$

$$= \frac{50 \times 8^2}{8} = 400 \text{ kNm} = 400 \times 10^6 \text{ Nmm}$$

$$\frac{\sigma_{max}}{y_{max}} = \frac{M}{I}$$

$$\therefore \ \sigma_{max} = \frac{400 \times 10^6}{1162 \times 10^6} \times 300 = 103.2 \frac{N}{mm^2} \quad \rightarrow \quad \text{Maximum stress}$$

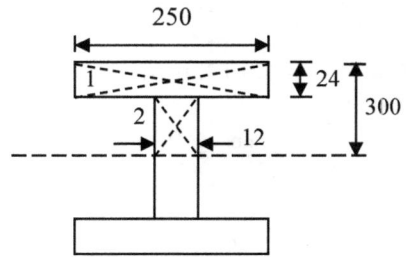

Fig. 4.8a

Example 4.5 *A water main of 1.2 m internal diameter and 12 mm thick is running full. If the bending stress is not to exceed 55 MPa, find the greatest span on which the pipe may be freely supported. Specific weight of steel and water are 75300 N/m³ and 9810 N/m³ respectively.*

Solution:

Total weight = steel pipe weight + weight of water

This load is uniformly distributed.

Let l = length of span

$\gamma_{steel} = 75300 \text{ N/m}^3$, $\gamma_{water} = 9810 \text{ N/m}^3$

Steel pipe weight = volume × specific weight

$$= \frac{\pi}{4}(1.224^2 - 1.2^2) \times l \times 75300 = (3440)l$$

Weight of water = volume × specific weight

$$= \frac{\pi}{4}(1.2^2) \times l \times 9810 = (11094)l$$

Total weight = $(3440)l + (11094)l = (14534)l$ N

∴ Intensity of uniformly distributed load = 14534 N/m

$$I = \frac{\pi}{64}(1.224^4 - 1.2^4) = 8.39 \times 10^{-3} \, m^4$$

Maximum bending moment,

$$M = \frac{wl^2}{8} = \frac{14534 \times l^2}{8} = (1816.75)l^2$$

$$\frac{\sigma_{max}}{y_{max}} = \frac{M}{I} \quad \text{or} \quad \frac{55 \times 10^6}{\left(\frac{1.224}{2}\right)} = \frac{(1816.75)l^2}{8.39 \times 10^{-3}}$$

∴ $l = 20.37$ m   → Length of span

Example 4.6 *A steel beam having T section as shown (Fig. 4.9) is simply supported over a span of 5 m. If the maximum allowable stress is 50 N/mm², calculate the maximum uniformly distributed load the beam can carry over the entire length.*

Solution:

Refer Fig. 4.9a

To locate the centroidal axis

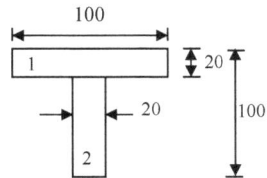

Fig. 4.9

$$y_L = \frac{A_1 y_1 + A_2 y_2}{A_1 + A_2} = \frac{100 \times 20 \times (100 - 10) + 20 \times 80 \times \frac{80}{2}}{(100 \times 20) + (20 \times 80)}$$

$$= 67.77 \text{ mm}$$

$\therefore y_u = 100 - 67.77 = 32.33$ mm

Moment of inertia of the cross section

$$I = \frac{100 \times 20^3}{12} + 100 \times 20 \times (33.33 - 10)^2 + \frac{20 \times 80^3}{12}$$
$$+ 20 \times 80 \times (67.77 - 40)^2$$

$$= 3.14 \times 10^6 \text{ mm}^4$$

Maximum stress occurs in the bottom fiber.

$$\frac{\sigma_L}{y_L} = \frac{M}{I}$$

$$\frac{50}{67.77} = \frac{w \times (5000)^2/8}{3.14 \times 10^6}$$

$$\therefore w = 0.741 \frac{N}{mm} = 741 \frac{N}{m}$$

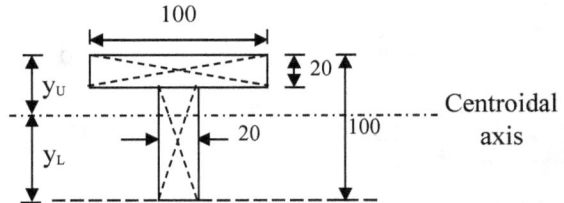

Fig. 4.9a

$\rightarrow$ Maximum uniformly distributed load

Example 4.7 *You are to cut a beam of rectangular section from a cylindrical log of wood of diameter 300 mm. Find the width and depth of the strongest beam (in respect of bending strength) which you can cut.*

Solution:

Refer Fig. 4.10

$$b^2 + h^2 = d^2$$

$$I = \frac{bh^3}{12}, \qquad y_{max} = \frac{h}{2}$$

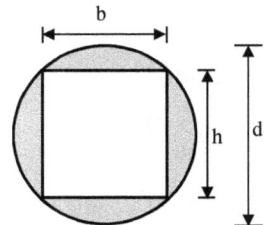

Fig. 4.10

bending strength, $M = \sigma_{all} \times Z$

Bending strength is maximum if section modulus Z is maximum.

$$Z = \frac{I}{y_{max}} = \frac{bh^2}{6}$$

For Z to be maximum, $\dfrac{dZ}{db} = 0$

$\rightarrow \quad \dfrac{d}{db}\left(\dfrac{b(d^2 - b^2)}{6}\right) = 0$

$\rightarrow \quad d^2 - 3b^2 = 0$

$\therefore \quad b = \dfrac{d}{\sqrt{3}} \quad$ and $\quad h = \sqrt{\dfrac{2}{3}}\, d$

given $d = 300$ mm

$\therefore \quad b = \dfrac{300}{\sqrt{3}} = 173.2$ mm, $\quad h = \sqrt{\dfrac{2}{3}} \times 300 = 245$ mm

These dimensions will give the best bending strength and the beam will be the strongest one.

Example 4.8 *Compute the maximum value of force P, the T beam can be subjected to (Fig. 4.11) if the allowable stresses due to bending are +150 MPa (tension) and -120 MPa (compression).*

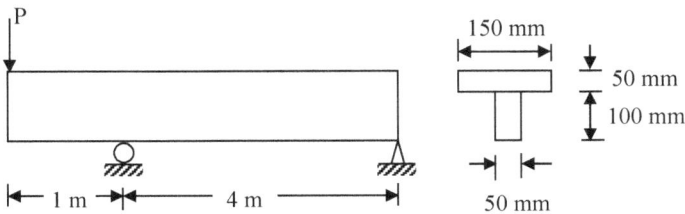

Fig. 4.11

Solution: (refer Fig. 4.11a)

$$\sum M_B = 0$$

$$(R_C \times 4) + (P \times 1) = 0$$

$$\therefore R_C = \frac{-P}{4}$$

$$\therefore R_B = P + \frac{P}{4} = \frac{5P}{4}$$

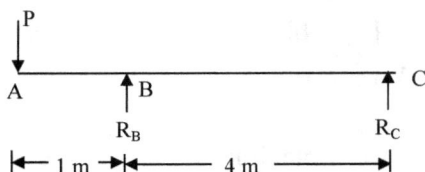

Fig. 4.11a

The bending moment diagram is shown in Fig. 4.11b

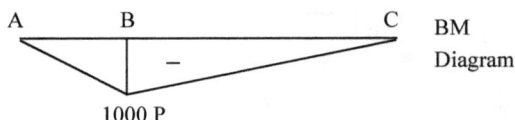

Fig. 4.11b

Maximum bending moment = $-1000\ P$ Nmm (Hogging bending moment)

Refer Fig. 4.11c

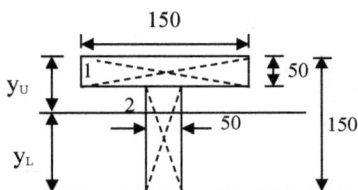

Fig. 4.11c

$$y_L = \frac{A_1 y_1 + A_2 y_2}{A_1 + A_2} = \frac{150 \times 50 \times (150 - 25) + 50 \times 100 \times \frac{100}{2}}{(150 \times 50) + (50 \times 100)}$$

$$= 94.4\ \text{mm}$$

$$\therefore y_u = 150 - 94.4 = 55.6\ \text{mm}$$

Moment of inertia of the cross section

$$I = \frac{150 \times 50^3}{12} + 150 \times 50 \times (55.6 - 25)^2 + \frac{50 \times 100^3}{12}$$
$$+ 50 \times 100 \times (94.4 - 50)^2$$

$$= 24 \times 10^6\ \text{mm}^4$$

Upper side of the beam is subjected to tension and lower side is subjected to compression.

$$\frac{\sigma_u}{y_u} = \frac{M}{I} \quad \rightarrow \quad \frac{150}{55.6} = \frac{1000\ P}{24 \times 10^6}$$

$\therefore$ P = 64.23×10³ N

$$\frac{\sigma_L}{y_L} = \frac{M}{I} \qquad \rightarrow \qquad \frac{120}{94.4} = \frac{1000P}{24×10^6}$$

$\therefore$ P = 30.36×10³ N

Safe load that can be applied is the lesser of the two values of P calculated.

$\therefore$ Safe load, P = 30.36×10³ N = 30.36 kN

Example 4.9 *Bending takes place around the horizontal axis for a square member in the two orientations shown (Fig 4.12). If the material is linearly elastic and maximum stress is same in both the cases, find the ratio of bending moments in the two cases.*

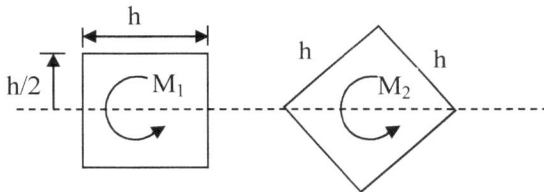

Fig. 4.12

Solution:

(a) Square bent about axis parallel to the side.

$$\frac{M}{I} = \frac{\sigma}{y} \qquad\qquad (I = \frac{h^4}{12})$$

$$\frac{M_1}{\left(\frac{h^4}{12}\right)} = \frac{\sigma}{\left(\frac{h}{2}\right)}$$

$$\therefore M_1 = \sigma \frac{h^3}{6}$$

(b) Square bent about diagonal

$$\frac{M}{I} = \frac{\sigma}{y}$$

$$\frac{M_2}{\left(\frac{h^4}{12}\right)} = \frac{\sigma}{\left(\frac{h}{\sqrt{2}}\right)}$$

$$\therefore M_2 = \sigma \frac{h^3}{6\sqrt{2}}$$

$$\therefore \text{Ratio of bending moments} = \frac{M_1}{M_2} = \frac{\sigma \dfrac{h^3}{6}}{\sigma \dfrac{h^3}{6\sqrt{2}}} = \sqrt{2}$$

Note:

$$M_1 = \sqrt{2} \, M_2$$

$\therefore$ *Square element about the diagonal is weaker.*

Example 4.10 *Compare the bending strengths of two beams one of square cross section and other of circular cross section made of same material and having same cross sectional area.*

Solution:

For same material, permissible stress is same.

Given that area of square = area of circle.

Let a = side of square   and d = diameter of circle

$$\therefore a^2 = \frac{\pi}{4} d^2$$

$$\rightarrow a = (0.886) \, d$$

Consider square section:

$$\frac{M}{I} = \frac{\sigma}{y} \quad \rightarrow \quad \frac{M_1}{\left(\frac{a^4}{12}\right)} = \frac{\sigma}{\left(\frac{a}{2}\right)} \quad \rightarrow \quad M_1 = \sigma \frac{a^3}{6}$$

Consider circular cross section:

$$\frac{M}{I} = \frac{\sigma}{y} \quad \rightarrow \quad \frac{M_2}{\left(\frac{\pi d^4}{64}\right)} = \frac{\sigma}{\left(\frac{d}{2}\right)} \quad \rightarrow \quad M_2 = \sigma \frac{\pi d^3}{32}$$

$$\therefore \frac{M_1}{M_2} = \frac{\sigma \frac{a^3}{6}}{\sigma \frac{\pi d^3}{32}} = \frac{16a^3}{3\pi d^3} = \frac{16(0.886\, d)^3}{3\pi d^3} = 1.18$$

∴ Square beam is stronger than circular beam by 18% for same material and same cross sectional area.

Example 4.11 *Compare the bending strength of two beams, one of solid circular cross section and other of hollow circular cross section having the inner diameter equal to ¾ of outer diameter, if both are made of same material and having same cross sectional area.*

Solution:

Consider solid circular beam (of diameter D_1):

$$\frac{M}{I} = \frac{\sigma}{y} \quad \rightarrow \quad \frac{M_1}{\left(\frac{\pi D_1^4}{64}\right)} = \frac{\sigma}{\left(\frac{D_1}{2}\right)} \quad \rightarrow \quad M_1 = \sigma \frac{\pi D_1^3}{32}$$

Consider hollow circular beam (of outer diameter D):

$$\frac{M}{I} = \frac{\sigma}{y} \quad \rightarrow \quad \frac{M_2}{\left(\frac{\pi[D^4 - \left(\frac{3}{4}D\right)^4]}{64}\right)} = \frac{\sigma}{\left(\frac{D}{2}\right)} \quad \rightarrow \quad M_2$$

$$= \sigma(0.067)D^3$$

As both beams are made of same material, permissible stress is same.

Given that area of solid beam=Area of hollow beam

$$\frac{\pi D_1^2}{4} = \frac{\pi[D^2 - \left(\frac{3}{4}D\right)^2]}{4} \quad \rightarrow \quad D_1 = (0.66)D$$

$$\therefore \frac{M_1}{M_2} = \frac{\sigma \frac{\pi D_1^3}{32}}{\sigma\,(0.067)D^3} = 0.423$$

Therefore, hollow beam is stronger than solid beam for same material and equal cross sectional area.

Example 4.12 *A simply supported beam is to have the inverted T section as shown (Fig 4.13). If the allowable stresses in tension and compression are* $\sigma_t = 2.8 \, N/mm^2$ *and* $\sigma_c = 5.6 \, N/mm^2$, *calculate the proper stem thickness t of the section.*

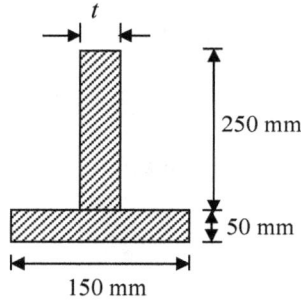

Fig. 4.13

Solution:

Refer Fig. 4.13

To locate centroid,

$$y_L = \frac{A_1 y_1 + A_2 y_2}{A_1 + A_2}$$

$$= \frac{(250\,t \times 175) + (150 \times 50 \times 25)}{250\,t + 7500} = \frac{175\,t + 750}{t + 30}$$

$$\therefore \quad y_u = (300 - y_L)$$

$$= 300 - \frac{175\,t + 750}{t + 30} = \frac{125t + 8250}{t + 30}$$

$$\frac{\sigma_L}{y_L} = \frac{M}{I} \quad \text{and} \quad \frac{\sigma_u}{y_u} = \frac{M}{I}$$

$$\therefore \quad \frac{\sigma_L}{y_L} = \frac{\sigma_u}{y_u} \quad \rightarrow \quad \frac{\sigma_t}{y_L} = \frac{\sigma_c}{y_u}$$

$$\rightarrow \quad \frac{y_L}{y_u} = \frac{\sigma_t}{\sigma_c} = \frac{2.8}{5.6} = \frac{1}{2}$$

$$\rightarrow \quad y_u = 2\,y_L$$

$$\rightarrow \quad \frac{125\,t + 8250}{t + 30} = 2 \times \frac{175\,t + 750}{t + 30}$$

$\therefore$ t = 30 mm

Example 4.13 *For the channel beam shown (Fig. 4.14), it is desired to have the ratio of extreme fiber bending stresses* $\sigma_t : \sigma_c = 3 : 7$. *Determine the proper wall thickness* t *to realize this condition.*

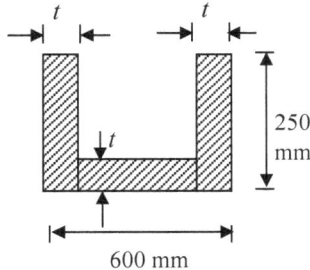

Fig. 4.14

Solution:

$$y_L = \frac{\Sigma\,Ay}{A} = \frac{\left(2 \times 250\,t \times \frac{250}{2}\right) + (600 - 2\,t)\,t\left(\frac{t}{2}\right)}{250\,t + (600 - 2\,t)t + 250\,t}$$

$$= \frac{t^2 - 300\,t - 62500}{2t - 1100}$$

$$y_u = (250 - y_L) = 250 - \frac{t^2 - 300\,t - 62500}{2t - 1100}$$

$$= \frac{-t^2 + 800\,t - 212500}{2t - 1100}$$

$$\frac{\sigma_t}{\sigma_c} = \frac{y_L}{y_u} \quad \text{but} \quad \frac{\sigma_t}{\sigma_c} = \frac{3}{7} \quad \text{given}$$

$$\therefore \quad \frac{y_L}{y_u} = \frac{3}{7} \quad \rightarrow \quad 7y_L = 3y_u$$

$$7 \times \frac{t^2 - 300\,t - 62500}{2t - 1100} = 3 \times \frac{-t^2 + 800\,t - 212500}{2t - 1100}$$

$$7t^2 - 2100\,t - 437500 = -3t^2 + 2400\,t - 637500$$

$$t^2 - 450\,t + 20000 = 0$$

$$\therefore t = \frac{450 \pm \sqrt{450^2 - 4 \times 1 \times 20000}}{2 \times 1}$$

$$= \frac{450 \pm 350}{2} = 400, 50$$

Therefore, $t = 50$ mm   (t can not be 400 mm)

Example 4.14 *Calculate the two section moduli Z_1 and Z_2 for the inverted T section shown (Fig. 4.15).*

Solution:

To locate the centroid G (Refer Fig. 4.15)

$$\bar{y} = \frac{A_1 y_1 + A_2 y_2}{A_1 + A_2}$$

$$= \frac{30 \times 250 \times \left(\frac{250}{2} + 50\right) + 150 \times 50 \times \frac{50}{2}}{(30 \times 250) + (150 \times 50)}$$

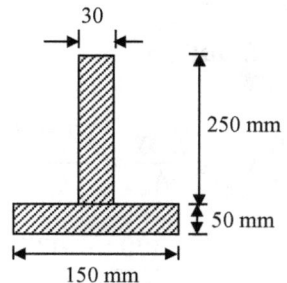

Fig. 4.15

$$= 100 \text{ mm}$$

Moment of inertia I of the cross section

$$I = \frac{30 \times 250^3}{12} + 30 \times 250 \left(200 - \frac{250}{2}\right)^2 + \frac{150 \times 50^3}{12}$$
$$+ 150 \times 50 \left(100 - \frac{50}{2}\right)^2$$

$$= 125 \times 10^6 \text{ mm}^4$$

∴ Section modulus

$$Z_u = \frac{I}{y_u} = \frac{125 \times 10^6}{200} = 62.5 \times 10^4 \text{ mm}^3$$

$$Z_L = \frac{I}{y_L} = \frac{125 \times 10^6}{100} = 125 \times 10^4 \text{ mm}^3$$

Example 4.15 *A long rod of uniform rectangular section and thickness t originally straight is bent into the form of a circular arc and the displacement d of the mid point of a length L is measured by means of dial gauge. If d is regarded small as compared with L show that the longitudinal surface strain ε in the rod is given by* $\varepsilon = \frac{4td}{L^2}$.

Solution:

Refer Fig. 4.16

$$R^2 = (R - d)^2 + \left(\frac{L}{2}\right)^2 = R^2 + d^2 - 2Rd + \frac{L^2}{4}$$

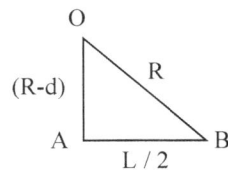

Fig. 4.16

$$\therefore 2Rd = \frac{L^2}{4} \qquad (d^2 \cong 0)$$

$$\rightarrow \quad R = \frac{L^2}{8d}$$

$$\frac{\sigma}{y} = \frac{E}{R} \quad \rightarrow \quad \frac{\sigma}{E} = \frac{y}{R} \qquad \text{but} \qquad \frac{\sigma}{E} = \varepsilon$$

$$\therefore \varepsilon = \frac{y}{R} = \frac{\frac{t}{2}}{R} = \frac{t}{2R}$$

$$\therefore \varepsilon = \frac{t}{2\frac{L^2}{8d}} = \frac{4td}{L^2} \qquad \text{(Proved)}$$

Example 4.16 *A wood beam 1.8 m long is simply supported at its ends has a cross section 150 mm wide by 600 mm deep and carries a uniformly distributed load of intensity w = 8 kN/m over the full span. Calculate the bending stress at a point 200 mm above the bottom of the beam and 600 mm from the left support.*

Solution:

Refer Fig. 4.17

Reactions at A and B:

$$\Sigma M_A = 0$$

$$R_B \times 1.8 = 8 \times 1.8 \times \frac{1.8}{2}$$

$$R_B = 7.2 \text{ kN}, \quad R_A = 7.2 \text{ kN}$$

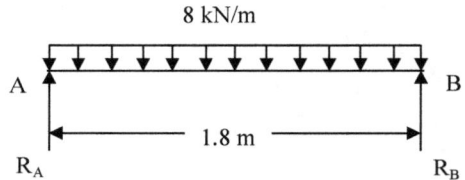

Fig. 4.17

The point where bending stress is required is at 600 mm from the left support and 200 mm above the bottom of beam (Fig. 4.17a)

$$y = 300 - 200 = 100 \text{ mm} = 0.1 \text{ m}$$

Bending moment

$$M = 7200 \times 0.6 - 8000 \times 0.6 \times \frac{0.6}{2}$$

$$= 2880 \text{ Nm}$$

$$I = \frac{0.15 \times 0.6^3}{12} = 0.0027 \text{ m}^4$$

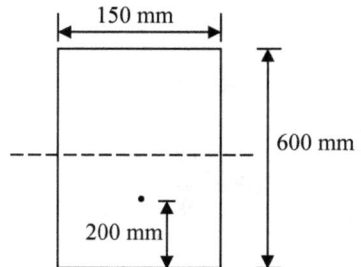

Fig. 4.17a

Bending stress equation: $\dfrac{\sigma}{y} = \dfrac{M}{I}$

$$\therefore \frac{\sigma}{0.1} = \frac{2880}{0.0027}$$

$$\therefore \sigma = 1.067 \times 10^5 \text{ N/m}^2$$

Example 4.17 *A simply supported beam of span 3.6 m is to carry a uniformly distributed load of 1.6 kN/m. The cross section of the beam is to be rectangular with depth h and width h/2. If the allowable bending stress in tension or compression is 0.84 N/mm², calculate the required depth h for the cross section.*

Solution:

Refer Fig. 4.18

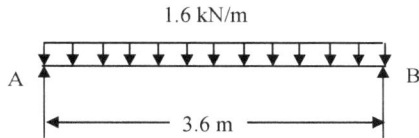

Fig. 4.18

Reactions at supports:

$$R_A = 1600 \times \frac{3.6}{2} = 2880 \text{ N}, \qquad R_B = 2880 \text{ N}$$

Maximum bending moment (at mid-span)

$$M_{max} = \frac{wl^2}{8} = \frac{1600 \times 3.6^2}{8} = 2592 \text{ Nm}$$

$$y = \frac{h}{2}, \qquad I = \frac{\frac{h}{2} \times h^3}{12} = \frac{h^4}{24}$$

$$\frac{\sigma}{y} = \frac{M}{I}$$

$$\therefore \sigma_{max} = (\frac{M_{max}}{I}) y_{max}$$

$$0.84 \times 10^6 = \frac{2592}{\left(\frac{h^4}{24}\right)} \times \frac{h}{2}$$

$$\therefore h = 0.333 \text{ m} = 333 \text{ mm}$$

Example 4.18   *A cantilever beam of length 4 m has a circular cross section. The diameter of beam cross section varies from 300 mm at fixed end to 200 mm at free end. It carries a load of 10 kN at the free end. Find the maximum bending stress.*

Solution:

Refer Fig. 4.19

Bending moment is zero at free end and is maximum at fixed end.

Suppose maximum bending stress occurs at a distance x from free end (Fig. 4.19).

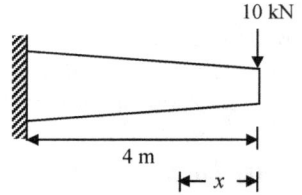

Fig. 4.19

Bending moment at a distance x from free end , $M_x = -10000\,x$

Diameter at a distance x ,

$$D_x = 200 + \frac{(300 - 200)x}{4000} = 200 + 0.025\,x$$

Moment of Inertia, $I_x = \dfrac{\pi D_x^{\,4}}{64} = \dfrac{\pi(200 + 0.025\,x)^4}{64}$

$$\frac{M}{I} = \frac{\sigma}{y}$$

$\rightarrow$ $\dfrac{10000\,x}{\dfrac{\pi(200 + 0.025\,x)^4}{64}} = \dfrac{\sigma}{\dfrac{200 + 0.025\,x}{2}}$

$\rightarrow$ $\sigma = \dfrac{32 \times 10^4\,x}{\pi(200 + 0.025\,x)^3}$

For maximum bending stress

$$\frac{d\sigma}{dx} = 0$$

$$\therefore \frac{32 \times 10^4}{\pi} \left[\frac{(200 + 0.025\, x)^3 \times 1 - x \times 3 \times (200 + 0.025\, x)^2 \times 0.025}{(200 + 0.025\, x)^6} \right]$$
$$= 0$$

$$(200 + 0.025\, x)^3 - 3\, x\, (200 + 0.025\, x)^2 \times 0.025 = 0$$

$$\therefore x = 4000 \text{ mm}$$

$$\therefore \sigma_{max} = \frac{32 \times 10^4 \times 4000}{\pi(200 + 0.025 \times 4000)^3} = 15.09 \text{ N/mm}^2$$

Example 4.19 *A flag pole is of circular cross section. The diameter is varying from 200 mm at ground to 100 mm at top. The height of pole is 3000 mm. it is subjected to a pull of 1000 N at top. Find maximum bending stress.*

Solution:

Let us assume that maximum bending stress occurs at a distance x from top.

Bending moment, $M_x = -1000\, x$

Diameter, $\quad D_x = 100 + \left(\dfrac{200 - 100}{3000}\right) x = 100 + \dfrac{1}{30} x$

Bending stress equation $\dfrac{M}{I} = \dfrac{\sigma}{y}$

$$\frac{1000\, x}{\dfrac{\pi(100 + \dfrac{1}{30} x)^4}{64}} = \frac{\sigma}{\dfrac{100 + \dfrac{1}{30} x}{2}} \quad \rightarrow \quad \sigma = \frac{32000\, x}{\pi(100 + \dfrac{1}{30} x)^3}$$

for maximum bending stress, $\quad \dfrac{d\sigma}{dx} = 0$

$$\therefore \frac{32000}{\pi} \left[\frac{\left(100 + \dfrac{1}{30} x\right)^3 \times 1 - x \times 3 \times \left(100 + \dfrac{1}{30} x\right)^2 \times \dfrac{1}{30}}{\left(100 + \dfrac{1}{30} x\right)^6} \right] = 0$$

$$\left(100 + \frac{1}{30}x\right)^3 - \frac{x}{10}\left(100 + \frac{1}{30}x\right)^2 = 0$$

$\therefore x = 1500$ mm

$\therefore$ Maximum bending stress occurs at mid-length.

$$\therefore \sigma_{max} = \frac{32000}{\pi}\left[\frac{1500}{\left(100 + \frac{1}{30} \times 1500\right)^3}\right] = 4.527 \frac{N}{mm^2}$$

Bending stress at fixed end can be calculated as

$$\sigma_{fix} = \frac{32000}{\pi}\left[\frac{3000}{\left(100 + \frac{1}{30} \times 3000\right)^3}\right] = 3.819 \frac{N}{mm^2}$$

Example 4.20 *A beam of I section of moment of inertia 9.54×10^6 mm^4 and depth of 140 mm is freely supported at its ends. Over what span can a uniform load of 5 kN/m run be carried if the maximum stress is 60 N/mm^2. What additional central load can be carried when the maximum stress is 90 N/mm^2.*

Solution:

Let l = length of span (mm)

Uniformly distributed load, $w = 5$ kN/m = 5 N/mm

Maximum bending moment

$$M_{max} = \frac{wl^2}{8} = \frac{5 \times l^2}{8} = 0.625\, l^2$$

$$\frac{\sigma_{max}}{y_{max}} = \frac{M_{max}}{I}$$

$\therefore$ Maximum stress, $\qquad \sigma_{max} = \dfrac{M_{max}}{I}\, y_{max}$

or $60 = \dfrac{0.625 \; l^2}{9.54 \times 10^6} \times 70$

$\therefore l = 3617$ mm $= 3.617$ m

Now central concentration load is applied in addition to uniformly distributed load.

Let P = Additional central load

Maximum bending moment, $\quad M_{max} = \dfrac{wl^2}{8} + \dfrac{Pl}{4}$

$= \dfrac{5(3617)^2}{8} + \dfrac{P \times 3617}{4} = 8.17 \times 10^6 + 904.25 \; P$

new maximum stress $= 90$ N/mm^2

$\therefore \dfrac{\sigma}{y} = \dfrac{M}{I} \quad \rightarrow \quad \dfrac{90}{70} = \dfrac{8.17 \times 10^6 + 904.25 \; P}{9.54 \times 10^6}$

$\therefore$ P = 4529 N

Example 4.21 *A rail road as shown (Fig. 4.20) is 2.4 m long and has a 300 mm × 250 mm rectangular cross section with the 300 mm faces horizontal. The maximum load transmitted to the tie by the rails is P = 25000 N each and the ballast is assumed to exert a uniformly distrib-uted reactive load on the bottom of the tie as shown. Calculate the maximum bending stress in the tie if l = 1425 mm and a = 487.5 mm.*

Solution:

Given P = 25000 N

$l = 1425 \; mm$, $\quad a = 487.5 \; mm$

Let w = uniformly distributed load
(Reactive load from bottom of tie)

Refer Fig. 4.20

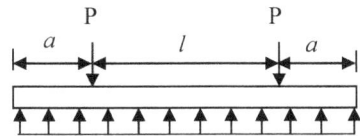

Fig. 4.20

w × (1425+2×487.5) = 25000 + 25000

$\therefore$ w = 20.83 N/mm

$$I = \frac{300 \times 250^3}{12} = 3.906 \times 10^8 \text{mm}^4$$

Bending stress equation $\dfrac{\sigma}{y} = \dfrac{M}{I}$

Maximum bending moment will be developed at mid span.

$$M_{max} = 20.83 \times 1200 \times \frac{1200}{2} - 25000 \times \frac{1425}{2}$$

$$= -2.82 \times 10^6 \text{ Nmm}$$

$$\frac{\sigma}{y} = \frac{M}{I}$$

or $\quad \dfrac{\sigma}{125} = \dfrac{2.82 \times 10^6}{3.906 \times 10^8}$

$\therefore \quad \sigma = 0.902 \text{ N/mm}^2 \quad \rightarrow$ Maximum stress in the tie

4.8 SHEAR STRESSES IN BEAMS

The shear force at any cross section of a beam will set up shear stress on transverse section. Due to the shear stress in transverse section, there will be a complementary shear stress of equal intensity on longitudinal planes parallel to the neutral plane. Hence to know the intensity of shear stress, the intensity of longitudinal shear stress can be determined.

4.8.1 Derivation of Shear Stress Formula

Consider a beam subjected to nonuniform bending i.e. say subjected to bending moment M at section AB and bending moment M+dM at section CD as shown (Fig. 4.21a).

Sections AB and CD are at a distance dx from each other. Consider left face AB of the element ABCD (Fig. 4.21b).

Consider elemental area dA of thickness dy at a distance y from neutral axis as shown (Fig. 4.21c).

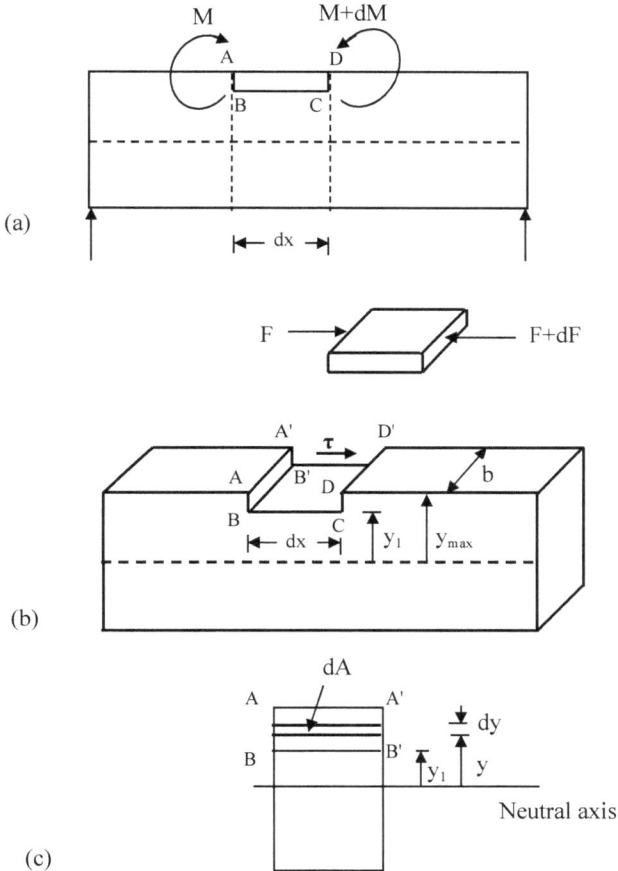

Fig. 4.21

Bending stress (normal stress) $\qquad \sigma = \dfrac{M}{I} y$

Normal force acting on left face of element

$$F = \int_{y_1}^{y_{max}} \sigma \, dA = \int_{y_1}^{y_{max}} \frac{M}{I} y \, dA$$

Similarly consider right face CD of element ABCD

Bending stress (normal stress) $\sigma' = \dfrac{M + dM}{I}\, y$

Normal force acting on right face of element

$$F + dF = \int_{y_1}^{y_{max}} \frac{M + dM}{I}\, y\, dA$$

∴ Unbalanced normal force acting on element

$$dF = \int_{y_1}^{y_{max}} \frac{dM}{I}\, y\, dA = \frac{dM}{I} \int_{y_1}^{y_{max}} y\, dA = \frac{dM}{I}\, A\,\bar{y}$$

where $\int y\, dA$ = Moment of area ABB'A' about neutral axis

A= Area above the line on which shear stress is found

$\bar{y}$ = Distance of C.G. of area A from neutral axis

This unbalanced normal force causes shear stress. Unbalanced normal force acting on the element is equal to horizontal shear force developed at the bottom face of element.

Let τ = Shear stress at a distance y_1 from neutral axis

$$\tau\, bdx = dF = \frac{dM}{I}\, A\,\bar{y} \qquad (b = \text{width of beam})$$

$$\rightarrow \quad \tau = \frac{dM}{dx}\left(\frac{A\,\bar{y}}{I\, b}\right)$$

$$\boxed{\tau = \frac{V\, A\,\bar{y}}{I\, b}} \longrightarrow \textbf{Shear Formula}$$

$V = \dfrac{dM}{dx}$ is the shear force at the cross section

I = moment of inertia of entire cross section

b = width of beam at the level where shear stress is calculated

A= area above the level where shear stress is calculated

$\bar{y}$ = distance of C.G. of area A from neutral axis

Computation of shear stress for various cross sections is discussed below:

(a) Rectangular cross section

Consider a rectangular section of width b and depth d subjected to shear force V.

Let x-x be at a distance y_1 from neutral axis as shown (Fig. 4.22a).

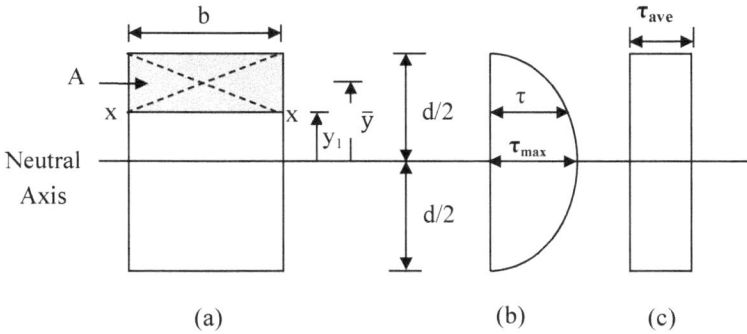

(a)   (b)   (c)

Fig. 4.22

Refer Fig. 4.22a

$$A = b\left(\frac{d}{2} - y_1\right)$$

$$\bar{y} = y_1 + \frac{1}{2}\left(\frac{d}{2} - y_1\right) = \frac{y_1}{2} + \frac{d}{4}$$

Shear stress at x-x (at a distance y_1 from neutral axis) is

$$\tau = \frac{V A \bar{y}}{I b}$$

$$= \frac{V}{I b}\left[b\left(\frac{d}{2} - y_1\right)\left(\frac{y_1}{2} + \frac{d}{4}\right)\right] = \frac{V}{2 I}\left(\frac{d}{2} - y_1\right)\left(\frac{d}{2} + y_1\right)$$

$$= \frac{V}{2I} \left[\left(\frac{d}{2}\right)^2 - y_1{}^2 \right]$$

$$\boxed{\tau = \frac{V}{8I} (d^2 - 4y_1{}^2)}$$

Shear stress variation is parabolic and this is shown in Fig. 4.22b. Putting $y_1 = 0$ maximum shear stress (τ_{max}) can be obtained.

$$\tau_{max} = \frac{V}{8I} (d^2) = \frac{V\,d^2}{8 \left(\dfrac{bd^3}{12}\right)} = \frac{3V}{2bd} = \frac{3}{2} \left(\frac{V}{A}\right)$$

However, average shear stress $\tau_{ave} = \dfrac{V}{A}$ (Fig. 4.22c)

$$\therefore \quad \tau_{max} = \frac{3}{2} \times \tau_{ave}$$

Maximum shear stress is greater than average shear stress by 50 %. Shear stress is zero at extreme fibers (i.e. when $y_1 = d/2$)

(b) I section

Refer Fig. 4.23a and 4.23b

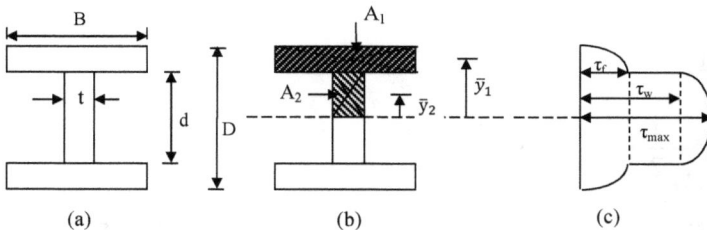

(a)   (b)   (c)

Fig. 4.23

Shear stress in flange $= \dfrac{V\,A\,\bar{y}}{I\,B}$

Shear stress in web $= \dfrac{V\,A\,\bar{y}}{I\,t}$

Shear stress at flange web junction: (refer Fig. 4.23c)

(i) in the flange

$$\tau_f = \frac{V A_1 \bar{y}_1}{I B} = \frac{V B \left(\frac{D-d}{2}\right)\left(\frac{d}{2}+\frac{D-d}{4}\right)}{I B} = \frac{V}{4 I}(D-d)\left(\frac{D}{2}+\frac{d}{2}\right)$$

$$= \frac{V}{8 I}(D^2 - d^2)$$

(ii) in the web

$$\tau_w = \frac{V A_1 \bar{y}_1}{I t} = \frac{V B}{8 I t}(D^2 - d^2)$$

Maximum shear stress:

$$\tau_{max} = \frac{V A \bar{y}}{I t} = \frac{V (A_1 \bar{y}_1 + A_2 \bar{y}_2)}{I t} = \frac{V A_1 \bar{y}_1}{I t} + \frac{V A_2 \bar{y}_2}{I t}$$

$$= \tau_w + \frac{V}{I t}\left(t \frac{d}{2}\right)\frac{d}{4}$$

$$= \frac{V B}{8 I t}(D^2 - d^2) + \frac{V d^2}{8 I}$$

(c) Circular cross section

Consider a circular section of diameter D as shown (Fig. 4.24a).

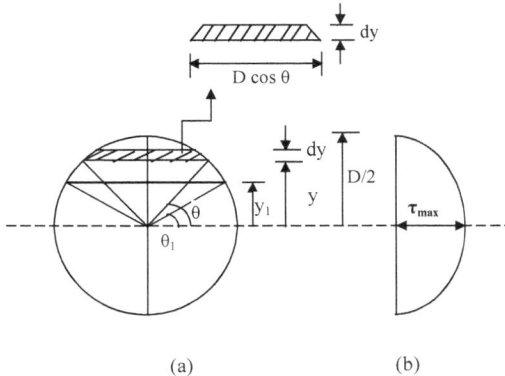

(a)   (b)

Fig. 4.24

A section at a distance y from neutral axis is also at an angular distance θ from neutral axis (Fig. 4.24a). On this section shear stress is to be calculated. Instead of y, it is convenient in this case to take θ as the variable.

D = Diameter of circle

$$y = \frac{D}{2} \sin \theta \quad \text{and} \quad dy = \frac{D}{2} \cos \theta \, d\theta$$

Elemental area dA= D cos θ dy

$$\therefore \, dA = (D \cos \theta) \, \frac{D}{2} \cos \theta \, d\theta = \frac{D^2}{2} \cos^2 \theta \, d\theta$$

$$A\bar{y} = \int_{y_1}^{D/2} y \, dA = \int_{\theta_1}^{\pi/2} \left(\frac{D}{2} \sin \theta \right) \frac{D^2}{2} \cos^2 \theta \, d\theta$$

$$= \frac{D^3}{4} \int_{\theta_1}^{\pi/2} \cos^2 \theta \sin \theta \, d\theta = \frac{D^3}{4} \left[-\frac{\cos^3 \theta}{3} \right]_{\theta_1}^{\pi/2} = \frac{D^3}{4} \left(\frac{\cos^3 \theta_1}{3} \right)$$

$$= \frac{D^3 \cos^3 \theta_1}{12}$$

Shear stress, $\quad \tau = \dfrac{V A \bar{y}}{I b} = \dfrac{V \left(\dfrac{D^3 \cos^3 \theta_1}{12} \right)}{I (D \cos \theta_1)} = \dfrac{V D^2}{12 \, I} \cos^2 \theta_1$

$$= \frac{V D^2}{12 \, I} (1 - \sin^2 \theta_1) = \frac{V D^2}{12 \, I} \left\{ 1 - \left(\frac{y_1}{D/2} \right)^2 \right\}$$

$$\therefore \quad \tau = \frac{V}{12 \, I} (D^2 - 4y_1{}^2)$$

Therefore, shear stress varies parabolically and its variation is explained in Fig. 4.24b.

Maximum shear stress ($\tau_{\max}$) occurs when $y_1 = 0$ (i.e. at neutral axis).

$$\therefore \tau_{max} = \frac{V D^2}{12\ I} = \frac{V D^2}{12\left(\dfrac{\pi D^4}{64}\right)} = \frac{16\ V}{3\ \pi D^2}$$

$$\text{Also } \tau_{max} = \frac{16\ V}{3\ \pi D^2} = \frac{4}{3}\,\frac{V}{\dfrac{\pi D^2}{4}} = \frac{4}{3}\left(\frac{V}{A}\right)$$

$$\therefore \quad \tau_{max} = \frac{4}{3}\,\tau_{ave}$$

Maximum shear stress is greater than average shear stress by $33\dfrac{1}{3}\%$.

(d) Triangular cross section

Consider a triangular cross section ABC as shown (Fig. 4.25a).

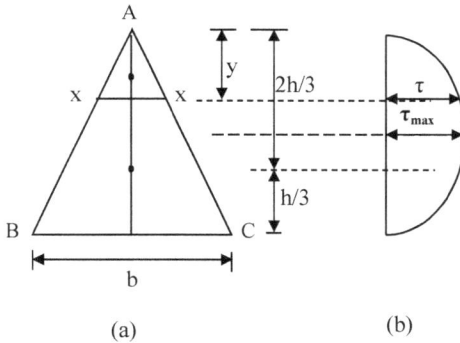

(a)   (b)

Fig. 4.25

Let x-x be a section at a distance y from the top fiber.

Here $\quad A = \dfrac{1}{2}\left(\dfrac{by}{h}\right) y = \dfrac{by^2}{2h} \quad$ and $\quad \bar{y} = \dfrac{2h}{3} - \dfrac{2y}{3} = \dfrac{2}{3}(h - y)$

Now shear stress,

$$\tau = \frac{V\left(\dfrac{by^2}{2h}\right)\left\{\dfrac{2}{3}(h - y)\right\}}{I\,\dfrac{by}{h}} = \frac{V}{3\ I}\,(h - y)y = \frac{V}{3\ I}\,(hy - y^2)$$

At neutral axis (passing through centroid), $\quad y = \dfrac{2h}{3}$

$$\tau_{NA} = \frac{V}{3I}\left\{h\left(\frac{2h}{3}\right)-\left(\frac{2h}{3}\right)^2\right\} = \frac{V}{3I}\left(\frac{2h^2}{3}-\frac{4h^2}{9}\right) = \frac{2Vh^2}{27I}$$

$$= \frac{2Vh^2}{27\left(\frac{bh^3}{36}\right)} = \frac{8V}{3bh} = \frac{4V}{3\left(\frac{bh}{2}\right)}$$

Average shear stress, $\quad \tau_{ave} = \dfrac{V}{\dfrac{bh}{2}}$

$$\therefore \quad \tau_{NA} = \frac{4}{3}\,\tau_{ave}$$

For maximum shear stress

$$\frac{d}{dy}(\tau) = 0$$

$$\frac{d}{dy}\left(\frac{V}{3I}(hy-y^2)\right) = 0$$

$$\frac{V}{3I}(h-2y) = 0 \qquad\qquad \therefore \quad y = \frac{h}{2}$$

$\therefore$ Maximum shear stress

$$\tau_{max} = \frac{V}{3I}\left[h\times\frac{h}{2}-\frac{h^2}{4}\right] = \frac{Vh^2}{12\times I} = \frac{Vh^2}{12\times\dfrac{bh^3}{36}}$$

$$= \frac{3V}{bh} = \frac{3V}{2\left(\frac{bh}{2}\right)}$$

$$\therefore \quad \tau_{max} = \frac{3}{2}\times\tau_{ave}$$

Example 4.22 *A square section with one diagonal horizontal is used as a beam. Draw the shear stress variation over the section. Find the magnitude of maximum shear stress and its location.*

Solution:

A square section under bending about its diagonal is as shown (Fig. 4.26a).

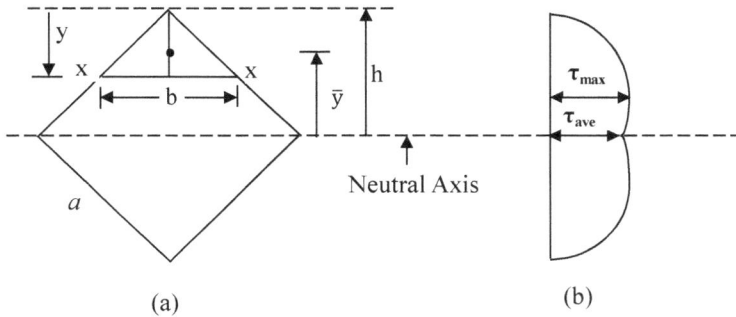

(a)   (b)

Fig. 4.26

a = side of square

Consider a section x-x at a distance y from top fiber and shear stress is required to be calculated here.

$$A = \frac{1}{2} b y \quad \text{and} \quad \bar{y} = h - \frac{2y}{3}$$

Moment of inertia of square about diagonal,

$$I = 2\left(a\sqrt{2} \times \frac{h^3}{12}\right) = \frac{a^4}{12}$$

∴ Shear stress, $\quad \tau = \dfrac{VA\bar{y}}{Ib} = \dfrac{V\left(\frac{1}{2} b y\right)\left(h - \frac{2y}{3}\right)}{\left(\frac{a^4}{12}\right)b} = \dfrac{6Vy}{a^4}\left(h - \frac{2y}{3}\right)$

∴ Shear stress variation is Parabolic.

At $y = 0$, $\tau = 0$

At neutral axis, $y = h$

$$\therefore \ \tau_{NA} = \frac{6Vh}{a^4}\left(h - \frac{2h}{3}\right) = \frac{2Vh^2}{a^4} = \frac{2V\left(\frac{a}{\sqrt{2}}\right)^2}{a^4} = \frac{V}{a^2}$$

But $\dfrac{V}{a^2} = \tau_{ave}$ (a^2 is the area of cross section)

$$\therefore \ \tau_{NA} = \tau_{ave}$$

For maximum shear stress $\quad \dfrac{d\tau}{dy} = 0$

$$\frac{d}{dy}\left[\frac{6Vy}{a^4}\left(h - \frac{2y}{3}\right)\right] = 0 \quad \text{or} \quad h - \frac{4}{3}y = 0$$

$$\therefore \ y = \frac{3}{4}\,h$$

Now maximum shear stress

$$\tau_{max} = \frac{6V}{a^4}\left(\frac{3}{4}h\right)\left(h - \frac{2}{3}\times\frac{3}{4}\,h\right) = \frac{9\,Vh^2}{4\,a^4} = \frac{9V}{4a^4}\left(\frac{a}{\sqrt{2}}\right)^2 = \frac{9}{8}\left(\frac{V}{a^2}\right)$$

$$\therefore \ \tau_{max} = \frac{9}{8}\times\tau_{ave}$$

The maximum shear stress occurs when $\ y = \dfrac{3}{4}\times h = 0.53\,a$

The variation of shear stress is explained in Fig. 4.26b.

Example 4.23 *A timber beam 100 mm wide × 150 mm deep carries u.d.l. over a span of 2m. If the permissible stresses are 28 N/mm² longitudinally and 2 N/mm² in transverse shear, calculate the maximum load per meter run which can safely be carried.*

Solution:

Longitudinal stress → Normal stress (Bending stress)
Transverse stress → Shear stress

For the rectangular cross section (100 mm wide × 150 mm deep)

$$I = \frac{100 \times 150^3}{12} = 2.812 \times 10^7 \text{ mm}^4$$

$$A = 100 \times 150 = 1.5 \times 10^4 \text{ mm}^2$$

Let w = uniformly distributed load (N/mm)

Maximum bending moment for the loaded beam is

$$M_{max} = \frac{w \times 2000^2}{8} = 5 \times 10^5 \text{ w} \quad \text{Nmm}$$

maximum shear force, $\quad V_{max} = \frac{wl}{2} = w \times \frac{2000}{2} = 1000 \text{ w} \quad \text{N}$

Maximum bending stress, $\quad \sigma_{max} = \frac{M}{I} y_{max}$

$\sigma_{max} \leq 28 \text{ N/mm}^2$

$$28 = \frac{5 \times 10^5 w}{2.812 \times 10^7} \times \frac{150}{2} \qquad \therefore \quad w = 20.99 \text{ N/mm}$$

Maximum shear stress $\quad \tau_{max} = 1.5 \frac{V}{A} \quad$ (for rectangular section)

$\tau_{max} \leq 2 \text{ N/mm}^2$

$$2 = 1.5 \times \frac{1000 \text{ w}}{1.5 \times 10^4} \qquad \therefore \quad w = 20 \text{ N/mm}$$

∴ Load that can safely be carried is 20 N/mm.

Example 4.24 *A laminated wooden beam is made up of three planks 100 mm × 50 mm glued together to form a solid cross section 100 mm wide and 150 mm deep. The allowable shear stress in the glued joint is 0.35 N/mm². If the beam is 1.8 m long and simply supported at the ends, what is the safe load P that can be carried at mid span?*

Solution:

Refer Fig. 4.27

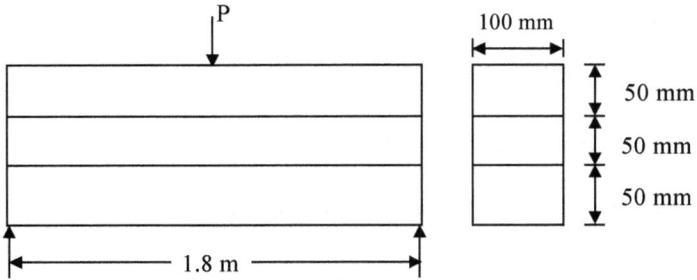

Fig. 4.27

Maximum shear stress for glued joint is 0.35N/mm²

$$I = \frac{100 \times 150^3}{12} = 28.12 \times 10^6 \ mm^4$$

Shear force, $V = \dfrac{P}{2}$

(P = safe load on midspan)

Shear stress induced in the glued joint, τ

$$= \frac{VA\bar{y}}{I\,b}$$

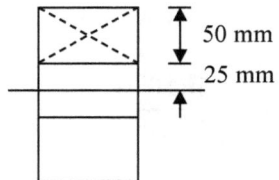

Fig. 4.27a

$$0.35 = \frac{\dfrac{P}{2}(100 \times 50) \times 50}{28.12 \times 10^6 \times 100}$$

$\therefore P = 7873 \ N$

$\therefore$ Safe load at mid span, P = 7873 N

Example 4.25 *A 50 mm × 120 mm I beam is subjected to a shearing force of 10 kN. Calculate the value of transverse shear stress at the neutral axis. Find what percentage of total shear force is carried by the web. Given I = 2.18×10⁶ mm⁴, A = 940 mm².*
Web thickness = 3.5 mm, Flange thickness = 5.5 mm.

Solution:

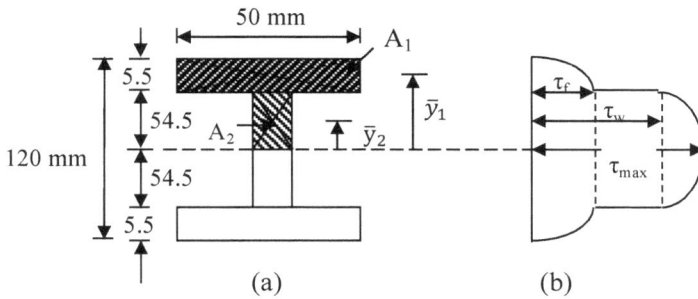

Fig. 4.28

Refer Fig. 4.28a

Shear stress in the flange, $\tau_f = \dfrac{VA_1\bar{y}_1}{I\,b}$

$$= \frac{(10000)(50 \times 5.5)(60 - \frac{5.5}{2})}{2.18 \times 10^6 \times 50} = 1.44 \text{ N/mm}^2$$

Shear stress in the web, $\tau_w = \dfrac{VA_1\bar{y}_1}{I\,t}$

$$= \frac{(10000)(50 \times 5.5)\left(60 - \frac{5.5}{2}\right)}{2.18 \times 10^6 \times 3.5} = 20.63 \text{ N/mm}^2$$

Maximum shear stress (occurs at the neutral axis)

$$\tau_{max} = \frac{VA\bar{y}}{I\,t} = \frac{V(A_1\bar{y}_1 + A_2\bar{y}_2)}{I\,t}$$

$$= \frac{(10000)\left[(50 \times 5.5)(60 - \frac{5.5}{2}) + (54.5 \times 3.5)\left(\frac{54.5}{2}\right)\right]}{2.18 \times 10^6 \times 3.5}$$
$$= 27.44 \text{ N/mm}^2$$

Shear stress distribution diagram is shown (Fig. 4.28b).

Total shear force (V)
$$= \text{ shear resisted by web}(V_w)$$
$$+ \text{ shear resisted by flange}(V_f)$$

Shear force resisted by web can be calculated as

$$V_w = 20.63(109 \times 3.5) + \frac{2}{3}(27.44 - 20.63)(109 \times 3.5) = 9602 \text{ N}$$

$$\therefore V_f = V - V_w = (10000 - 9602) = 398 \text{ N}$$

$$\therefore \text{Percentage shear force resisted by the web} = \frac{V_w}{V} \times 100$$

$$= \frac{9602}{10000} \times 100 = 96.02 \%$$

$$\text{Percentage shear force resisted by the flange} = \frac{V_f}{V} \times 100$$

$$= \frac{398}{10000} \times 100 = 3.98 \%$$

Note:

1. In case of I section beam, most of the shear about 96 % is carried by the web. But most of the bending moment is carried by flange.
2. Function of web is to resist shear force while that of flange is to resist bending moment.

Example 4.26 *The shear force acting at the section of beam shown (Fig. 4.29) is 40 kN. Show the shear stress distribution along the depth with the values at the neutral axis and at the junction of web and flanges.*

Solution:

To locate the centroidal axis (Refer Fig. 4.9a)

All dimensions are in mm

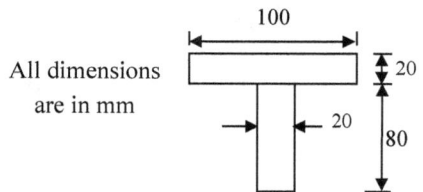

Fig. 4.29

$$y_L = \frac{A_1 y_1 + A_2 y_2}{A_1 + A_2}$$

$$= \frac{(100 \times 20)90 + (20 \times 80)40}{(100 \times 20) + (20 \times 80)} = 67.77 \text{ mm}$$

$\therefore y_u = 100 - 67.77 = 32.33$ mm

$$I = \frac{100 \times 20^3}{12} + (100 \times 20)\left(32.33 - \frac{20}{2}\right)^2 + \frac{20 \times 80^3}{12}$$
$$+ (80 \times 20)\left(67.77 - \frac{80}{2}\right)^2$$

$= 3.14 \times 10^6$ mm^4

At the flange web junction:

Shear stress in the flange, $\tau_f = \dfrac{VA_1\bar{y}_1}{I\,b}$

$$= \frac{(40 \times 10^3)(100 \times 20)\left(32.23 - \frac{20}{2}\right)}{3.14 \times 10^6 \times 100}$$

$= 5.66$ N/mm^2

Shear stress in the web, $\tau_w = \dfrac{VA_1\bar{y}_1}{I\,t}$

$$= \frac{(40 \times 10^3)(100 \times 20)\left(32.23 - \frac{20}{2}\right)}{3.14 \times 10^6 \times 20}$$

$= 28.81$ N/mm^2

Maximum shear stress,

$$\tau_{max} = \frac{V(A_1\bar{y}_1 + A_2\bar{y}_2)}{I.t}$$

$$= \frac{(100 \times 10^3)\left[(100 \times 20)\left(32.23 - \frac{20}{2}\right) + (20 \times 12.33)\left(\frac{12.23}{2}\right)\right]}{3.14 \times 10^6 \times 20}$$

$= 73.17$ N/mm^2

Example 4.27 *A T-section member has dimensions as shown in Fig. 4.30. The member is used as a simply supported beam of span 1.5 m, the flange being horizontal. Calculate the uniformly distributed load which can be applied over the entire span such that maximum shearing stress induced in the cross section is not to exceed 30 MN/mm². What is the maximum bending stress induced in the beam.*

Solution:

To find the location of centroid (refer Fig 4.30a)

$$\bar{y} = \frac{(150 \times 10)95 + (10 \times 90)45}{(150 \times 10) + (10 \times 90)}$$

$$= 76.25 \text{ mm}$$

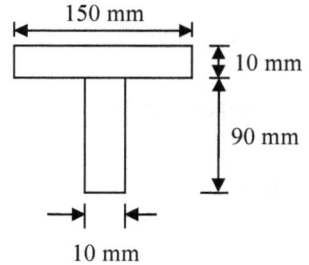

150 mm

10 mm

90 mm

10 mm

Fig. 4.30

$$I = \frac{150 \times 10^3}{12} + (150 \times 10)\left(23.75 - \frac{10}{2}\right)^2 + \frac{10 \times 90^3}{12}$$
$$+ (10 \times 90)\left(76.25 - \frac{90}{2}\right)^2$$
$$= 2.026 \times 10^6 \text{ mm}^4$$

Maximum bending moment for the beam

$$M_{max} = \frac{w \times 1500^2}{8} = (2.81 \times 10^5)w \text{ N mm}$$

Maximum shear force

$$V_{max} = \frac{wl}{2} = \frac{w \times 1500}{2} = 750 \text{ w N}$$

Maximum shear stress, $\tau_{max} = \dfrac{V A \bar{y}}{I t}$

$$30 = \frac{(750 \text{ w})(76.25 \times 10) \times \frac{76.25}{2}}{2.026 \times 10^6 \times 10}$$

$$\therefore \text{ w} = 27.88 \text{ N/mm}$$

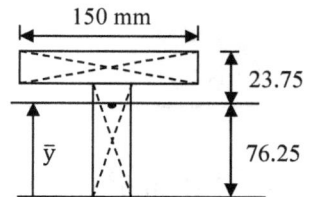

150 mm

23.75

$\bar{y}$

76.25

Fig. 4.30a

$\therefore$ Maximum bending moment

$$M_{max} = (2.81 \times 10^5)w = 2.81 \times 10^5 \times 27.88 = 7.83 \times 10^6 \text{ Nmm}$$

Maximum bending stress

$$\sigma_L = \frac{M}{I} y_L = \frac{7.83 \times 10^6}{2.026 \times 10^6} \times 76.25 = 294.8 \text{ N/mm}^2 \text{ (tensile)}$$

$[\sigma_L \text{ is } \sigma_{max} \text{ as } y_L > y_u]$

If required we can calculate

$$\sigma_u = \frac{M}{I} y_U = \frac{7.83 \times 10^6}{2.026 \times 10^6} \times 23.75 = 91.78 \text{ N/mm}^2 \text{(compressive)}$$

Example 4.28 *A simply supported beam of rectangular cross section and length l is subjected to a concentrated load W at distance of l/5 from the left support. The ratio of maximum allowable stress in bending and shear is 6:1. Find the ratio of length to depth of beam so that both bending and shear stresses reach the maximum allowable limits simultaneously.*

Solution:

Refer Fig. 4.31

Reactions at A and B

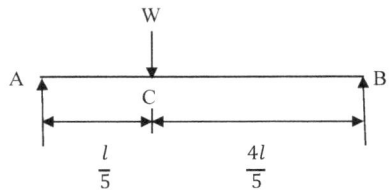

Fig. 4.31

$$\Sigma M_B = 0 \quad \rightarrow \quad R_A \times l = W \times \frac{4}{5} l$$

$$\therefore R_A = \frac{4}{5} W \quad , \quad R_B = \frac{W}{5}$$

Maximum bending moment (at C)

$$M_{max} = R_A \times \frac{l}{5} = \left(\frac{4}{5} W\right) \frac{l}{5} = \frac{4}{25} Wl$$

Maximum shear force, $V = R_A = \frac{4}{5} W$

Maximum bending stress $\sigma_{max} = \dfrac{M}{I}\, y = \dfrac{\left(\frac{4}{25}\, Wl\right)}{\left(\frac{bd^3}{12}\right)} \times \dfrac{d}{2} = \dfrac{24}{25}\left(\dfrac{Wl}{bd^2}\right)$

Maximum shear stress,

$$\tau_{max} = 1.5 \times \tau_{ave} = 1.5 \times \dfrac{V}{bd} \qquad \text{(for rectangular cross section)}$$

Given that $\dfrac{\sigma_{max}}{\tau_{max}} = 6$

$$\therefore \dfrac{\frac{24}{25}\left(\frac{Wl}{bd^2}\right)}{1.5 \times \frac{V}{bd}} = 6$$

$$\dfrac{24}{25}\left(\dfrac{Wl}{bd^2}\right) = 6 \times 1.5 \times \dfrac{\frac{4}{5}W}{bd}$$

$$\therefore \dfrac{l}{d} = \dfrac{15}{2} = 7.5 \quad \rightarrow \quad \text{Ratio of length to depth of beam.}$$

Example 4.29 *A 100 mm × 500 mm rectangular wooden beam supports a 40 kN load as shown (Fig. 4.32). At a section a-a, the grain of wood makes an angle of 20° with the axis of beam. Find the shear stresses along the grain of wood at the point Q caused by the applied concentrated load.*

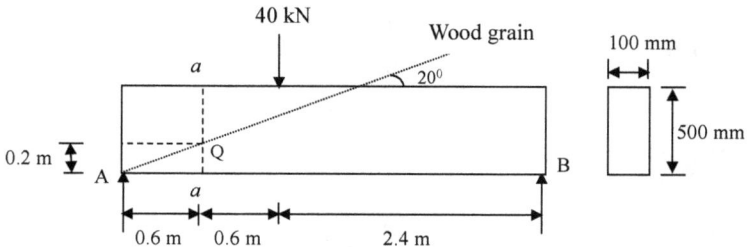

Fig. 4.32

Solution:

$$\Sigma M_B = 0 \quad \rightarrow \quad R_A \times 3.6 = 40 \times 2.4$$

$\therefore R_A = 26.67$ kN

Bending moment at section *a-a* is

$M = R_A \times 0.6 = 26.67 \times 0.6 = 16$ kN m $= 16 \times 10^6$ N mm

Shear force at section *a-a* is

$V = R_A = 26.67$ kN

$$I = \frac{100 \times 500^3}{12} = 1.04 \times 10^9 \text{ mm}^4$$

Bending stress at point Q

$$\sigma = \frac{M}{I} y = \frac{16 \times 10^6}{1.04 \times 10^9} \times 50 = 0.769 \text{ N/mm}^2 \text{ (Tensile)}$$

Shear stress at point Q (refer Fig. 4.32a)

$$\tau = \frac{VA\bar{y}}{I b}$$

$$= \frac{(26.67 \times 10^3) \times (100 \times 200)150}{1.04 \times 10^9 \times 100}$$

$= 0.769$ N/mm²

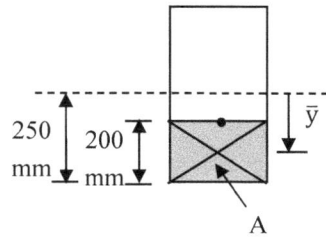

Fig. 4.32a

Consider element at point Q at section *a-a*

Refer Fig. 4.32b

$\sigma_x = 0.769$ N/mm²

$\sigma_y = 0$

$\tau = -0.769$ N/mm²

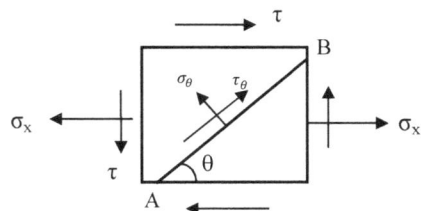

Fig. 4.32b

$$\tau_\theta = \left(\frac{\sigma_y - \sigma_x}{2}\right) \sin 2\theta + \tau \cos 2\theta$$

$$= \frac{0 - 0.769}{2} \sin(2 \times 20) + (-0.769) \cos(2 \times 20)$$

$$= -0.835 \text{ N/mm}^2$$

$\therefore$ Shear stress along the grain $= 0.835 \dfrac{\text{N}}{\text{mm}^2}$ (in magnitude)

4.9 BEAM OF UNIFORM STRENGTH

When a beam is designed by maintaining its cross section in such a way that the maximum bending stress developed at all the cross sections is same, then the beam is known as beam of uniform strength. This can be achieved by

- Maintaining constant width and varying depth.
- Maintaining constant depth and varying width
- Varying both width and depth

A beam of uniform strength is an ideal beam from economic considerations.

We know that safe moment $M = \sigma_{all} \times Z$

Here allowable stress (σ_{all}) is constant and section modulus (Z) can be computed after finding bending moment (M). For the required section modulus, different sections can be designed by varying the width or/and depth.

Example 4.30 *A simply supported beam of span 6 m is subjected to a point load of 20 kN at mid span. Find the cross section of the beam for uniform strength by keeping (i) depth of 300 mm throughout constant (ii) width of 200 mm throughout constant. Permissible stress due to bending is 8 N/mm².*

Solution:

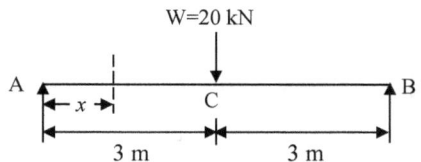

Fig. 4.33

The loaded beam is shown (Fig. 4.33)

Bending moment at a section at distance x from end A is

$$M_x = \frac{W}{2} x = \frac{20000}{2} x = 10000 x \quad N \, mm$$

Safe moment,
$$M = \sigma_{all} \times Z \qquad (\sigma_{all} \rightarrow \text{Allowable stress}, Z \rightarrow \text{section modulus})$$

$$\therefore 10000 \, x = \sigma_{all} \times \frac{bd^2}{6} = 8 \times \frac{bd^2}{6}$$

or $bd^2 = 7500 \, x$

(i) keeping depth of 300 mm constant

$$b = \frac{7500 \, x}{(300)^2} = (0.0833) \, x \qquad \text{(Linear variation of width)}$$

At $x = 0$, $b = 0$, At $x = 3000 \, mm$, $b = 250 \, mm$

The beam having uniform strength with constant depth is explained in Fig. 4.33a.

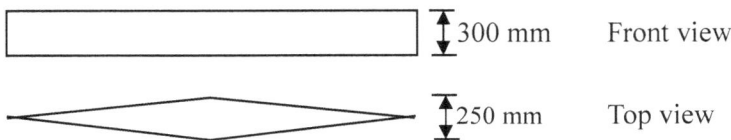

300 mm   Front view

250 mm   Top view

Fig. 4.33a

(ii) keeping width of 200 mm constant

$$bd^2 = 7500 \, x$$

$$d = \sqrt{\frac{7500 \, x}{200}} = 6.123\sqrt{x} \qquad \text{(Parabolic variation of width)}$$

At $x = 0$, $d = 0$

At $x = 3000$ mm, $d = 335.3$ mm

The beam of uniform strength having constant width and varying depth is explained in Fig. 4.33b.

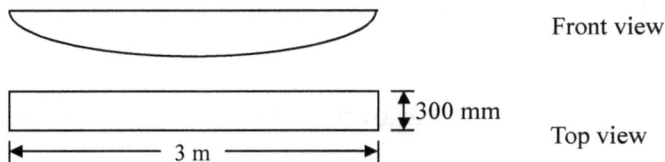

Front view

Top view

300 mm

3 m

Fig. 4.33b

4.10 COMPOSITE BEAMS (Flitched Beams)

Composite beam is a type of beam made of more than one material. The different materials are rigidly connected so that there is no slip at the common faces. Figure 4.34 presents a composite beam made of wood and steel as shown.

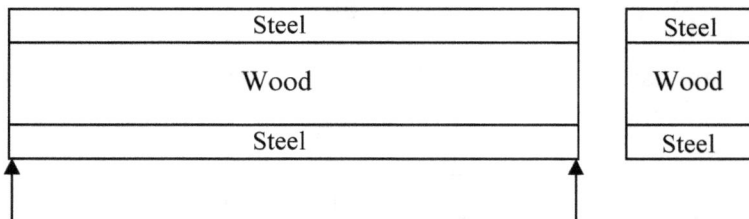

Fig. 4.34

In composite beams at the junction of the two materials, same strain occurs in both the materials.

Strain in steel = Strain in wood

$$\frac{\sigma_s}{E_s} = \frac{\sigma_w}{E_w}$$

$$\therefore \quad \sigma_s = \frac{E_s}{E_w} \sigma_w = m \, \sigma_w \quad \text{where } m = \frac{E_s}{E_w} = \text{Modular Ratio}$$

At the junction, Stress in steel $= m \times$ Stress in wood

Suppose the composite beam is replaced by an equivalent beam made of a single material, say steel is converted into equivalent wood (Fig. 4.35).

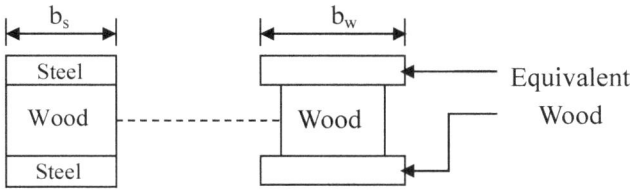

Fig. 4.35

Moment of resistance (moment carrying capacity) of original composite beam (M_c) and that of equivalent beam (M_e) shall be same.

$$M_o = M_e$$

Keeping the lever arm same, force carried by original beam and equivalent beam are same.

$$F_s = F_w \quad \text{(subscripts s and w denote steel and wood, respectively)}$$

$$\text{or} \quad \sigma_s \times A_s = \sigma_w \times A_w \quad \rightarrow \quad \frac{\sigma_s}{\sigma_w} \times A_s = A_w$$

$$\therefore \quad A_w = \text{modular ratio} \times A_s$$

$$b_w \times t_w = \text{Modular ratio} \times b_s \times t_s \quad \text{(b and t are width and depth)}$$

$$\therefore \quad b_w = \text{modular ratio} \times b_s$$

i.e. **width of equivalent wood = modular ratio × width of steel**

The stress distributions with the equivalent wood as well as for the original beam cross section are explained in Figs. 4.36a and 4.36b respectively.

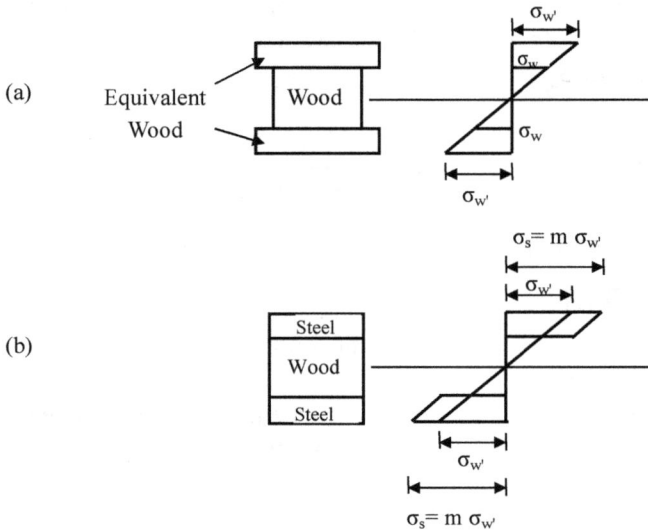

(a) Equivalent
     Wood

(b)

Fig. 4.36

Example 4.31 *A simply supported composite beam 3 m long carries a point load of 5 kN at the mid span. The beam is constructed with wood 100 mm wide and 150 mm deep reinforced on its lower side by a steel plate 8 mm thick and 100 mm wide. Find the maximum bending stresses in the steel and the wood. Given E_{wood}=10 GPa, E_{steel}=210 GPa.*

Solution:

Maximum bending moment,

$$M = \frac{wl}{4} = \frac{5 \times 3}{4} = 3.75 \text{ kNm}$$

Modular ratio, $m = \dfrac{E_s}{E_w} = \dfrac{210}{10} = 21$

∴ Width of equivalent wood = modular ratio × width of steel = 21 × 100 = 2100 mm

The beam cross section and the equivalent wood cross section are shown in Figs. 4.37a and 4.37b respectively.

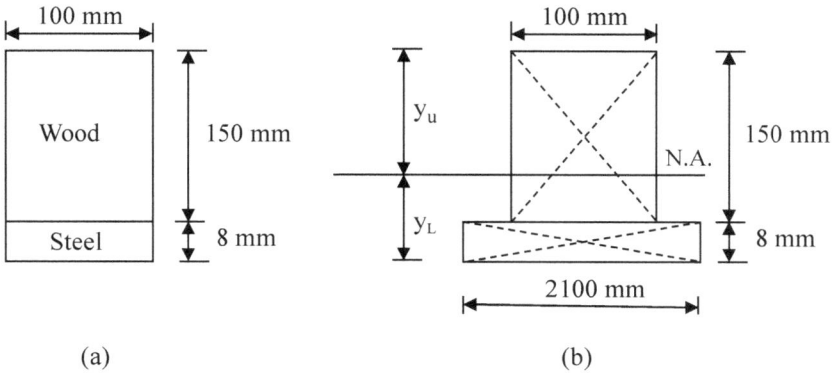

(a) (b)

Fig. 4.37

To locate the neutral axis (Refer Fig. 4.37b)

$$y_L = \frac{(2100 + 8)4 + (100 \times 150)(8 + 75)}{(2100 \times 8) + (100 \times 150)} = 41.26 \text{ mm}$$

$$\therefore \ y_U = 158 - 41.26 = 116.74 \text{ mm}$$

$$I = \frac{100 \times 150^3}{12} + (100 \times 150)(116.74 - 75)^2 + \frac{2100 \times 8^3}{12}$$
$$+ (2100 \times 8)(41.26 - 4)^2$$

$$= 78.3 \times 10^6 \text{ mm}^4$$

$$\sigma_w = \frac{M}{I_w} \, y_u = \frac{3.75 \times 10^6}{78.3 \times 10^6} \times 116.74 = 5.59 \text{ N/mm}^2$$

$$\sigma_{w\prime} = \frac{M}{I_w} \, y_L = \frac{3.75 \times 10^6}{78.3 \times 10^6} \times 41.26 = 1.97 \text{ N/mm}^2$$

The stress distribution in the beam cross section is presented in Fig. 4.38.

Maximum stress in wood, $\sigma_w = 5.59$ N/mm^2 (see Fig. 4.38a)

Maximum stress in steel, $\sigma_s =$ modular ratio $\times \sigma_{w\prime} = 21 \times 1.97 = 41.37$ N/mm^2 (see Fig. 4.38b)

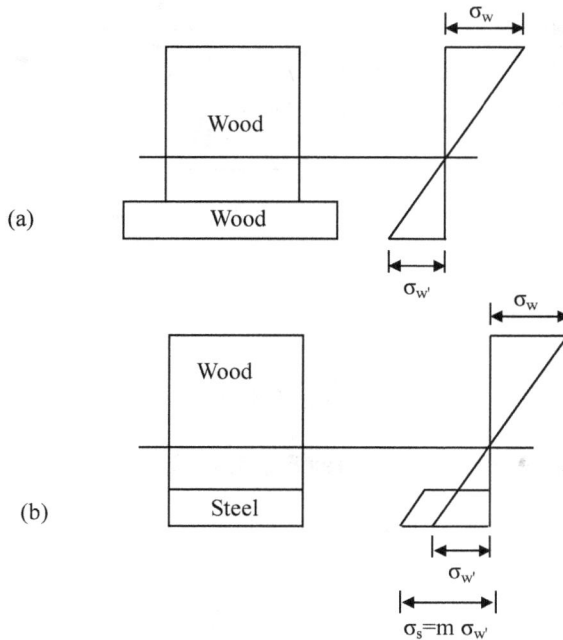

Fig. 4.38

Example 4.32 *Determine the allowable bending moment around horizontal neutral axis for the composite beam of wood and steel shown in Fig. 4.39. Materials are fastened to act as a unit. Young's modulus of wood and steel are 8×10^9Pa and 200×10^9Pa. Allowable bending stress in steel is 140 MPa and that in wood is 8 MPa.*

Solution:

Modular ratio $\quad m = \dfrac{E_s}{E_w} = \dfrac{200}{8} = 25$

Width of equivalent wood

= m × width of steel = 25 × 200

= 5000 mm

Refer Fig. 4.40a

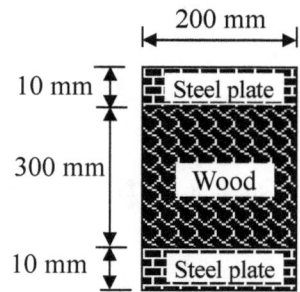

Fig. 4.39

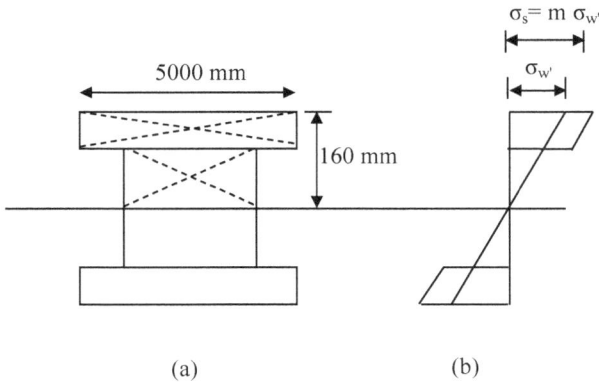

(a)   (b)

Fig. 4.40

$$I = 2\left[\frac{5000\times10^3}{12} + (5000 \times 10) \times 155^2\right] + \frac{2100\times300^3}{12}$$

$$= 2.853\times10^9 \text{mm}^4$$

Case-I : Stress in wood shall not exceed 8 N/mm^2 i.e. $\sigma_w \le$ 8 N/mm^2

$$\sigma_w = \frac{M}{I}y$$

$$\sigma_w = \frac{M}{2.853\times10^9}\times150$$

$$\therefore M = 152.17\times10^6 \text{ N mm}$$

Case – II : stress in steel shall not exceed 140 N/mm^2 i.e. $\sigma_s \le$ 140 N/mm^2

Stress in equivalent wood $\le \dfrac{140}{25}$ i. e. $\sigma_{w\prime} \le 5.6$ N/mm^2

$$5.6 = \frac{M}{2.853\times10^9}\times160$$

$$\therefore M = 99.85\times10^6 \text{ Nmm}$$

$$\therefore \text{ Safe moment} = 99.85\times10^6 \text{ Nmm} = 99.85 \text{ kNm}$$

Example 4.33 *Two rectangular bars one of brass other of steel are placed together to form a beam as shown (Fig 4.41). Determine maximum allowable load P which can be applied on the beam if the bars are (i) firmly secured to each other throughout the length (ii) separate and can bend independently.*

Permissible stresses in brass and steel are 75 N/mm² and 150 N/mm², respectively. $E_b = 0.8 \times 10^5$ N/mm², $E_s = 2 \times 10^5$ N/mm².

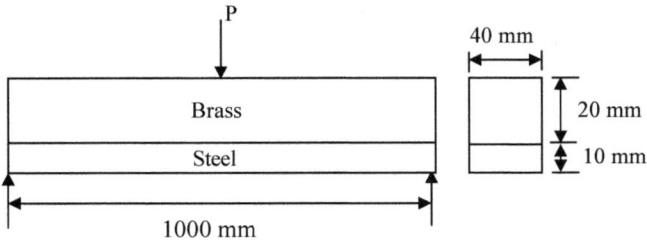

Fig. 4.41

Solution:

(i) Beams are rigidly connected:

Maximum bending moment

$$M = \frac{Pl}{4} = \frac{P \times 1000}{4} = 250\ P \ \text{N mm}$$

Modular ratio,　　$m = \dfrac{E_s}{E_b} = \dfrac{2 \times 10^5}{0.8 \times 10^5} = 2.5$

Width of equivalent brass = m × width of steel = 2.5 × 40 = 100 mm

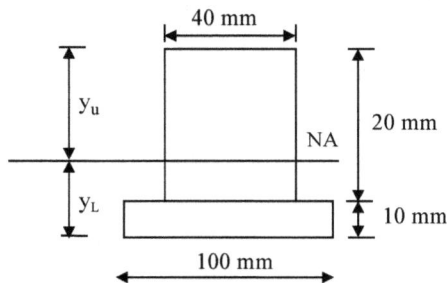

Fig. 4.42

Refer Fig. 4.42

$$y_L = \frac{(40 + 20)20 + (100 \times 10)5}{(40 \times 20) + (100 \times 10)} = 11.67 \text{ mm}$$

$$y_u = 30 - 11.67 = 18.33 \text{ mm}$$

$$I = \frac{40 \times 20^3}{12} + (40 \times 20)\left(18.33 - \frac{20}{2}\right)^2 + \frac{100 \times 10^3}{12}$$
$$+ (100 \times 10)\left(11.67 - \frac{10}{2}\right)^2$$

$$= 135 \times 10^3 \text{ mm}^4$$

Stress in brass shall not exceed 75 N/mm^2 :

$$\sigma = \frac{M}{I}y$$

$$75 = \frac{250 \, P}{135 \times 10^3} \times 18.33 \qquad \rightarrow \qquad P = 2.209 \times 10^3 \text{ N}$$

Stress in steel shall not exceed 150 N/mm^2 :

i.e. stress in equivalent brass , $\sigma_{b'} \leq 60$ N/mm^2

$$60 = \frac{250 \, P}{135 \times 10^3} \times 11.67 \quad \rightarrow \qquad P = 2.776 \times 10^3 \text{ N}$$

$\therefore$ Safe load, $P = 2.209 \times 10^3$ N \qquad (lesser of the two values)

(ii) Beams are separate and can bend independently :

Bending moment

$$M = \frac{Pl}{4} = \frac{P \times 1000}{4} = 250 \, P$$

Let M_s = bending moment resisted by steel

M_b = bending moment resisted by brass

$\therefore M_s + M_b = M = 250P$ — (1)

From the bending stress equation, radius of curvature $R = \dfrac{EI}{M}$

Assuming same radius of curvature for both steel and brass, $R_s = R_b$

$\therefore \dfrac{E_s I_s}{M_s} = \dfrac{E_b I_b}{M_b}$

$\therefore \dfrac{M_s}{M_b} = \dfrac{E_s I_s}{E_b I_b} = \dfrac{2 \times 10^5 \times 3.33 \times 10^3}{0.8 \times 10^5 \times 26.67 \times 10^3} \quad \rightarrow \quad \dfrac{M_s}{M_b} = 0.312$ — (2)

$\left[I_b = \dfrac{40 \times 20^3}{12} = 26.67 \times 10^3 \, \text{mm}^4, \quad I_s = \dfrac{40 \times 10^3}{12} = 3.33 \times 10^3 \, \text{mm}^4 \right]$

From (1) and (2)

$M_s = 59.52 \, P, \quad M_b = 190.48 \, P$

Consider brass :

$\sigma_b = \dfrac{M_b}{I_b} y_b \qquad \rightarrow \qquad 75 = \dfrac{190.48P}{26.67 \times 10^3} \times \dfrac{20}{2}$

$\therefore P = 1050.1 \, N$

Consider steel :

$\sigma_s = \dfrac{M_s}{I_s} y_s \qquad \rightarrow \qquad 150 = \dfrac{59.52P}{3.33 \times 10^3} \times \dfrac{10}{2}$

$\therefore P = 1680 \, N$

Safe load P = 1050.1 N (Lesser of the two value)

Note: Load carrying capacity of connected beam is more. Connected beam will be stronger and stiffer compared to that without connection.

4.11 COMBINED DIRECT AND BENDING STRESSES

Consider the case when structural members such as columns or pillars are loaded with vertical loads. If the load is acting through the centroid of cross section, then direct stress (compressive in nature) will be induced in the member (Fig. 4.43a). If the load is not acting through the centroid of cross section then the loading is known as *Eccentric loading*. The distance between the axis of member and the line of action of load is called eccentricity 'e' (Fig 4.43b).

Due to eccentric loading, bending moment is additionally developed in the member thereby causing additional bending stresses. Therefore due to eccentric loading, the cross section is subjected to both direct stress as well as bending stress.

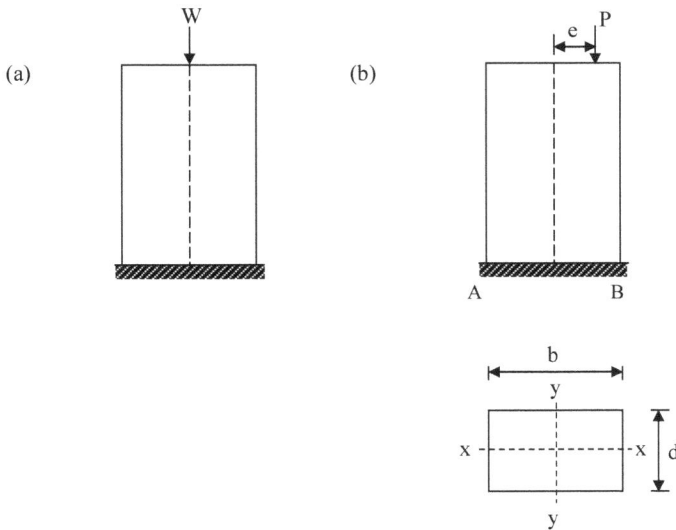

Fig. 4.43

Consider a short column of rectangular cross section as shown in Fig. 4.43b.

P = Eccentric point load

e = Eccentricity

Area of cross section, $A = b \times d$

Direct stress, $\qquad \sigma_d = \dfrac{P}{A}$

Due to eccentric loading, bending moment is developed.

Bending moment, $\qquad M = P\,e$

Bending stress, $\qquad \sigma_b = \dfrac{M}{Z} = \dfrac{P\,e}{Z}$

where the section modulus, $\qquad Z = \dfrac{db^2}{6} \qquad\qquad$ (Z is about y axis)

The bending stress may be compressive or tensile depending upon the location of the point of interest with respect to the neutral axis.

Resultant stress will be algebraic sum of direct and bending stress.

For the loading shown in Fig. 4.43b

Maximum stress at B, $\qquad \sigma_{max} = \sigma_d + \sigma_b = \dfrac{P}{A} + \dfrac{P\,e}{Z}$

Minimum stress at A, $\qquad \sigma_{min} = \sigma_d - \sigma_b = \dfrac{P}{A} - \dfrac{P\,e}{Z}$

The value of σ_{min} at A positive means the resultant stress is compressive and negative means the resultant stress is tensile in nature.

Cases:

(i) If $\sigma_d > \sigma_b$, then the stresses on the section are wholly compressive.

(ii) If $\sigma_d = \sigma_b$, then $\sigma_{max} = \sigma_d + \sigma_b = 2\sigma_d$ and $\sigma_{min} = \sigma_d - \sigma_b = 0$
This means the stresses on the section are compressive and the stress intensity varies from zero at one extremity to maximum value (compressive) at the other.

(iii) If $\sigma_d < \sigma_b$ then $\sigma_{max} = \sigma_d + \sigma_b$ and $\sigma_{min} = \sigma_d - \sigma_b = -(\sigma_b - \sigma_d)$

In this case, σ_{max} will be compressive but σ_{min} will be tensile.

4.11.1 Condition for the stresses to be wholly compressive

For the stresses to be wholly compressive (i.e. no tensile stress) the condition is

$$\sigma_d \geq \sigma_b$$

or $\quad \dfrac{P}{A} \geq \dfrac{Pe}{Z}$

$\therefore \ e \leq \dfrac{Z}{A}$

$\rightarrow \quad$ *Condition for the stresses to be wholly compressive*

For rectangular section:

Refer Fig. 4.44a

$$Z = \frac{db^2}{6}, \quad A = bd \quad \therefore \quad \frac{Z}{A} = \frac{b}{6}$$

Therefore, $\quad e \leq \dfrac{b}{6} \quad$ is the condition

For rectangular sections, eccentricity 'e' must be less than or equal to b/6 from y-y axis so that the stress becomes wholly compressive. This means if the load is applied at a distance less than or equal to b/6 from y-y axis on either side of the axis, then no tensile stress will be developed in the section. Hence the limit of eccentricity on either side of the y-y axis so as not to develop any tensile stress is within middle third of the base.

Similarly, if the load is eccentric with respect to x-x axis, then the limit of eccentricity on either side of x-x is d/6 so as not to develop any tensile stress anywhere in the section. Hence the range within which the load may be applied is within the middle third of the respective dimension. This is known as **middle third rule for rectangular section**. With reference to Fig. 4.44a, the condition for not to develop tensile stresses is when the load is applied anywhere within the diamond-shaped figure ABCD whose diagonals are AC = b/3 and

BD = d/3. This figure ABCD within which the load should be placed so as not to develop any tensile stress represents **Core or Kernel** of the section.

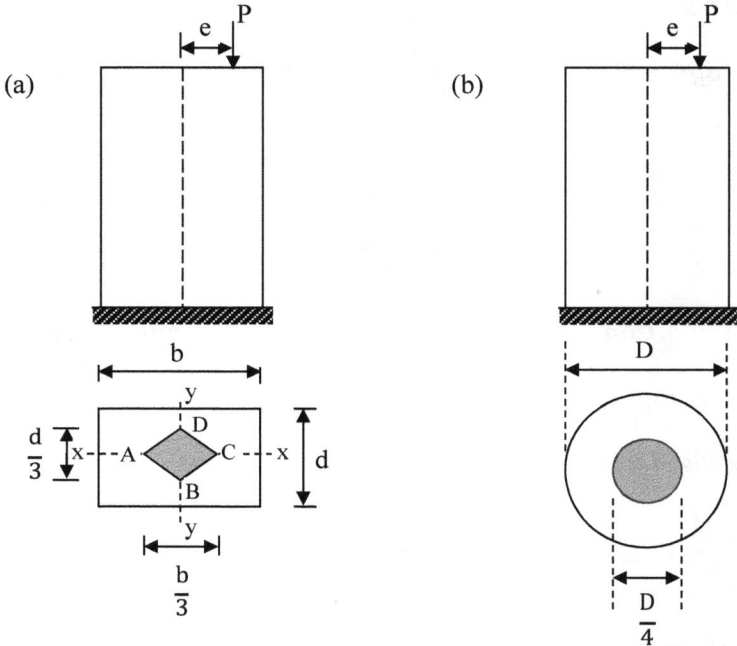

Fig. 4.44

For the rectangular section shown (Fig. 4.44a), the core or kernel is represented by the shaded portion ABCD.

$$\text{Core area} = 2 \times \frac{1}{2} \times \frac{b}{3} \times \frac{d}{6} = \frac{bd}{18}$$

For circular section:

Refer Fig 4.44b

$$Z = \frac{\pi D^3}{32}, \qquad A = \frac{\pi D^2}{4} \qquad \therefore \quad \frac{Z}{A} = \frac{D}{8}$$

Therefore, $e \le \dfrac{D}{8}$ is the condition

In other words, if the load be applied anywhere within a concentric circle of diameter D/4, the stress will be wholly compressive. This is known as **middle quarter rule for circular section**.

The core or kernel of the solid circular section is represented by a circle of diameter d/4 shown shaded in the section (Fig. 4.44b).

$$\text{Core area} = \frac{\pi}{4} \times \left(\frac{D}{4}\right)^2 = \frac{\pi D^2}{64}$$

HIGHLIGHTS

Definitions

1. **Pure bending**: In pure bending, beam is subjected to constant bending moment and no shear force. Due to this the beam bends as an arc of a circle.

2. **Neutral plane** (or neutral layer): It is a layer in which the longitudinal fibers do not change in length. At neutral plane, stress and strain are zero.

3. **Neutral Axis:** It is the line of intersection of neutral plane with the beam cross section. At neutral axis, stress and strain are zero.

4. **Section modulus**: It is the ratio of moment of inertia of beam cross section about the centroidal axis to the distance of extreme fiber from the centroidal axis.

5. **Moment of resistance** (or bending strength or moment carrying capacity): It is the product of allowable bending stress and the section modulus. $(M = \sigma_{all} \times Z)$

6. **Beam of uniform strength**: A beam in which the maximum bending stress developed at all the cross sections is the same is known as beam of uniform strength.

7. **Composite beam** (or Flitched beam): A beam composed of two or more materials rigidly connected to each other and preventing any

slip between them during bending is known as composite beam.

Concepts and Formulae

1. Bending stress equation (Flexure formula) $\dfrac{\sigma}{y} = \dfrac{M}{I} = \dfrac{E}{R}$

   where
   σ = Bending stress
   y = Distance from neutral axis
   M = Bending moment
   I = Moment of inertia of beam cross section about neutral axis
   E = Young's modulus
   R = Radius of curvature

2. Section modulus, $Z = \dfrac{I}{y_{max}}$

   (i) Rectangular cross section: $Z = \dfrac{bd^2}{6}$

   (ii) Hollow rectangular cross section: $Z = \dfrac{(BD^3 - bd^3)}{6D}$

   (iii) Circular cross section: $Z = \dfrac{\pi d^3}{32}$

   (iv) Hollow circular cross section: $Z = \dfrac{\pi}{32}\left(\dfrac{D^4 - d^4}{D}\right)$

   (v) Triangular cross section: $Z = \dfrac{bh^2}{24}$

3. Maximum bending stress:

For symmetrical section:

$$\sigma_{max} = \frac{M}{I}\, y_{max} = \frac{M}{Z}$$

For unsymmetrical section:

$$(\sigma_{max})_U = \frac{M}{I} \ (y_{max})_U = \frac{M}{Z_U} \quad \text{and} \quad (\sigma_{max})_L = \frac{M}{I} \ (y_{max})_L = \frac{M}{Z_L}$$

4. Moment carrying capacity or Bending strength: $M = \sigma_{all} \times Z$

5. Shear stress formula, $\quad \tau = \dfrac{V A \bar{y}}{I \, b}$

V= shear force at the cross section
A= area above the level where shear stress is calculated
$\bar{y}$ = distance of C.G. of area A from neutral axis
I = moment of inertia of entire cross section
b = width of beam at the level where shear stress is calculated

6. Maximum Shear stress (τ_{max})

(i) Rectangular cross section: $\quad \tau_{max} = \dfrac{3}{2} \times \tau_{ave}$

(ii) Circular cross section: $\quad \tau_{max} = \dfrac{4}{3} \times \tau_{ave}$

(iii) Triangular cross section (isosceles triangle):

$$\tau_{max} = \frac{3}{2} \times \tau_{ave}$$

However, $\quad \tau_{NA} = \dfrac{4}{3} \times \tau_{ave}$

(iv) Square cross section bending about diagonal: $\quad \tau_{max} = \dfrac{9}{8} \times \tau_{ave}$

7. Beam of uniform strength can be obtained by

(i) maintaining constant width and varying depth
(ii) maintaining constant depth and varying width
(iii) varying both width and depth

The most common way of obtaining a beam of uniform strength is by keeping the width constant and varying the depth

8. In eccentric loading, for no tension in the sections (or the stress to be wholly compressive),

Eccentricity, $e \leq \dfrac{Z}{A}$

where Z = section modulus, A = area of cross section

For *rectangular section* the stress will be wholly compressive if the load line lies within the middle third of the respective dimension of the section (*middle third rule*).

For *circular section* the stress will be wholly compressive if the load line lies within the middle quarter of the diameter of the circle (*middle quarter rule*).

SHORT TYPE QUESTIONS

1. A mild steel beam develops a bending stress of 80 N/mm² at a distance of 80 mm from the neutral layer. If E = 200 GPa, what is the radius of curvature?

[Ans. 200 m]

2. When a beam is said to be loaded in pure bending?
[Ans. under constant bending moment]

3. The elastic curve of a beam subjected to pure bending is a
(a) parabolic   (b) elliptic (c) circular (d) none of these

[Ans (c)]

4. A timber and a steel beam having identical dimensions are subjected to identical loads. The stress in the timber beam will be ____ the stress in the steel beam.
(a) less than   (b) equal to   (c) greater than   (d) none of these

[Ans. (b)]

5. Two prismatic bars A and B have same length. The one having larger ____ will be stronger in flexure.
(a) moment of inertia   (b) section modulus   (c) area of cross section

[Ans. (b)]

6. The strength of a beam of square cross section placed with its diagonal horizontal is ____ times the strength when it is placed with its sides horizontal.
(a) 1/2   (b) $\sqrt{2}$ (c) $1/\sqrt{2}$

[Ans. (c)]

7. The strength of a beam mainly depends upon
(a) bending moment   (b) centroid of section   (c) section modulus
(d) its weight

[Ans. (c)]

8. The most common way of keeping the beam of uniform strength is by
(a) keeping the width uniform and varying the depth
(b) keeping the depth uniform and varying the width
(c) varying both the width and the depth
(d) none of these

[Ans. (a)]

9. A beam cross section is used in two different orientations as shown (Fig. 4.45). Bending moments applied to the beam in both the cases are same. The maximum bending stresses induced in cases A and B are related as
(a) $\sigma_A = \sigma_B$ (b) $\sigma_A = 2\,\sigma_B$ (c) $\sigma_A = \sigma_B\,/2$ (d) $\sigma_A = \sigma_B\,/4$

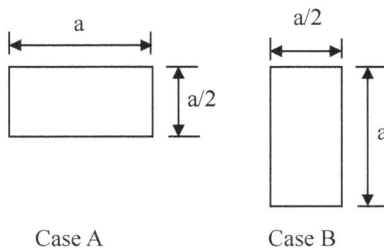

Case A   Case B

Fig. 4.45

[Ans. (b)]

10. In a cross section of a beam subjected to lateral loads where the shear stress will be maximum?

[Ans. at or near the neutral axis]

11. The ratio of maximum shear stress to average shear stress is 1.5 in a beam of
(a) circular cross section (b) rectangular cross section (c) any cross section

[Ans. (b)]

12. The ratio of maximum shear stress to average shear stress is 4/3 in a beam of
(a) circular cross section (b) rectangular cross section (c) any cross section

[Ans. (a)]

13. The diameter of kernel of circular cross section of diameter d is
(a) $\dfrac{d}{2}$ (b) $\dfrac{d}{3}$ (c) $\dfrac{d}{4}$ (d) $\dfrac{d}{8}$

[Ans. (c)]

14. The variation of shear stress on the transverse plane of a normal beam section caused by a transverse shear force is usually
(a) linear   (b) parabolic   (c) nonlinear

[Ans. (b)]

15. In an I section beam subjected to transverse shear force, the maximum shear stress is developed
(a) at the centre of web   (b) at top edge of top flange
(c) at bottom edge of top flange   (d) none of these

[Ans. (a)]

16. In a rectangular section, the stress will be of the same sign throughout the section if the load lies within the ____ of the section.
(a) middle third (b) middle half (c) either of the above (d) none

[Ans. (a)]

17. A beam of square section with side x is subjected to shear force V, the magnitude of shear stress at the top edge of square is
(a) $\dfrac{15}{x^2}$ V   (b) $\dfrac{1}{x^2}$ V   (c) $\dfrac{0.55}{x^2}$ V   (d) zero

[Ans. (d)]

18. In case of circular section, the maximum shear stress is ____ percent more than the mean shear stress.

[Ans. 33.33]

EXERCISE PROBLEMS

1. Calculate the maximum stress induced in a cast iron pipe of external diameter 40 mm, internal diameter 20 mm and length 4 meter when the pipe is supported at its ends and carries a central point load of 80 N.

[Ans. 13.58 N/mm²]

2. A rectangular beam 200 mm deep by 100 mm wide is subjected to maximum bending moment of 500 kNm. Determine the maximum stress in the beam. If E = 200 GPa, find the radius of curvature for that portion of the beam where the bending moment is maximum.

[Ans. 750 MPa, 26.67 m]

3. A wooden beam 3 m long is simply supported at its ends and has a cross section 200 mm × 400 mm deep. It carries a uniformly distributed load of 40 kN/m over the entire span. Calculate the ebnding stress at a point 100 mm above bottom and 1 m from left support.

[Ans. 3.7 N/mm²]

4. A rolled steel joist of I section has the following dimensions:
Flange: width = 250 mm, thickness = 25 mm, overall depth =600 mm, web thickness = 12 mm.
Calculate the safe uniformly distributed load per meter length of the beam if the effective span is 8 m and the maximum stress in steel is 103 N/mm².

[Ans. 7321 N/m]

5. Find the dimensions of the strongest rectangular beam that can be cut out of a log of wood of 180 mm diameter.

[Ans. b = 103.9 mm, d = 147 mm]

6. A wooden beam 100 mm wide and 200 mm deep is reinforced at the top and bottom by steel plates 100 mm wide and 10 mm thick. If the allowable stresses in steel and wood are 200 Mpa and 10 Mpa, find the moment of resistance of the beam. Take $E_S = 15\ E_W$.

[Ans. 39.77 kNm]

7. A timber beam 10 mm wide and 220 mm deep is strengthened by two steel plates 110 mm wide and 20 mm thick at top and bottom of the timber section. Find the moment of resistance of the section. Allowable stress in timber = 7 MPa and E_S = 20 E_W.

[Ans. 87035.7 Nm]

8. A beam of I section 400 mm×200 mm has a web and flange thickness 20 mm. Calculate the maximum intensity of shear stress across the section and sketch the shear stress distribution across the section of the beam if it carries a shear force of 300 kN at a section.

[Ans. τ_{max} = 44.42 N/mm²]

9. A T-section beam of flange 200 mm × 20 mm and web 250 mm × 25 mm is subjected to a shear force of 35 kN. Find the maximum shear intensity and draw the shear stress distribution diagram across the section.

[Ans. τ_{max} = 10.94 N/mm²]

Chapter 5

Deflection of Beams

Learning Objectives

After going through this chapter, the reader will be able to
- understand the differential equation of elastic curve.
- apply double integration method to find slope and deflection in beams.
- apply Macaulay's method of double integration to find slope and deflection in beams.
- state and derive Mohr's moment-area theorems.
- apply moment-area theorems to find slope and deflection in beams.

5.1 INTRODUCTION

Materials used for beams are elastic in nature and hence the beam axis is deflected under the action of loads. In addition to stresses in beams (bending stress, shear stress), this deflection plays an important role in many applications. The beam may be sufficiently strong against the stresses but it may fail due to excessive deflection beyond the permissible limit. Say for example, the beams supporting the plaster ceiling should not get deflected excessively in order to avoid cracks. Again elements of machines must be sufficiently rigid to prevent misalignments and to maintain dimensional accuracy under loading conditions. The machine members designed solely on the basis of stresses may not function properly due to excessive deflection of parts. The knowledge of deflection characteristics of machine members is also essential in the study of their vibration characteristics. In this Chapter, deflection of statically determinate beams will be taken up.

5.2 DEFINITIONS

Refer Fig. 5.1 which represents a simply supported beam AB before and after the application of a point load W at C.

Deflection curve (or elastic curve): It is the longitudinal centroidal axis of the beam after loading.

Deflection (y): It is the linear displacement of the longitudinal centroidal axis measured in lateral directions.

Angle of rotation (θ): It is the rotational displacement of the cross section of the beam.

Slope: It is the tangent to angle of rotation.

slope = tan θ   (θ → angle of rotation)
For small values of θ , tan θ ≈ θ and
slope = θ   or   slope = angle of rotation

Slope is also defined as the rate of change of deflection.

$$\text{slope,} \qquad \theta = \frac{dy}{dx} \qquad\qquad (y = \text{deflection})$$

Curvature: It is the rate of change of slope.

$$\text{Curvature} = \frac{d\theta}{dx} = \frac{d}{dx}\left(\frac{dy}{dx}\right) = \frac{d^2y}{dx^2}$$

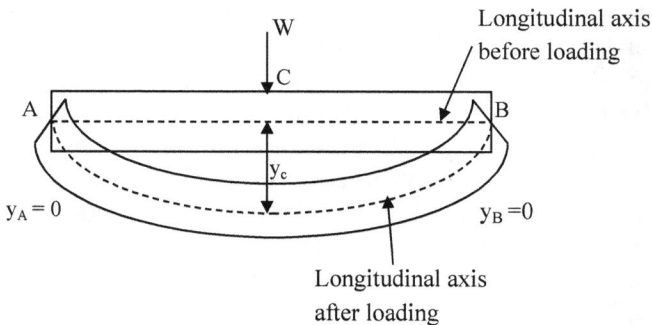

Fig. 5.1

5.3 DIFFERENTIAL EQUATION OF ELASTIC CURVE

Consider a beam which is subjected to pure bending as shown in Fig.

5.2. The beam bends as an arc of a circle. In Fig. 5.2, O is the center of the circle (curvature) and R is the radius of curvature.

Let us consider a small beam element PQ of length dx and at a distance x from left end A as shown (Fig. 5.2).

$d\theta$ = Angular displacement between P and Q.

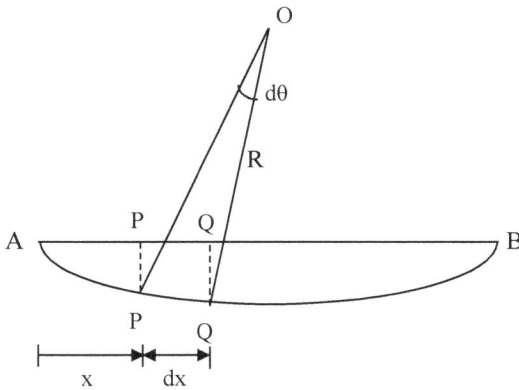

Fig. 5.2

For the element PQ, arc length ds = dx

$$\therefore \ dx \ = \ Rd\theta \qquad or \qquad \frac{d\theta}{dx} = \frac{1}{R}$$

From bending stress equation, $\qquad \dfrac{1}{R} = \dfrac{M}{E\,I}$

$$\therefore \quad \frac{d\theta}{dx} = \frac{M}{E\,I}$$

slope, $\quad \tan\theta = \dfrac{dy}{dx} \qquad$ and $\ \tan\theta \approx \theta \qquad$ (for small values of θ)

$$\therefore \ \theta = \frac{dy}{dx}$$

Now $\qquad \dfrac{d\theta}{dx} = \dfrac{d^2y}{dx^2} = \dfrac{M}{E\,I}$

$$EI \frac{d^2y}{dx^2} = M$$

→ Differential Equation of Elastic curve

or **Curvature Equation**

On integrating the curvature equation,

$$\frac{dy}{dx} = \int \frac{M}{E\,I}\, dx + C_1$$

→ **Slope Equation**

Again integrating

$$y = \iint \frac{M}{E\,I}\, dx\ dx + C_1\, x + C_2$$

→ **Deflection Equation**

The constants of integration C_1 and C_2 are obtained from boundary conditions and from the continuity of the deflection curve.

The product EI is called the *flexural rigidity* in which
E = Young's modulus of elasticity
I = Moment of inertia of beam cross section

5.3.1 Sign convention for deflection and slope

It is important to adopt uniform sign convention for all quantities.

(i) For deflection: upward deflection is positive and downward deflection is negative (Fig. 5.3a).

(ii) For slope: The positive slope occurs only on the part of the curve upto maximum deflection say point P (Fig. 5.3b). Beyond point P, the slope is negative. If origin O is shifted to right side and x is measured +ve to left as shown (Fig. 5.3c), the slope is positive for part OP and negative for part towards left of P. Opposite conditions apply if the curve is below x axis as shown (Figs. 5.3d and 5.3e)

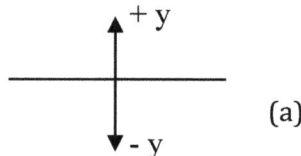

(a)

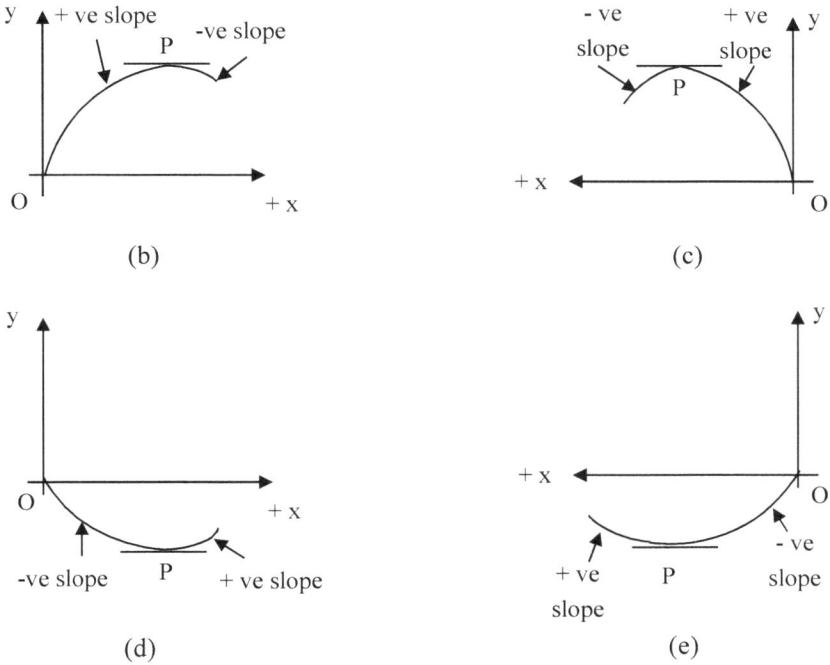

(b)

(c)

(d)

(e)

Fig. 5.3

5.4 DOUBLE INTEGRATION METHOD OF FINDING SLOPE AND DEFLECTION

The double integration method of finding slope and deflection in beams is explained considering suitable examples as below:

5.4.1 Cantilever subjected to point load at free end

A cantilever beam of span *l* carries a point load W at its free end (Fig. 5.4).

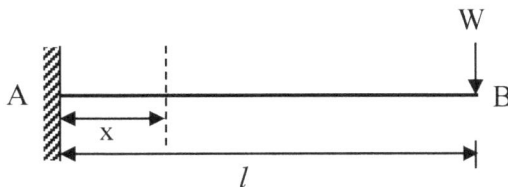

Fig. 5.4

Bending moment at a section at distance x from end A is

$$M = -W(l - x)$$

Curvature equation: $\qquad EI \dfrac{d^2y}{dx^2} = M$

$$EI \dfrac{d^2y}{dx^2} = -W(l - x)$$

Integrating $\quad EI \dfrac{dy}{dx} = \dfrac{W(l - x)^2}{2} + C_1 \qquad\qquad ---- (1)$

Again integrating $\quad EI\ y = \dfrac{-W(l - x)^3}{6} + C_1 x + C_2 \qquad ---- (2)$

Boundary condition:

At fixed end A, slope and deflection are zero.

(i) Put $x = 0$, $\qquad \dfrac{dy}{dx} = 0$ in equation (1)

$$0 = \dfrac{W\,l^2}{2} + C_1 \qquad \therefore\ C_1 = -\dfrac{W\,l^2}{2}$$

(ii) Put $x = 0$, $\qquad y = 0$ in equation (2)

$$0 = \dfrac{-W\,l^3}{6} + C_1 \times 0 + C_2 \qquad\qquad \therefore\ C_2 = \dfrac{W\,l^3}{6}$$

$$\therefore\ EI \dfrac{dy}{dx} = \dfrac{W(l - x)^2}{2} - \dfrac{W\,l^2}{2} \qquad \rightarrow\ \text{Slope Equation}$$

$$EI\ y = \dfrac{-W(l - x)^3}{6} - \dfrac{W\,l^2}{2}x + \dfrac{W\,l^3}{6} \qquad \rightarrow\ \text{Deflection Equation}$$

To find maximum slope and deflection put x=*l*

$$EI\ \left(\dfrac{dy}{dx}\right)_B = \dfrac{W(l - l)^2}{2} - \dfrac{W\,l^2}{2} = -\dfrac{W\,l^2}{2}$$

$$\therefore \left(\frac{dy}{dx}\right)_B = -\frac{W\,l^2}{2EI}$$

In magnitude $\quad \theta_B = \dfrac{W\,l^2}{2EI}$

Similarly, $\quad$ EI $y_B = \dfrac{-W(l-l)^3}{6} - \dfrac{W\,l^2}{2} \times l + \dfrac{W\,l^3}{6}$

$$y_B = -\frac{W\,l^3}{3\,EI}$$

$$\therefore \quad y_B = \frac{W\,l^3}{3EI} \quad \text{(Downward)}$$

5.4.2 Cantilever subjected to uniformly distributed load over whole span

A cantilever beam of span l is carrying uniformly distributed load of intensity w per unit length over whole of its span as shown (Fig. 5.5).

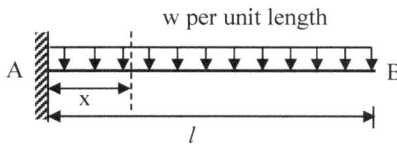

Fig. 5.5

Bending moment at a section at distance x from end A is

$$M = -w(l-x)\frac{(l-x)}{2} = -\frac{w(l-x)^2}{2}$$

Curvature equation: $\quad$ EI $\dfrac{d^2y}{dx^2} = M = -\dfrac{w(l-x)^2}{2}$

Integrating $\quad$ EI $\dfrac{dy}{dx} = \dfrac{w(l-x)^3}{6} + C_1 \qquad\qquad ----(1)$

Again integrating $\quad$ EI $y = \dfrac{-w(l-x)^4}{24} + C_1x + C_2 \qquad ----(2)$

Boundary condition:

At fixed end A, slope and deflection are zero.

(i) At $x = 0$, $\dfrac{dy}{dx} = 0$

(1) $\Rightarrow$ $0 = \dfrac{w \, l^6}{6} + C_1$ $\therefore C_1 = -\dfrac{w \, l^3}{6}$

(ii) At $x = 0$, $y = 0$

(2) $\Rightarrow$ $0 = \dfrac{-w \, l^4}{24} + C_1 \times 0 + C_2$ $\therefore C_2 = \dfrac{w \, l^4}{24}$

$\therefore$ $EI \dfrac{dy}{dx} = \dfrac{w(l - x)^3}{6} - \dfrac{w \, l^3}{6}$ $\rightarrow$ Slope Equation

$EI \, y = \dfrac{-w(l - x)^4}{24} - \dfrac{w \, l^3}{6} x + \dfrac{w \, l^4}{24}$ $\rightarrow$ Deflection Equation

To find maximum slope and maximum deflection (which occur at free end) put $x = l$

$EI \left(\dfrac{dy}{dx}\right)_B = \dfrac{w(l - l)^3}{6} - \dfrac{w \, l^3}{6} = -\dfrac{w \, l^3}{6}$

$\left(\dfrac{dy}{dx}\right)_B = -\dfrac{w \, l^3}{6 \, EI}$

In magnitude $\theta_B = \dfrac{w \, l^3}{6 \, EI}$

Now putting $x = l$ in deflection equation

$EI \, y_B = \dfrac{-w(l - l)^4}{24} - \dfrac{w \, l^3}{6} \times l + \dfrac{w \, l^4}{24}$

$\therefore y_B = -\dfrac{w \, l^4}{8 \, EI}$

In magnitude $\quad y_B = \dfrac{w\, l^4}{8\, EI} \qquad$ (Downward deflection)

5.4.3 Cantilever subjected to couple at free end

A cantilever beam of span l is subjected to a couple M at its free end as shown (Fig. 5.6).

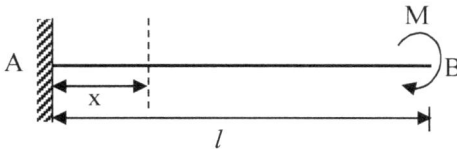

Fig. 5.6

Bending moment at a section at distance x from end A is

$$M_x = -M$$

Curvature equation: $\qquad EI\, \dfrac{d^2y}{dx^2} = -M$

Integrating $\quad EI\, \dfrac{dy}{dx} = -Mx + C_1 \qquad\qquad ----(1)$

Again integrating $\quad EI\, y = \dfrac{-Mx^2}{2} + C_1x + C_2 \qquad ----(2)$

Boundary condition:

At fixed end A, slope and deflection are zero.

(i) At x = 0, $\quad \dfrac{dy}{dx} = 0$

(1) $\quad \Rightarrow \quad 0 = -M \times 0 + C_1 \qquad \therefore\ C_1 = 0$

(ii) At x = 0, $\quad y = 0$

(2) $\quad \Rightarrow \quad 0 = \dfrac{-M \times 0}{2} + 0 + C_2 \qquad\qquad \therefore\ C_2 = 0$

$$\therefore \ EI\ \frac{dy}{dx} = -Mx \qquad \rightarrow \ \text{Slope Equation}$$

$$EI\ y = \frac{-Mx^2}{2} \qquad \rightarrow \ \text{Deflection Equation}$$

At free end, x=*l*

$$EI\ \left(\frac{dy}{dx}\right)_B = -Ml$$

$$\therefore \ \left(\frac{dy}{dx}\right)_B = -\frac{Ml}{EI}$$

$$EI\ y_B = \frac{-Ml^2}{2}$$

$$\therefore \ y_B = -\frac{Ml^2}{2EI}$$

In this case, variation of slope along the length of beam will be straight line and that of deflection will be parabolic in nature.

5.4.4 Cantilever subjected to uniformly varying load

A cantilever beam of span *l* is subjected to uniformly varying load from zero at free end to w per unit length at fixed end as shown (Fig. 5.7a).

Bending moment at a section at distance x from end A is

$$M = -\frac{1}{2}\left(\frac{l-x}{l}\right) w\ (l-x)\frac{(l-x)}{3}$$

$$= \frac{-w(l-x)^3}{6\ l}$$

(Refer Fig. 5.7b)

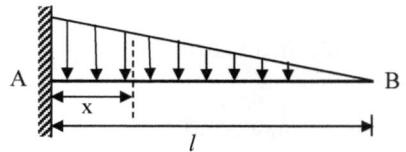

(a)

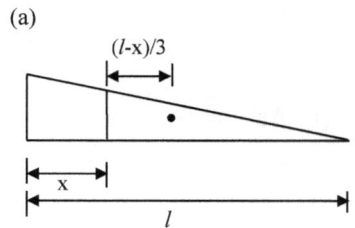

(b)

Fig. 5.7

Curvature equation: $EI \dfrac{d^2y}{dx^2} = M = \dfrac{-w(l-x)^3}{6\,l}$

Integrating $EI \dfrac{dy}{dx} = \dfrac{w(l-x)^4}{24\,l} + C_1$ $\quad\quad$ $----$ (1)

Again integrating $EI\ y = \dfrac{-w(l-x)^5}{120\,l} + C_1 x + C_2$ $\quad$ $----$ (2)

Boundary condition:

(i) At x = 0, $\quad \dfrac{dy}{dx} = 0$

(1) $\Rightarrow$ $C_1 = \dfrac{-wl^3}{24}$

(ii) At x = 0, $\quad$ y = 0

(2) $\Rightarrow$ $C_2 = \dfrac{wl^4}{120}$

$\therefore$ $EI \dfrac{dy}{dx} = \dfrac{w(l-x)^4}{24\,l} - \dfrac{wl^3}{24}$ $\quad\quad \rightarrow$ Slope Equation

$EI\ y = \dfrac{-w(l-x)^5}{120\,l} - \dfrac{wl^3}{24}x + \dfrac{wl^4}{120}$ $\quad \rightarrow$ Deflection Equation

At free end, x=*l*

$EI \left(\dfrac{dy}{dx}\right)_B = 0 - \dfrac{wl^3}{24}$

$\left(\dfrac{dy}{dx}\right)_B = \dfrac{-wl^3}{24\ EI}$

In magnitude $\quad \theta_B = \dfrac{wl^3}{24\ EI}$

$EI\ y_B = -\dfrac{wl^4}{24} + \dfrac{wl^4}{120} = \dfrac{-wl^4}{30}$

$$y_B = \frac{-wl^4}{30\ EI}$$

In magnitude $y_B = \dfrac{w\ l^4}{30\ EI}$ (Downward deflection)

Example 5.1 *A cantilever beam 4 m long carries a point load of 100 kN at a distance of 3m from the fixed end. Calculate the slope and deflection at the free end and at the point of loading. Given I = 4×10 ⁻⁴ m⁴, E =2×10 ¹¹ N/m².*

Solution:

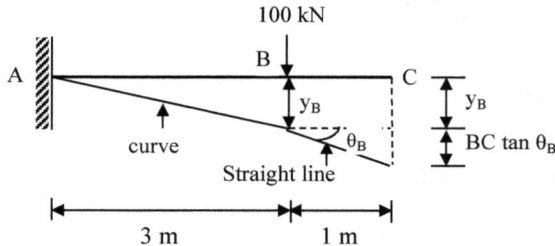

Fig. 5.8

$EI = 2 \times 10^{11} \times 4 \times 10^{-4} = 8 \times 10^7 \ Nm^2$

Refer Fig. 5.8

$$\theta_B = \frac{Wl^2}{2\ EI} = \frac{100 \times 10^3 \times 3^2}{2 \times 8 \times 10^7} = 5.625 \times 10^{-3}\ rad \quad \rightarrow \quad \text{Slope at B}$$

$$y_B = \frac{Wl^3}{3\ EI} = \frac{100 \times 10^3 \times 3^3}{3 \times 8 \times 10^7} = 0.0112\ m \quad \rightarrow \quad \text{Deflection at B}$$

Bending moment is zero and slope is constant in the segment BC of the beam.

$\therefore \theta_B = \theta_C = 5.625 \times 10^{-3}\ rad \quad \rightarrow \quad$ Slope at free end C

$y_C = y_B + BC \tan \theta_B$

$= 0.0112 + (1 \times 5.625 \times 10^{-3}) = 0.0168\ m$
$\qquad\qquad \rightarrow \quad$ Deflection at free end C

5.4.5 Simply supported beam subjected to point load

A simply supported beam of span l is subjected to point load W at mid span as shown (Fig. 5.9).

Reactions at supports:

$$R_A = R_B = \frac{W}{2}$$

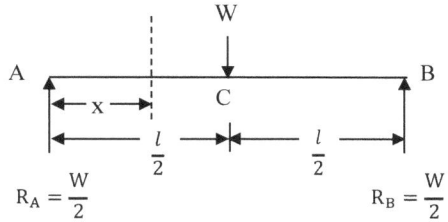

Fig. 5.9

Bending moment at a distance x from end A is

$$M = \frac{W}{2} x \quad \text{(valid from A to C)}$$

Curvature equation: $\quad EI \dfrac{d^2y}{dx^2} = M$

$$EI \frac{d^2y}{dx^2} = \frac{W}{2} x$$

Integrating $\quad EI \dfrac{dy}{dx} = \dfrac{Wx^2}{4} + C_1 \qquad\qquad ---- (1)$

Again integrating $\quad EI\, y = \dfrac{Wx^3}{12} + C_1 x + C_2 \qquad ---- (2)$

Boundary condition:

At $x = 0$, $y = 0$ and at $x = l$, $y = 0$

(deflection is zero at point of support)

At $x = \dfrac{l}{2}$, $\quad \dfrac{dy}{dx} = 0$

(i) Put $x = 0$, $\quad y = 0$ in equation (2)

$0 = 0 + 0 + C_2 \qquad \therefore C_2 = 0$

(ii) Put $x = \dfrac{l}{2}$, $\dfrac{dy}{dx} = 0$ in equation (1)

$$0 = \frac{W(\frac{l}{2})^2}{4} + C_1 \quad \therefore C_1 = \frac{-Wl^2}{16}$$

$$\therefore \ EI \ \frac{dy}{dx} = \frac{Wx^2}{4} - \frac{Wl^2}{16} \qquad \rightarrow \ \text{Slope Equation}$$

$$EI \ y = \frac{Wx^3}{12} - \frac{Wl^2}{16} \ x \qquad \rightarrow \ \text{Deflection Equation}$$

To find slope at A   (where x = 0)

$$EI \left(\frac{dy}{dx}\right)_A = -\frac{Wl^2}{16}$$

In magnitude $\theta_A = \dfrac{Wl^2}{16 \ EI}$

Maximum deflection occurs at mid span (where x=l/2)

$$\therefore \ \ EI \ y_C = \frac{W}{12}\left(\frac{l}{2}\right)^3 - \frac{Wl^2}{16}\left(\frac{l}{2}\right) = -\frac{Wl^3}{48}$$

$$y_C = -\frac{Wl^3}{48 \ EI}$$

In magnitude $y_C = \dfrac{Wl^3}{48 \ EI}$ (Downward deflection)

$$\boxed{y_{max} = \frac{Wl^3}{48 \ EI}} \longrightarrow \ \textbf{Maximum deflection}$$

By symmetry   Slope at B, $\theta_B = -\theta_A = \dfrac{Wl^2}{16 \ EI}$

Slope and deflection are shown in Fig. 5.9a.

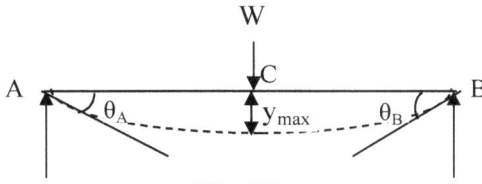

Fig. 5.9a

5.4.6 Simply supported beam subjected to uniformly distributed load

A simply supported beam of span l is carrying uniformly distributed load of intensity w per unit length over its span as shown (Fig. 5.10).

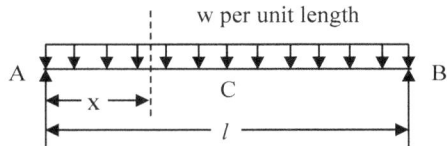

Fig. 5.10

Reactions at supports:

$$R_A = R_B = \frac{wl}{2}$$

Bending moment at a distance x from end A is

$$M = \frac{wl}{2}\, x - (w\, x)\left(\frac{x}{2}\right) = \frac{wlx}{2} - \frac{wx^2}{2} \qquad \text{(valid from A to B)}$$

Curvature equation: $\qquad EI\, \dfrac{d^2y}{dx^2} = M$

$$EI\, \frac{d^2y}{dx^2} = \frac{wlx}{2} - \frac{wx^2}{2}$$

Integrating $\quad EI\, \dfrac{dy}{dx} = \dfrac{wlx^2}{4} - \dfrac{wx^3}{6} + C_1 \qquad\qquad ----(1)$

Again integrating $\quad EI\, y = \dfrac{wlx^3}{12} - \dfrac{wx^4}{24} + C_1 x + C_2 \qquad ----(2)$

Boundary condition:

At A, $x = 0$, $y = 0$

At B, $x = l$, $y = 0$

At C, $x = \dfrac{l}{2}$, $\dfrac{dy}{dx} = 0$

(i) At $x = 0$, $y = 0$

(2) $\rightarrow$ $C_2 = 0$

(ii) At $x = \dfrac{l}{2}$, $\dfrac{dy}{dx} = 0$

(1) $\rightarrow$ $0 = \dfrac{wl}{4}\left(\dfrac{l}{2}\right)^2 - \dfrac{w}{6}\left(\dfrac{l}{2}\right)^3 + C_1$ $\therefore C_1 = \dfrac{-wl^3}{24}$

$\therefore$ EI $\dfrac{dy}{dx} = \dfrac{wlx^2}{4} - \dfrac{wx^3}{6} - \dfrac{wl^3}{24}$ $\rightarrow$ Slope Equation

EI $y = \dfrac{wlx^3}{12} - \dfrac{wx^4}{24} - \dfrac{wl^3}{24}x$ $\rightarrow$ Deflection Equation

Maximum deflection:

Maximum deflection occurs at mid span (where $x=l/2$)

$\therefore$ EI $y_{max} = \dfrac{wl}{12}\left(\dfrac{l}{2}\right)^3 - \dfrac{w}{24}\left(\dfrac{l}{2}\right)^4 - \dfrac{wl^3}{24}\left(\dfrac{l}{2}\right) = -\dfrac{5}{384}wl^4$

$y_{max} = -\dfrac{5}{384}\dfrac{wl^4}{EI}$

In magnitude   (Downward deflection)

$$\boxed{y_{max} = \dfrac{5}{384}\left(\dfrac{wl^4}{EI}\right)}$$ $\longrightarrow$ **Maximum deflection**

Slope at ends:

To find slope at A, put x=0 in slope equation

$$EI \left(\frac{dy}{dx}\right)_A = 0 - 0 - \frac{wl^3}{24} = -\frac{wl^3}{24}$$

In magnitude $\quad \theta_A = \dfrac{wl^3}{24\ EI}$

By symmetry $\quad$ Slope at B, $\quad \theta_B = -\theta_A = -\dfrac{wl^3}{24\ EI}$

5.4.7 Simply supported beam subjected to uniformly varying load

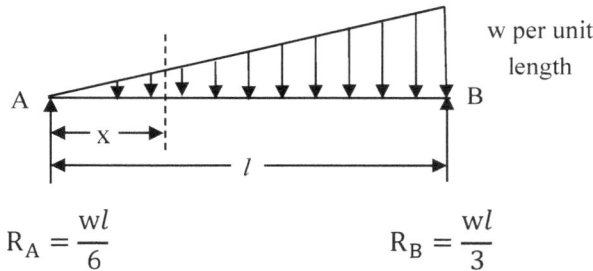

Fig. 5.11

A simply supported beam of span l is subjected to uniformly varying load from zero at A to w per unit length at B as shown (Fig. 5.11).

Reactions at supports:

$$\Sigma M_A = 0$$

$$R_B \times l = \left(\frac{1}{2} \times l \times w\right)\frac{2l}{3}$$

$$\therefore \quad R_B = \frac{wl}{3} \quad \text{and} \quad R_A = \frac{wl}{2} - \frac{wl}{3} = \frac{wl}{6}$$

Bending moment at a section at distance x from A is

$$M = \frac{wl}{6}\,x - \left\{\frac{1}{2}(x)\left(\frac{wx}{l}\right)\right\}\left(\frac{x}{3}\right) = \frac{wlx}{6} - \frac{wx^3}{6\,l} = \frac{w}{6l}(l^2x - x^3)$$

Curvature equation: $EI \dfrac{d^2y}{dx^2} = M$

$$EI \dfrac{d^2y}{dx^2} = \dfrac{w}{6l}(l^2x - x^3)$$

Integrating $EI \dfrac{dy}{dx} = \dfrac{w}{6l}\left(\dfrac{l^2x^2}{2} - \dfrac{x^4}{4}\right) + C_1$ $----(1)$

Again integrating $EI \ y$
$$= \dfrac{w}{6l}\left(\dfrac{l^2x^3}{6} - \dfrac{x^5}{20}\right) + C_1x + C_2 \qquad ----(2)$$

Boundary condition:

(i) At $x = 0$, $y = 0$

(2) $\rightarrow$ $C_2 = 0$

(ii) At $x = l$, $y = 0$

(2) $\rightarrow$ $0 = \dfrac{w}{6l}\left(\dfrac{l^5}{6} - \dfrac{l^5}{20}\right) + C_1 \times l$ $\therefore C_1 = \dfrac{-7wl^3}{360}$

$\therefore \ EI \dfrac{dy}{dx} = \dfrac{w}{6l}\left(\dfrac{l^2x^2}{2} - \dfrac{x^4}{4}\right) - \dfrac{7wl^3}{360}$ $\rightarrow$ Slope Equation

$EI \ y = \dfrac{w}{6l}\left(\dfrac{l^2x^3}{6} - \dfrac{x^5}{20}\right) - \dfrac{7wl^3}{360}x$ $\rightarrow$ Deflection Equation

Maximum deflection:

Maximum deflection occurs where slope is zero.

From slope equation $\dfrac{w}{6l}\left(\dfrac{l^2x^2}{2} - \dfrac{x^4}{4}\right) = \dfrac{7wl^3}{360}$

$\dfrac{2l^2x^2 - x^4}{4 \times 6l} = \dfrac{7l^3}{360}$ $\rightarrow$ $x^4 - 2l^2x^2 + \dfrac{7}{15}l^4 = 0$

$$x^2 = \frac{+2l^2 \pm \sqrt{4l^44 \times 1 \times \frac{7}{15} l^4}}{2} = l^2 \pm \sqrt{l^4 - \frac{7}{15} l^4} = l^2 \pm \sqrt{\frac{8}{15}} l^2$$

$$x^2 = l^2 - \sqrt{\frac{8}{15}} l^2$$

(for positive sign, x > l which is not possible)

x = (0.5193) l

From deflection equation we get

$$EI \; y_{max} = \frac{w}{6l} \left(\frac{l^2}{6} (0.5193)^3 \; l^3 - \frac{(0.5193)^5 l^5}{20} \right) - \frac{7wl^3}{360} \times 0.5193 \times l$$

$$= \frac{-wl^4}{153}$$

$$\therefore \; y_{max} = \frac{-wl^4}{153 \; EI}$$

In magnitude

$$\boxed{y_{max} = \frac{wl^4}{135 \; EI}}$$ (Downward deflection)

Slope at ends:

To find slope at A put x=0 in slope equation

$$EI \left(\frac{dy}{dx} \right)_A = -\frac{7wl^3}{360}$$

In magnitude $$\boxed{\theta_A = \frac{7wl^3}{360 \; EI}}$$

To find slope at B put x=*l* in slope equation

$$\text{EI} \left(\frac{dy}{dx}\right)_B = \frac{w}{6l}\left(\frac{l^2}{2}\ l^2 - \frac{l^4}{4}\right) - \frac{7wl^3}{360} = \frac{wl^4}{24\ l} - \frac{7wl^3}{360} = \frac{wl^3}{45}$$

∴   Slope at B,

$$\theta_B = \frac{wl^3}{45\ \text{EI}}$$

5.4.8 Simply supported beam with couple at one end

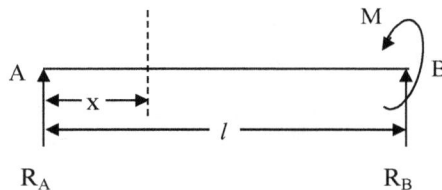

Fig. 5.12

A simply supported beam of span l is subjected to couple M as shown (Fig. 5.12).

Reactions at supports:

$$\sum M_A = 0$$

$$R_B \times l + M = 0 \qquad \Rightarrow \qquad R_B = \frac{-M}{l} \quad \text{and} \quad R_A = \frac{M}{l}$$

Bending moment at a distance x from A is

$$M_1 = \frac{M}{l}\ x$$

Curvature equation: $\qquad \text{EI}\ \dfrac{d^2y}{dx^2} = M_1$

$$\text{EI}\ \frac{d^2y}{dx^2} = \frac{M}{l}\ x$$

Integrating $\text{EI}\ \dfrac{dy}{dx} = \dfrac{Mx^2}{2l} + C_1$ $\qquad\qquad\qquad - - - -(1)$

Again integrating $EI\ y = \dfrac{Mx^3}{6\,l} + C_1 x + C_2$ $----(2)$

Boundary condition:

(i) At $x = 0$, $y = 0$

$(2) \rightarrow C_2 = 0$

(ii) At $x = l$, $y = 0$

$(1) \rightarrow 0 = \dfrac{Ml^3}{6\,l} + C_1 l$ $\therefore C_1 = \dfrac{-Ml}{6}$

$\therefore EI\ \dfrac{dy}{dx} = \dfrac{Mx^2}{2l} - \dfrac{Ml}{6}$ $\rightarrow$ Slope Equation

$EI\ y = \dfrac{Mx^3}{6\,l} - \dfrac{Ml}{6}\,x$ $\rightarrow$ Deflection Equation

Maximum deflection:

Maximum deflection occurs where slope is zero.

From slope equation $\dfrac{Mx^2}{2l} - \dfrac{Ml}{6} = 0$

$\rightarrow \dfrac{Mx^2}{2l} = \dfrac{Ml}{6}$ $\rightarrow x = \dfrac{l}{\sqrt{3}}$

From deflection equation we get

$EI\ y_{max} = \dfrac{M}{6\,l}\left(\dfrac{l}{\sqrt{3}}\right)^3 - \dfrac{Ml}{6}\left(\dfrac{l}{\sqrt{3}}\right) = \dfrac{-Ml^2}{9\sqrt{3}}$

$y_{max} = \dfrac{-Ml^2}{9\sqrt{3}\ EI}$

In magnitude

$$\boxed{y_{max} = \dfrac{Ml^2}{9\sqrt{3}\ EI}}$$

(Downward deflection)

Slope at ends:

To find slope at A   put x=0 in slope equation

$$EI \left(\frac{dy}{dx}\right)_A = -\frac{Ml}{6}$$

In magnitude

$$\boxed{\theta_A = \frac{Ml}{6\ EI}}$$

To find slope at B   put x=*l* in slope equation

$$EI \left(\frac{dy}{dx}\right)_B = \frac{Ml^2}{2l} - \frac{Ml}{6} = \frac{Ml}{3}$$

∴   Slope at B, $$\boxed{\theta_B = \frac{Ml}{3EI}}$$

Example 5.2 *A beam consists of symmetrical rolled steel joists. The beam is simply supported at its ends and carries a point load at the centre of the span (Fig. 5.13). If the maximum stress due to bending is 140 MPa, find the ratio of depth of beam to span in order that the central deflection may not exceed $\frac{1}{480}$ of the span. Take E=200 GPa.*

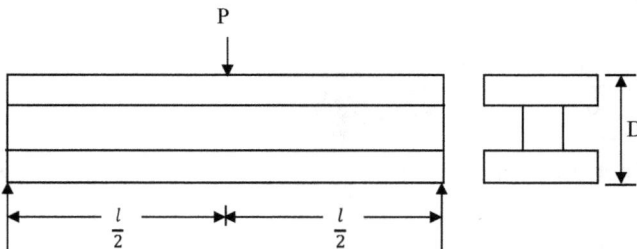

Fig. 5.13

Solution:

$$\sigma = 140 \text{ MPa} = 140 \times 10^6 \text{ Pa}, \qquad E = 200 \times 10^9 \text{ Pa}$$

$$\frac{M}{I} = \frac{\sigma_{max}}{y_{max}} \quad \text{(Bending stress equation)}$$

$$\frac{Pl/4}{I} = \frac{140 \times 10^6}{D/2} \quad \rightarrow \quad \frac{PlD}{I} = 1.12 \times 10^9$$

Deflection $\quad y_{max} = \dfrac{1}{480} \times l$

$$\frac{Pl^3}{48\,EI} = \frac{1}{480} \times l \quad \text{or} \quad \frac{Pl}{I} \times \frac{l}{48\,E} = \frac{1}{480}$$

or $\quad \dfrac{1.12 \times 10^9}{D} \times \dfrac{l}{48 \times 200 \times 10^9} = \dfrac{1}{480}$

$\therefore \quad \dfrac{D}{l} = 0.056$

Example 5.3 *If a cantilever is loaded at the mid span by a concentrated load W, find the slope and deflection at its free end.*

Solution:

Refer Fig. 5.14.

The slope and deflection at free end for cantilever subjected to concentrated load at free end are already computed earlier.

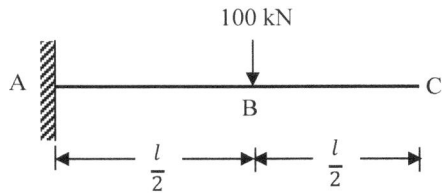

Fig. 5.14

$$\therefore \; \theta_B = -\frac{W\left(\frac{l}{2}\right)^2}{2EI} = -\frac{Wl^2}{8EI}$$

$$y_B = -\frac{W\left(\frac{l}{2}\right)^3}{3EI} = -\frac{Wl^3}{24EI}$$

Bending moment equation in part BC is $M = 0$

$$\therefore \; EI\frac{d^2y}{dx^2} = 0$$

Integrating $EI\dfrac{dy}{dx} = C_1$ $------ (1)$

Again integrating $EI\,y = C_1\,x + C_2$ $----- (2)$

(i) At $x = \dfrac{l}{2}$, $\dfrac{dy}{dx} = -\dfrac{Wl^2}{8EI}$

$(1) \rightarrow EI\left(-\dfrac{Wl^2}{8EI}\right) = C_1$ $\therefore C_1 = -\dfrac{Wl^2}{8}$

(ii) At $x = \dfrac{l}{2}$, $y = -\dfrac{Wl^3}{24EI}$

$(2) \rightarrow EI\left(-\dfrac{Wl^3}{24EI}\right) = -\dfrac{Wl^2}{8} \times \dfrac{l}{2} + C_2$ $\therefore C_2 = -\dfrac{Wl^3}{24} + \dfrac{Wl^3}{16}$

$= \dfrac{Wl^3}{48}$

$\therefore EI\dfrac{dy}{dx} = -\dfrac{Wl^2}{8}$ $\rightarrow$ Slope equation

$EI\,y = -\dfrac{Wl^2}{8}x + \dfrac{Wl^3}{48}$ $\rightarrow$ Deflection equation

Slope at free end $(x = l)$

$\therefore \left(\dfrac{dy}{dx}\right)_C = -\dfrac{Wl^2}{8EI}$

Deflection at free end $(x = l)$

$\therefore y_C = y_{max} = \dfrac{1}{EI}\left(-\dfrac{Wl^2}{8} \times l + \dfrac{Wl^3}{48}\right) = \dfrac{1}{EI}\left(\dfrac{-6Wl^3 + Wl^3}{48}\right)$

$\therefore y_{max} = \dfrac{-5Wl^3}{48EI}$

Example 5.4 *A simply supported beam carrying point load W at mid span is having a rectangular cross section. Maximum deflection is given* y_{max}. *If the length of the beam is doubled and depth is halved with no other change, then what will be the maximum deflection?*

Solution:

For the given condition, maximum deflection

$$y_{max} = \frac{Wl^3}{48EI}$$

Now new length of beam $l' = 2l$

New depth of beam $d' = \frac{d}{2}$

$$\therefore \text{New moment of inertia } I' = \frac{b(d')^3}{12} = \frac{b\left(\frac{d}{2}\right)^3}{12} = \frac{bd^3}{96} = \frac{I}{8}$$

$$\left(\text{Initial moment of inertia, } I = \frac{bd^3}{12}\right)$$

$$\text{New maximum deflection} = \frac{W(l')^3}{48EI'} = \frac{W(2l)^3}{48E \times \left(\frac{I}{8}\right)} = 64 \times \left(\frac{Wl^3}{48EI}\right)$$

$$= 64 \, y_{max}$$

5.5 MACAULAY'S METHOD OF FINDING SLOPE AND DEFLECTION

This method is used to find slope and deflection in a beam subjected to asymmetrical point load, number of point loads and uniformly distributed load applied over part of the span.

In this method, a generalized bending moment equation is written. The bending moment equation is valid throughout the span by considering or neglecting certain terms. The constants of integration are valid throughout the span. The procedure of finding slope and deflection in beams using Macaulay's method is explained in the following few examples.

Example 5.5 *A simply supported beam of span l carries a point load W at a distance a from the left end as shown (Fig. 5.15a). Find the deflection under load and the maximum deflection.*

Solution:

Reactions at supports:

$\Sigma M_B = 0$

$R_A \times l = W \times b$

$\therefore R_A = \dfrac{W \times b}{l}$

Similarly $\Sigma M_A = 0$

$R_B \times l = W \times a$

$\therefore R_B = \dfrac{W \times a}{l}$

Considering AC part: (refer Fig. 5.15b)

Fig. 5.15

Bending moment at a section at a distance x from A is

$$M = \frac{Wb}{l}\, x$$

(valid from A to C of the beam i.e. from x=0 to x=a)

Considering CB part: (refer Fig. 5.15c)

Bending moment at a section at a distance x from A is

$$M = \frac{Wb}{l} x - W(x - a) \qquad \text{(valid from C to B)}$$

It can be observed that bending moment equation on AC part is not valid in CB part but that on CB part is valid also in AC part by neglecting the second term.

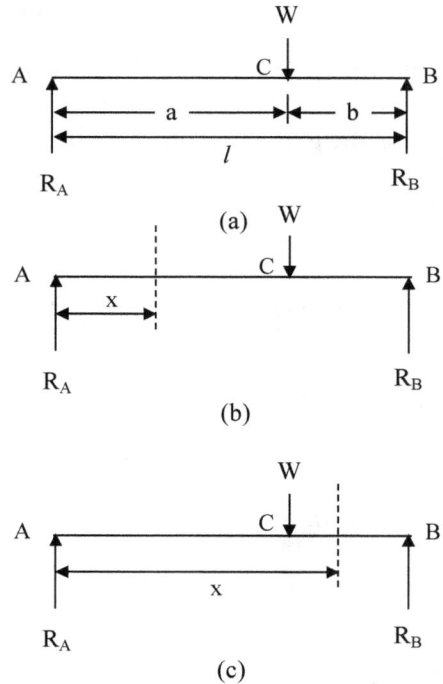

(when the value of the term in bracket becomes negative that term will be neglected)

$$\therefore M = \frac{Wb}{l}x - W(x - a)$$

$\rightarrow$ Generalised bending moment equation

Curvature equation $\quad EI\dfrac{d^2y}{dx^2} = M$

$$EI\frac{d^2y}{dx^2} = \frac{Wb}{l}x - W(x - a)$$

Integrating $\quad EI\dfrac{dy}{dx} = \dfrac{Wb}{2l}x^2 - \dfrac{W(x-a)^2}{2} + C_1 \qquad ---(1)$

Again integrating $EI\ y = \dfrac{Wb}{6l}x^3 - \dfrac{W(x-a)^3}{6} + C_1x + C_2 \ ---(2)$

(i) 1st boundary condition $\rightarrow x = 0, \qquad y = 0$

$$(2) \rightarrow EI \times 0 = \frac{Wb}{6l} \times 0 - \text{neglect} + C_1 \times 0 + C_2$$

$\therefore C_2 = 0$

(ii) 2nd boundary condition $\rightarrow x = l, y = 0$

$$(2) \rightarrow EI \times 0 = \frac{Wb}{6l} \times l^3 - \frac{W(l-a)^3}{6} + C_1l + 0$$

$\therefore \ C_1 = -\dfrac{Wb(l^2 - b^2)}{6l}$

(While considering AC part, 2nd term of equation (2) is neglected but in CB part, all the terms are considered)

Now (1) $\rightarrow \quad EI\dfrac{dy}{dx} = \dfrac{Wb}{2l}x^2 - \dfrac{W(x-a)^2}{2} - \dfrac{Wb(l^2-b^2)}{6l}$

$\rightarrow$ Slope equation

$$(2) \rightarrow \quad EIy = \frac{Wb}{6l}x^3 - \frac{W(x-a)^3}{6} - \frac{Wb(l^2-b^2)}{6l}x$$

$$\rightarrow \quad \text{Deflection equation}$$

To find deflection under load, Put x = a in deflection equation

$$\therefore EI\, y_c = \frac{Wb}{6l}a^3 - 0 - \frac{Wb(l^2-b^2)}{6l} \times a$$

$$= \frac{Wb}{6l}(a^3 - (l^2-b^2)a) = \frac{Wb}{6l}[a^3 - \{(a+b)^2 - b^2\}a]$$

$$= \frac{Wb}{6l}(a^3 - a^3 - 2a^2b) = \frac{-2Wa^2b^2}{6l} = \frac{-Wa^2b^2}{3\,l}$$

In magnitude $\quad y_c = \dfrac{Wa^2b^2}{3\,EI\,l} \quad\quad \rightarrow \quad$ Deflection under load

Maximum deflection here will not occur at the mid point of beam because loading is asymmetrical.

Assuming $a > b$

Maximum deflection occurs in AC part at the point where slope is Zero.

Slope equation $\quad \rightarrow \quad EI \times 0 = \dfrac{Wbx^2}{2l} - \text{Neglect} - \dfrac{Wb(l^2-b^2)}{6l}$

$$\therefore x = \sqrt{\frac{l^2-b^2}{3}}$$

Putting this value of x in deflection equation, maximum deflection can be found out.

$$\therefore EI\, y_c = \frac{Wb}{6l}\left(\frac{l^2-b^2}{3}\right)^{\frac{3}{2}} - \text{Neglect} - \frac{Wb(l^2-b^2)}{6l}\left(\frac{l^2-b^2}{3}\right)^{\frac{1}{2}}$$

$$= \frac{Wb}{6l}(l^2-b^2)^{\frac{3}{2}}\left[\frac{1}{3\sqrt{3}} - \frac{1}{\sqrt{3}}\right] = \frac{Wb}{6l}(l^2-b^2)^{\frac{3}{2}}\left(\frac{-2}{3\sqrt{3}}\right)$$

$$= -\frac{Wb}{9\sqrt{3}\,l}(l^2 - b^2)^{\frac{3}{2}}$$

In magnitude $y_{max} = \dfrac{Wb}{9\sqrt{3}\,EI\,l}(l^2 - b^2)^{\frac{3}{2}}$ (Downward)

$\rightarrow$ Maximum Deflection

Example 5.6 *A simply supported beam of span 6 m carries three point loads 25 KN, 50 KN and 25 KN acting at 1m, 3m and 5m respectively from left end support. Find the maximum deflection and maximum slope.*

Solution:

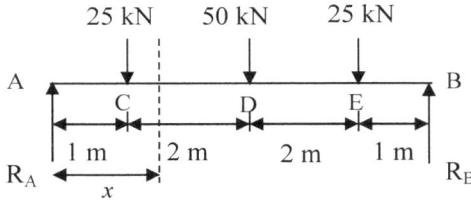

Fig. 5.16

Refer Fig. 5.16

This is a problem of symmetrical loading.

Reaction at supports:

$R_A = 50$ KN, $\qquad R_B = 50$ KN

Consider a section in CD part at a distance x from A as shown (Fig.5.16)

$M = 50\,x - 25(x - 1)$
(Here bending moment equation which is valid for half of span is sufficient)

$$EI\frac{d^2y}{dx^2} = M = 50\,x - 25(x - 1)$$

$$EI\frac{dy}{dx} = 50\left(\frac{x^2}{2}\right) - \frac{25(x - 1)^2}{2} + C_1 \qquad \text{-----(1)}$$

$$EI\ y = 50\left(\frac{x^3}{6}\right) - \frac{25(x-1)^3}{6} + C_1 x + C_2 \qquad ----- (2)$$

(i) At $x = 0$, $y = 0$

(2) $\rightarrow$ $\quad 0 = 0 - \text{neglect} + 0 + C_2 \qquad \therefore C_2 = 0$

(ii) At $x = 3$, $\dfrac{dy}{dx} = 0$ (As symmetrical loading)

(1) $\rightarrow$ $\quad 0 = 50\left(\dfrac{3^2}{2}\right) - \dfrac{25(3-1)^2}{2} + C_1$

$\therefore C_1 = -175$

$\therefore EI\ \dfrac{dy}{dx} = 50\left(\dfrac{x^2}{2}\right) - \dfrac{25(x-1)^2}{2} - 175 \quad \rightarrow$ Slope equation

$EI\ y = 50\left(\dfrac{x^3}{6}\right) - \dfrac{25(x-1)^3}{6} - 175\ x \quad \rightarrow$ Deflection equation

To find maximum slope, put x=0 in slope equation

$EI\ \theta_A = \dfrac{50 \times 0}{2} - \text{neglect} - 175 \quad \rightarrow \quad \theta_A = \dfrac{-175}{EI}$

$\therefore$ In magnitude $\quad \theta_A = \dfrac{175}{EI} \qquad$ (maximum slope)

To find maximum deflection, $\qquad$ put $x = 3m$ in deflection equation

$EI\ y_{max} = 50 \times \dfrac{3^3}{6} - \dfrac{25(3-1)^3}{6} - 175 \times 3 \qquad$ or $\quad y_{max}$

$= \dfrac{-333.33}{EI}$

In magnitude $y_{max} = \dfrac{333.33}{EI} \qquad$ (Downward)

Example 5.7 *A simply supported beam of span 5 m carries two concentrated loads 10 kN at 2m and 5 kN at 4 m from left end. Calculate the maximum deflection of beam. Given* $E = 200\ GPa$, $I = 50 \times 10^6\ mm^4$.

Solution:

$E = 200 \times 10^6\ \text{kN/m}^2$

$I = 50 \times 10^{-6}\ \text{m}^4$

$\therefore EI = 200 \times 10^6 \times 50 \times 10^{-6}$

$= 10000\ \text{kNm}^2$

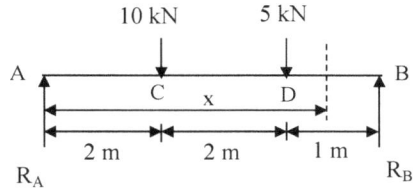

Fig. 5.17

Refer Fig. 5.17

Reactions at supports:

$\Sigma M_B = 0$

$R_A \times 5 = (10 \times 3) + (5 \times 1)$

$\therefore R_A = 7\ \text{kN}$ and $R_B = 10 + 5 - 7 = 8\ \text{kN}$

In AC Part:

Bending moment, $M = 7x$ (valid for $x = 0$ to 2 m)

In CD Part:

Bending moment, $M = 7x - 10(x - 2)$ (valid for $x = 2$ m to 4m)

In DB Part:

Bending moment, $M = 7x - 10(x - 2) - 5\ (x - 4)$ (valid for $x = 4$ m to 5m)

Now generalised bending moment equation is

$M = 7x - 10(x - 2) - 5(x - 4)$

The bending moment equation can be applied for AC by neglecting the 2nd and 3rd term and for CD by neglecting the 3rd term only.

Curvature equation: $\quad EI \dfrac{d^2y}{dx^2} = M$

$$EI \dfrac{d^2y}{dx^2} = 7x - 10(x-2) - 5(x-4)$$

Integrating

$$EI \dfrac{dy}{dx} = 7\dfrac{x^2}{2} - 10\dfrac{(x-2)^2}{2} - 5\dfrac{(x-4)^2}{2} + C_1 \qquad\qquad -(1)$$

Again integrating

$$EI\, y = 7\dfrac{x^3}{6} - 10\dfrac{(x-2)^3}{6} - 5\dfrac{(x-4)^3}{6} + C_1 x + C_2 \qquad\qquad -(2)$$

(i) At $x = 0$, $\quad y = 0$

$(2) \rightarrow EI \times 0 = 0 - \text{neglect} - \text{neglect} + 0 + C_2 \qquad \therefore C_2 = 0$

(ii) At $x = 5$, $\quad y = 0$

$$(2) \rightarrow EI \times 0 = 7\dfrac{5^3}{6} - 10\dfrac{(5-2)^3}{6} - 5\dfrac{(5-4)^3}{6} + C_1 x + 0$$

$\therefore C_1 = -20$

Putting the values of constant C_1 and C_2 in equations (1) and (2), the slope equation and the deflection equation can be obtained.

$$EI \dfrac{dy}{dx} = 7\dfrac{x^2}{2} - 10\dfrac{(x-2)^2}{2} - 5\dfrac{(x-4)^2}{2} - 20 \rightarrow \quad \text{Slope equation}$$

$$EI\, y = 7\dfrac{x^3}{6} - 10\dfrac{(x-2)^3}{6} - 5\dfrac{(x-4)^3}{6} - 20x \rightarrow \quad \text{Deflection equation}$$

Maximum deflection occurs in CD part where slope will be zero.

Equating slope at the section in CD part to zero we get

$$EI\frac{dy}{dx} = 7\frac{x^2}{2} - 10\frac{(x-2)^2}{2} - 5\frac{(x-4)^2}{2} - 20$$

$$0 = 7\frac{x^2}{2} - 10\frac{(x-2)^2}{2} - \text{neglect} - 20$$

$$7x^2 - 10(x^2 + 4 - 4x) - 40 = 0$$

$$3\,x^2 - 40\,x + 80 = 0$$

$$\therefore x = \frac{40 \pm \sqrt{1600 - 4 \times 3 \times 80}}{2 \times 3} = 2.45\ m$$

Put x = 2.45 m in deflection equation (neglect 3rd term)

$$EI\,y_{max} = 7\frac{(2.45)^3}{6} - 10\frac{(2.45-2)^3}{6} - \text{Neglect} - 20 \times 2.45$$

$$= -32.17$$

$$y_{max} = \frac{-32.17}{EI} = \frac{-32.17}{10000} = -0.003217\ m$$

In magnitude $y_{max} = 0.003217\ m$ (Downward Deflection)

Example 5.8 *A cantilever beam of length 2 m carries a uniformly distributed load of 2.5 kN/m for a length of 1.25 m from the fixed end and a point load of 1 kN at the free end. If moment of inertia I of the section is 24³ ×10⁻⁸ m⁴ and E=10¹⁰ N/m², find the slope and deflection at free end.*

Solution:

$$I = 24^3 \times 10^{-8}\ m^4, \qquad E = 10^{10}\ N/m^2$$

$$EI = (10^{10} \times 10^{-3}) \times (24^3 \times 10^{-8}) = 1382.4\ kN/m^2$$

Refer Fig. 5.18

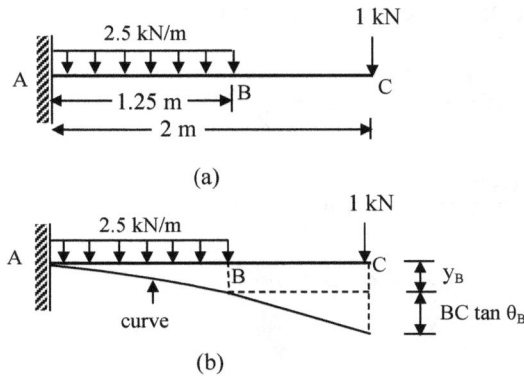

Fig. 5.18

(a) Due to point load at free end

$$\theta_c = \frac{Wl^2}{2EI} = \frac{1 \times 2^2}{2 \times 1382.4} = 1.44 \times 10^{-3} \text{ rad}$$

$$y_c = \frac{Wl^3}{3EI} = \frac{1 \times 2^3}{3 \times 1382.4} = 1.929 \times 10^{-3} \text{m}$$

(b) Due to uniformly distributed load

$$\theta_B = \frac{wl^2}{6EI} = \frac{2.5 \times 1.25^2}{6 \times 1382.4} = 5.88 \times 10^{-4} \text{ rad}$$

$$y_B = \frac{wl^4}{8EI} = \frac{2.5 \times 1.25^4}{8 \times 1382.4} = 5.51 \times 10^{-4} \text{m}$$

In BC part, slope is constant (Fig. 5.18b)

$$\therefore \ \theta_c = \theta_B = 5.88 \times 10^{-4} \text{ rad} \qquad \text{(Slope at free end)}$$

$$y_c = y_B + (\text{BC} \tan \theta_B) = y_B + (\text{BC} \times \theta_B)$$

$$= 5.51 \times 10^{-4} + (0.75 \times 5.88 \times 10^{-4}) = 9.95 \times 10^{-4} \text{ m}$$

Combined point load and uniformly distributed load:

$$\theta_c = (\theta_c)_{\text{Point load}} + (\theta_c)_{\text{udl}} = (1.44 \times 10^{-3}) + (5.88 \times 10^{-4})$$

$= 2.028 \times 10^{-3}$ rad

$y_c = (y_c)_{\text{Point load}} + (y_c)_{\text{udl}} = (1.929 \times 10^{-3}) + (9.95 \times 10^{-4})$

$= 2.915 \times 10^{-3}$ m   (Deflection at free end)

Example 5.9 *A rectangular beam 100 mm wide and 200 mm deep is freely supported over a span of 2 m. A load of 10 kN is dropped on to the middle of beam from a height of 10 mm. Find the maximum instantaneous deflection and stress induced in the beam. E = 2×10^5 N/mm².*

Solution:

$$I = \frac{100 \times 200^3}{12} = 6.67 \times 10^7 \text{mm}^4$$

$W = $ Impact load, $P = $ Equivalent gradual load

$$y_c = \frac{Pl^3}{48EI} = \frac{P \times 2000^3}{48 \times 2 \times 10^5 \times 6.67 \times 10^7} = 1.25 \times 10^{-5} \, P$$

workdone by impact load = work done by equivalent gradual load

$$W(h + y_c) = \frac{1}{2} \times P \times y_c \quad \text{(load W is dropped from a height of h)}$$

$$\therefore (10 \times 10^3)(10 + 1.25 \times 10^{-5}P) = \frac{1}{2} \times P \times 1.25 \times 10^{-5} \, P$$

$$0.625 \times 10^{-5} \, P^2 - 0.125 \, P - 10^{-5} = 0$$

$$\therefore P = \frac{0.125 \pm \sqrt{(0.125)^2 - 4 \times 0.625 \times 10^{-5} \times (-10^{-5})}}{2 \times 0.625 \times 10^{-5}}$$

$= 136.8 \times 10^3$ N   (Taking the + ve sign)

$\therefore P = 136.8 \times 10^3$ N $= 136.8$ kN

10 kN weight falling from a height of 10 mm will cause the same deflection as caused by equivalent gradual load of 136.8 kN.

$\therefore \qquad y_C = 1.25 \times 10^{-5} \times 136.8 \times 10^3 = 1.71 \text{mm}$

Bending stress equation:

$$\frac{\sigma_{max}}{y_{max}} = \frac{M}{I}$$

$$\therefore \sigma_{max} = \frac{M}{I} \times y_{max} = \frac{\left(\frac{Pl}{4}\right)}{I} \times y_{max}$$

$$= \frac{136.8 \times 10^3 \times 2000}{4 \times 6.67 \times 10^7} \times \frac{200}{2} = 102.5 \text{ N/mm}^2$$

Example 5.10 *A beam ABCD (with constant EI) is loaded as shown (Fig. 5.19a). Determine (i) slope at end A (ii) Deflection at B (iii) Deflection at end D.*

Solution:

Refer Fig. 5.19b

Reactions at supports:

$\sum M_c = 0$

$R_A \times 2a + P \times a = P \times a$

$\therefore \quad R_A = 0 \quad$ and $\quad R_C = 2P$

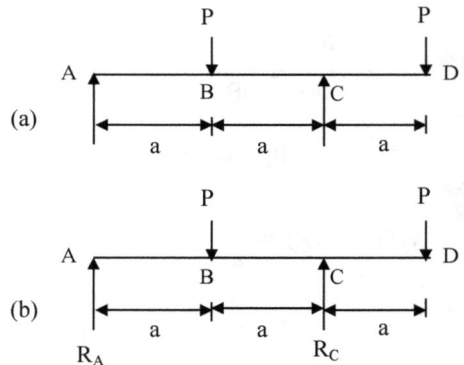

Fig. 5.19

Bending moment at a section in CD part at a distance x from A is

$M = R_A x - P(x - a) + R_C(x - 2a)$

$\rightarrow \quad M = 0 - P(x - a) + 2P(x - 2a)$

This is the *generalized bending moment equation* which is also valid in AB part by neglecting the second and third terms and is valid in BC part by neglecting the third term only.

Curvature equation : $EI \dfrac{d^2y}{dx^2} = M$

$$EI \dfrac{d^2y}{dx^2} = 0 - P(x - a) + 2P(x - 2a)$$

Integrating $EI \dfrac{dy}{dx} = -P \dfrac{(x-a)^2}{2} + \dfrac{2P(x-2a)^2}{2} + C_1$ $- - - (1)$

Again integrating $EI\, y$

$$= -P \dfrac{(x-a)^3}{6} + \dfrac{2P(x-2a)^3}{6} + C_1 x + C_2 \quad - - - (2)$$

(i) At $x = 0$, $y = 0$

$(2) \rightarrow C_2 = 0$

(ii) At $x = 2a$, $y = 0$

$(2) \rightarrow 0 = -P \dfrac{(2a-a)^3}{6} + \text{Neglect} + C_1 \times 2a + 0$

$\therefore C_1 = \dfrac{Pa^2}{12}$

Putting value of C_1 and C_2 in equations (1) and (2)

$$EI \dfrac{dy}{dx} = -P \dfrac{(x-a)^2}{2} + \dfrac{2P(x-2a)^2}{2} + \dfrac{Pa^2}{12} \quad \rightarrow \text{Slope equation}$$

$$EIy = -P \dfrac{(x-a)^3}{6} + \dfrac{2P(x-2a)^3}{6} + \dfrac{Pa^2}{12} x \quad \rightarrow \text{Deflection equation}$$

Slope at A (θ_A) :

Putting $x = 0$ in slope equation

$$EI \left(\dfrac{dy}{dx} \right)_A = -\text{Neglect} + \text{Neglect} + \dfrac{Pa^2}{12}$$

$\therefore \theta_A = \dfrac{Pa^2}{12}$ (anti clockwise w. r. t. x axis)

Deflection at B (y_B) :

Putting x = a in deflection equation

$$EI\, y_B = 0 + Neglect + \frac{Pa^2}{12} \times a = \frac{Pa^3}{12}$$

$$\therefore y_B = \frac{Pa^3}{12\,EI} \qquad \text{(upward deflection)}$$

Deflection at D (y_D) :

Put x =3a in deflection equation

$$EI\, y_D = -P\frac{(3a-a)^3}{6} + \frac{2P\,(3a-2a)^3}{6} + \frac{Pa^3}{12} \times 3a$$

$$= \frac{-8Pa^3}{6} + \frac{2Pa^3}{6} + \frac{3Pa^3}{12} = -\frac{9Pa^3}{12} = -\frac{3Pa^3}{4}$$

$$\therefore y_D = \frac{3Pa^3}{4EI} \qquad (\,\text{Downward deflection})$$

Note: In part AB, bending moment is zero, curvature $\left(\frac{M}{EI}\right)$ is zero, slope is constant and deflection curve is straight line.

Example 5.11 *A beam ABC of uniform flexural rigidity (EI) is supported by a fulcrum at B and a steel wire AD at A as shown (Fig. 5.20). The steel wire is 3 m long and has a diameter of 3 mm. The beam is subjected to a vertical load of P= 550 N at the free end C. Find the vertical deflection of point C. Take EI for beam as 90×10⁹ Nmm² and E$_s$=200 GPa.*

Solution:

Refer Fig. 5.20a

$$\Sigma M_A = 0$$

$$R_B \times 0.5 = 500 \times 1.25$$

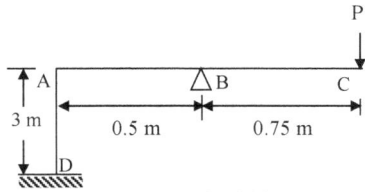

Fig. 5.20

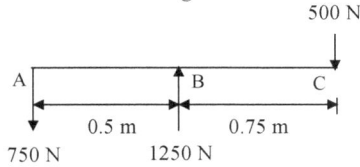

Fig. 5.20a

$\therefore R_B = 1250 \text{ N}$

$R_A + R_B = 500 \text{ N}$

$\therefore R_A = -750 \text{ N}$

Bending moment at a section in BC part at distance x from A is

$M = -750\,x + 1250(x - 0.5)$

$EI \dfrac{d^2 y}{dx^2} = M = -750\,x + 1250(x - 0.5)$

Integrating,

$EI \dfrac{dy}{dx} = -750\left(\dfrac{x^2}{2}\right) + 1250\,\dfrac{(x - 0.5)^2}{2} + C_1 \qquad --(1)$

Integrating again,

$EIy = -750\left(\dfrac{x^3}{6}\right) + 1250\,\dfrac{(x - 0.5)^3}{6} + C_1 x + C_2 \qquad --(2)$

(i) At $x = 0$, $y = 0$

$(2) \rightarrow 0 = 0 + \text{Neglect} + 0 + C_2 \qquad\qquad \therefore C_2 = 0$

(ii) At x = 0.5,   y = 0

(2) → $0 = -750\dfrac{0.5^3}{6} + 0 + C_1 x$ $\therefore C_1 = 31.25$

$\therefore EI\dfrac{dy}{dx} = -750\left(\dfrac{x^2}{2}\right) + 1250\dfrac{(x-0.5)^2}{2} + 31.25$

→   Slope equation

$EIy = -750\left(\dfrac{x^3}{6}\right) + 1250\dfrac{(x-0.5)^3}{6} + 31.25\,x$

→   Deflection equation

Deflection of free end C is the sum total of deflections at C due to beam bending (y_{c1}) and due to the extension of wire at A (y_{c2}).

To find the beam deflection at C, put x=1.25 in deflection equation.

$\therefore EI\,y_{C_1} = -750\dfrac{(1.25)^3}{6} + 1250\dfrac{(1.25-0.5)^3}{6} + (31.25 \times 1.25)$

$y_{C_1} = -1.31 \times 10^{-3}$ m

In magnitue, $y_{C_1} = 1.31 \times 10^{-3}$ m (downward)

Let y_{c2} = Deflection at C due to extension of wire

Elongation of wire AD is

$AA' = \left(\dfrac{Pl}{AE}\right)_{wire}$

$= \dfrac{750 \times 3}{\left(\frac{\pi}{4} \times (0.003^2)\right) \times 200 \times 10^9}$

$= 1.59 \times 10^{-3}$ m

Refer Fig. 5.20b

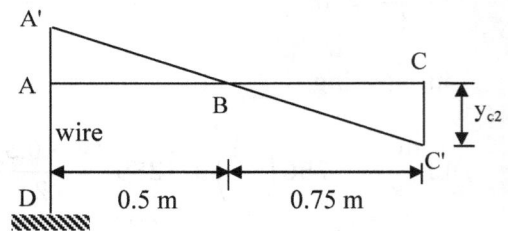

Fig. 5.20b

$$\frac{CC'}{BC} = \frac{AA'}{AB}$$

$$\therefore \frac{y_{c2}}{0.75} = \frac{1.59 \times 10^{-3}}{0.5} \qquad \Rightarrow \quad y_{c2} = 2.38 \times 10^{-3} \text{m}$$

$\therefore$ Total deflection at C,

$$y_C = y_{c1} + y_{c2} = 1.31 \times 10^{-3} + 2.38 \times 10^{-3}$$

$$y_C = 3.69 \times 10^{-3} \text{ m} \quad \text{(downward)}$$

Note: The combined deflection (y_c) can also be worked out by considering boundary condition as follows:

At $x = 0$, $\quad y = 1.59 \times 10^{-3}$ $\quad$ and $\quad$ at $x = 0.5$, $\quad y = 0$

Example 5.12 *A beam ABC has a simply supported span AB =4 m and overhanging span BC=2 m. It is subjected to uniformly distributed load of intensity 3 kN/m in the simply supported portion and a point load 5 kN at free end C. Find the slopes θ_A, θ_B and θ_C.*

Solution:

Refer Fig. 5.21

Reactions of supports:

$\Sigma M_A = 0$

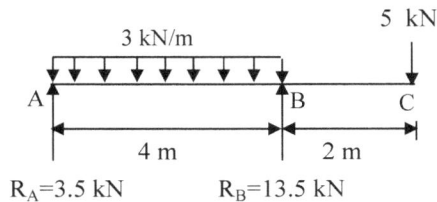

Fig. 5.21

$$R_B \times 4 = \left(3 \times 4 \times \frac{4}{2}\right) + (5 \times 6)$$

$$\therefore R_B = 13.5 \text{ kN}$$

$$\therefore R_A = (3 \times 4) + 5 - 13.5 = 3.5 \text{ kN}$$

Bending moment equation for AB part (at a section at distance x from A) is

$$M = 3.5 \, x - \frac{3x^2}{2}$$

Bending moment equation for BC part (at a section at x distance from A)

$$M = 3.5\,x - (3 \times 4)\left(x - \frac{4}{2}\right) + 13.5\,(x - 4)$$

Here both the equations are not generalized.

To get the generalized bending moment equation continue the uniformly distributed load of 3 kN/m upto right end C and balance the excess uniformly distributed load of same magnitude in BC part as explained (Fig. 5.21a).

Bending moment equation for a section in BC part at distance x from A can be written as (see Fig. 5.21a)

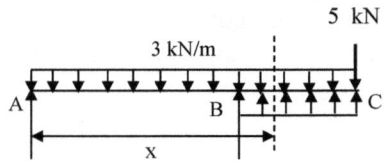

Fig. 5.21a

$$M = 3.5\,x - \frac{3x^2}{2} + 13.5(x - 4) + \frac{3(x - 4)^2}{2}$$
$$\rightarrow \quad \text{Generalized BM equation}$$

$$EI\frac{d^2y}{dx^2} = M = 3.5x - \frac{3x^2}{2} + 13.5(x - 4) + \frac{3(x - 4)^2}{2}$$

$\therefore$ Integrating,

$$EI\frac{dy}{dx} = 3.5\frac{x^2}{2} - \frac{3x^3}{6} + 13.5\frac{(x - 4)^2}{2} + \frac{3(x - 4)^3}{6} + C_1 \tag{1}$$

Integrating again,

$$EIy = 3.5\frac{x^3}{6} - \frac{3x^4}{24} + 13.5\frac{(x - 4)^3}{6} + \frac{3(x - 4)^4}{24} + C_1x + C_2 \tag{2}$$

(i) At x=0 , y=0

$(2) \rightarrow 0 = 0 - 0 + \text{Neglect} + \text{Neglect} + 0 + C_2 \quad \therefore\ C_2 = 0$

(ii) At x=4 , y=0

$$(2) \rightarrow 0 = 3.5 \times \frac{4^3}{6} - \frac{3 \times 4^4}{24} + 0 + 0 + C_1 \times 4 + 0 \quad \therefore C_1 = -1.33$$

Putting the value of constant C_1 and C_2 in equations (1) and (2) we get

$$EI\frac{dy}{dx} = 3.5 \ \frac{x^2}{2} - \frac{3x^3}{6} + 13.5\frac{(x-4)^2}{2} + \frac{3(x-4)^3}{6} - 1.33$$
$$\rightarrow \ \text{Slope equation}$$

$$EI \ y = 3.5 \ \frac{x^3}{6} - \frac{3x^4}{24} + 13.5\frac{(x-4)^3}{6} + \frac{3(x-4)^4}{24} - 1.33 \ x$$
$$\rightarrow \ \text{Deflection equation}$$

To find θ_A, put x=0 in slope equation

$$EI \ \theta_A \ = 0 - 0 + \text{Neglect} + \text{Neglect} - 1.33$$

$$\therefore \ \theta_A = \frac{-1.33}{EI}$$

In magnitude, $\qquad \theta_A = \frac{1.33}{EI}$

To find θ_B, put x=4 in slope equation

$$EI \ \theta_B = 3.5 \times \frac{4^2}{2} - \frac{3 \times 4^3}{6} + 0 + 0 - 1.33 = -5.33$$

$$\theta_B = \frac{-5.33}{EI}$$

In magnitude, $\qquad \theta_B = \frac{5.33}{EI}$

To find θ_C, put x = 6 in slope equation

$$EI\theta_c = 3.5 \times \frac{6^2}{2} - \frac{3 \times 6^3}{6} + 13.5 \times \frac{(6-4)^2}{2} + \frac{3(6-4)^3}{6} - 1.33$$

$$= -15.33$$

$$\theta_c = \frac{-15.33}{EI}$$

In magnitude, $\theta_C = \frac{15.33}{EI}$

Assuming that maximum deflection occurs in AB part equate slope equation to zero

$$3.5 \times \frac{x^2}{2} - \frac{3x^3}{6} - 1.33 = 0$$

$$\rightarrow \ 3.5 \times x^2 - x^3 - 2.66 = 0$$

$$\rightarrow \ x = 3.25 \ m$$

$$\therefore EI \ y_{max} = 3.5 \frac{(3.25)^3}{6} - \frac{3(3.25)^4}{24} + neglect + neglect$$
$$- 1.33 \times (3.25) = 1.756$$

$$\rightarrow \ y_{max} = \frac{1.756}{EI}$$

Example 5.13 *A double overhanging beam ABCD as shown (Fig. 5.22) is subjected to concentrated loads W_1 at free ends and to uniformly distributed load of $\frac{W_2}{l}$ per unit length in the portion BC. Determine the ratio of W_1 to W_2 such that the deflection at free end A equal to the deflection at E where E is the midpoint of BC.*

Solution:

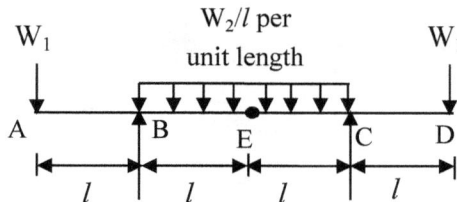

Fig. 5.22

In this case to write the generalized bending moment equation, the uniformly distributed load (udl) is extended to part CD and upward udl of same magnitude is applied in CD part as explained in Fig. 5.22a.

Due to symmetry, Reactions at B and C are

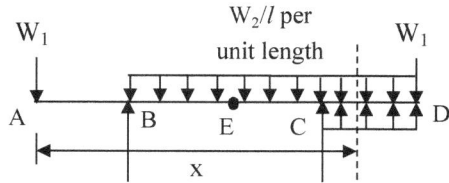

Fig. 5.22a

$$R_B = R_c = W_1 + W_2$$

Bending moment at a distance x from A   (Section in CD part)

$$M = -W_1x + (W_1 + W_2)(x - l) - \frac{W_2}{l}(x - l)\frac{(x - l)}{2}$$
$$+ (W_1 + W_2)(x - 3l) + \frac{W_2}{l}(x - 3l)\frac{(x - 3l)}{2}$$

$$EI\frac{d^2y}{dx^2} = M = -W_1x + (W_1 + W_2)(x - l) - \frac{W_2}{l}\frac{(x - l)^2}{2}$$
$$+ (W_1 + W_2)(x - 3l) + \frac{W_2}{l}\frac{(x - 3l)^2}{2}$$

$$EI\frac{dy}{dx} = -W_1\frac{x^2}{2} + (W_1 + W_2)\frac{(x - l)^2}{2} - \left(\frac{W_2}{l}\right)\frac{(x - l)^3}{6}$$
$$+ (W_1 + W_2)\frac{(x - 3l)^2}{2} + \left(\frac{W_2}{l}\right)\frac{(x - 3l)^3}{6} + C_1 \qquad (1)$$

$$EI\, y = -W_1\frac{x^3}{6} + (W_1 + W_2)\frac{(x - l)^3}{6} - \left(\frac{W_2}{l}\right)\frac{(x - l)^4}{24}$$
$$+ (W_1 + W_2)\frac{(x - 3l)^3}{6} + \left(\frac{W_2}{l}\right)\frac{(x - 3l)^4}{24} + C_1x$$
$$+ C_2 \qquad (2)$$

Boundary condition:

At midpoint E ($x=2l$), slope is zero due to symmetry of loading.

$$EI \times 0 = -W_1\frac{(2l)^2}{2} + (W_1 + W_2)\frac{(2l - l)^2}{2} - \frac{W_2}{l}\frac{(2l - l)^3}{6} + \text{neglect}$$
$$+ \text{neglect} + C_1$$

$$0 = -2W_1l^2 + (W_1 + W_2)\frac{l^2}{2} - \frac{W_2l^2}{6} + C_1$$

$$\therefore C_1 = \frac{3}{2}W_1 l^2 - \frac{1}{3}W_2 l^2 = \left(\frac{3}{2}W_1 - \frac{1}{3}W_2\right)l^2$$

Given that $y_A = y_E$ (Deflection at A and E should be same)

y_A will be obtained by putting x=0 and y_E will be obtained by putting x=2l in deflection equation.

$$\therefore C_2 = \frac{-W_1(2l)^3}{6} + (W_1 + W_2)\frac{(2l-l)^3}{6} - \frac{W_2}{l}\frac{(2l-l)^4}{24} + \text{Neglect}$$
$$+ \text{Neglect} + \left(\frac{3}{2}W_1 - \frac{1}{3}W_2\right)l^2 \times 2l + C_2$$

$$\rightarrow \quad 0 = -\frac{8}{6}W_1 l^3 + (W_1 + W_2)\frac{l^3}{6} - \frac{W_2 l^3}{24} + 3W_1 l^3 - \frac{2}{3}W_2 l^3$$

$$\rightarrow \quad 0 = -\frac{8}{6}W_1 + \frac{W_1}{6} + \frac{W_2}{6} - \frac{W_2}{24} + 3W_1 - \frac{2}{3}W_2$$

$$\rightarrow \quad W_1\left(-\frac{8}{6} + \frac{1}{6} + 3\right) + W_2\left(\frac{1}{6} - \frac{1}{24} - \frac{2}{3}\right) = 0$$

$$\rightarrow \quad W_1\left(\frac{11}{6}\right) + W_2\left(-\frac{13}{24}\right) = 0$$

$$\therefore \quad \frac{W_1}{W_2} = \frac{13}{44}$$

5.6 MOMENT AREA METHOD OF FINDING SLOPE AND DEFLECTION

Consider a beam AB subjected to pure bending which bends as an arc of a circle (Fig. 5.23).

O = Center of circle (center of curvature)

R= Radius of circle (radius of curvature)

θ = Angular displacement between A and B (change of slope)

$d\theta$ = Angular displacement between P and Q

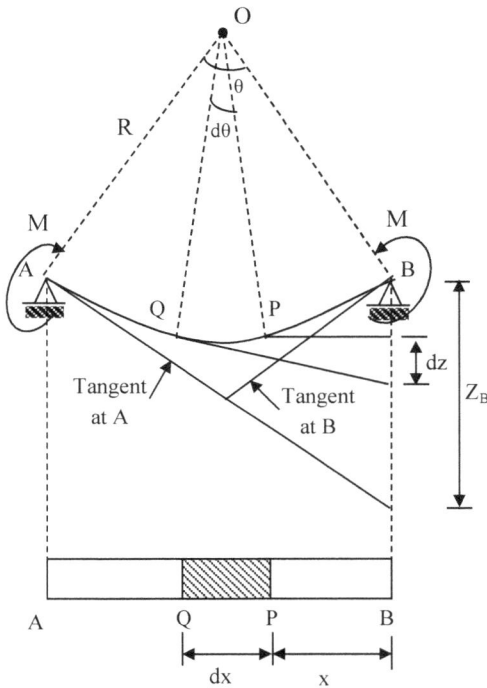

Fig. 5.23

Z_B = Vertical displacement of point B from tangent drawn at A (or deviation of point B from tangent drawn at A.

$PQ = ds = R\, d\theta \quad \text{and} \quad ds \approx dx$

$$dx = R\, d\theta \quad \rightarrow \quad \frac{1}{R} = \frac{d\theta}{dx}$$

Also $\dfrac{1}{R} = \dfrac{M}{EI}$

$$\therefore \quad \frac{d\theta}{dx} = \frac{M}{EI} \quad \rightarrow \quad d\theta = \frac{M}{EI}\, dx$$

Total angular displacement can be calculated as

$$\theta = \int_B^A \frac{M}{EI}\, dx$$

Angular displacement between A and B

$= $ Area of $\dfrac{M}{EI}$ diagram (curvature diagram)between A and B

Since deflection of beam is very small, the tangents drawn are almost horizontal lines.

Therefore, $dz = x\,d\theta = x\,\dfrac{M}{EI}\,dx$

(refer Fig. 5.23a)

$\therefore \ \ Z_B = \displaystyle\int_B^A \dfrac{M}{EI}\,x\,dx$

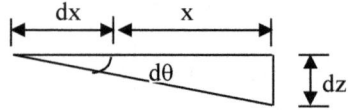

Fig. 5.23a

$\dfrac{M}{EI}\,x\,dx \ \rightarrow \ $ moment of area about B

Therefore, vertical displacement of point B from tangent drawn at A

$= $ moment of area of $\dfrac{M}{EI}$ diagram between A and B about B

5.6.1 Moment area Theorem (Mohr's Theorem)

Theorem -1

Angular displacement between two points A and B on the elastic curve is equal to the area of M/EI diagram between A and B.

$\theta = \displaystyle\int_B^A \dfrac{M}{EI}\,dx$

Theorem -2

Vertical displacement of point B from the tangent drawn at point A is equal to the moment of area of M/EI diagram between A and B about B.

$Z_B = \displaystyle\int_B^A \dfrac{M}{EI}\,x\,dx$

5.6.2 Application of Moment area Theorem to simply supported beam

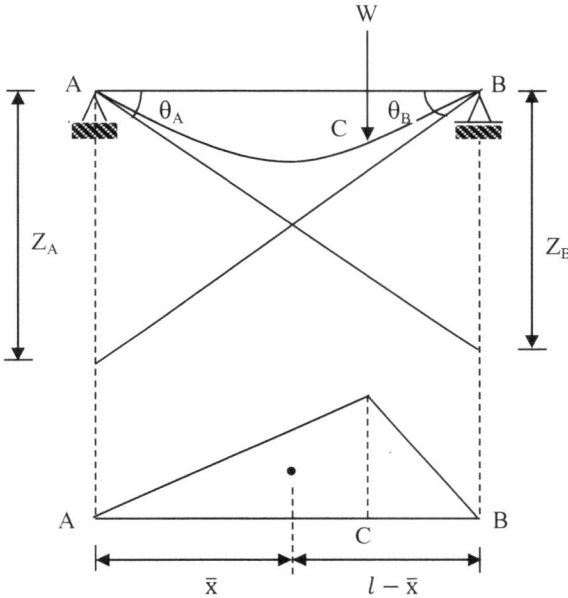

Fig. 5.24

Consider a simply supported beam loaded as shown in Fig. 5.24.

Applying moment area theorem 2,

Z_B = Moment of area of $\dfrac{M}{EI}$ diagram between A and B about B

$$= A(l - \bar{x})$$

where A = area of $\dfrac{M}{EI}$ diagram

$l - \bar{x}$ = Distance of centroid of area from B

Again $\tan \theta_A = \dfrac{Z_B}{l}$ (refer Fig. 5.24)

$$\boxed{\theta_A = \dfrac{A(l - \bar{x})}{l}}$$

Z_A = Moment of area of $\dfrac{M}{EI}$ diagram between A and B about A

$\quad\quad = A\,\bar{x}$

$\tan\theta_B = -\dfrac{Z_A}{l}$ (refer Fig. 5.24)

$$\boxed{\theta_B = \dfrac{-A\,\bar{x}}{l}}$$

where $\bar{x}$ = distance of centroid of area from A

Slope at any point:

Slope at C = Slope at A − angular displacement between A and C

$\theta_C = \theta_A - A_{AC} = \theta_A -$ area of $\dfrac{M}{EI}$ diagram between A and C

$$\boxed{\theta_C = \theta_A - A_{AC}}$$

Slope at D = Slope at B − angular displacement between B and D

$$\boxed{\theta_D = \theta_B + A_{BD}}$$

A_{BD} = area of $\dfrac{M}{EI}$ diagram between B and D

Therefore,
slope at any right side point
$\quad\quad\quad\quad$ = slope at left side point
$\quad\quad\quad\quad$ − area between the two points

slope at any left side point = slope at right side point + area between the two points

$\theta_R = \theta_L - A_{RL}$

$\theta_L = \theta_R + A_{RL}$

Deflection at any point:

Deflection at C

$$y_C = AC \ \tan \theta_A - Z_C$$

$= AC \times \theta_A$

$-$ moment of area of $\dfrac{M}{EI}$ diagram between C and A about C

$$\boxed{y_C = \theta_A AC - A_{AC} \ \bar{x}_C}$$

Deflection at D

$$y_D = BD \ \tan \theta_B - Z_D$$

$= -BD \times \theta_B$

$-$ moment of area of $\dfrac{M}{EI}$ diagram between B and D about D

$$\boxed{y_D = -\theta_B \ BD - A_{BD} \ \bar{x}_D}$$

5.6.3 Application of Moment area Theorem to cantilever beam

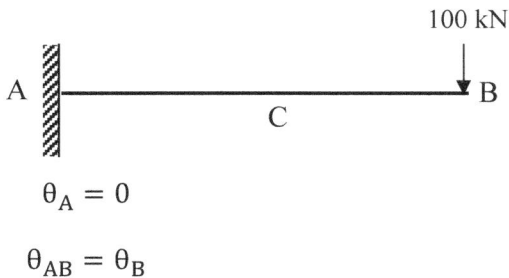

$\theta_A = 0$

$\theta_{AB} = \theta_B$

Fig. 5.25

Consider a cantilever beam AB as shown in Fig. 5.25.

$\theta_{AB} = $ Change of slope between A and B $=$ slope of B

In cantilever beam

$\theta_{AB} = 0 - \theta_B$

Slope at any right side point, $\theta_R = \theta_L - A_{LR}$

$$\boxed{\theta_B = -A_{AB}}$$

Similarly, slope at C, $\theta_C = \theta_A - A_{AC} = 0 - A_{AC}$

$$\boxed{\theta_C = -A_{AC}}$$

Deflection at B, $y_B = \theta_A AB - A_{AB}\,\bar{x}_B$

$$\boxed{y_B = -A_{AB}\,\bar{x}_B}$$

Deflection at C, $y_C = \theta_A AC - A_{AC}\,\bar{x}_C$

$$\boxed{y_C = -A_{AC}\,\bar{x}_C}$$

Example 5.14 *A simply supported beam of span 6 m carries a point load of 40 kN acting at a distance of 4 m from left end (Fig. 5.26a).*
Find (i) slope at the supports   (ii) deflection under load   (iii) maximum deflection

Solution:

$$R_A = \frac{40 \times 2}{6} = \frac{40}{3}\ kN$$

$$R_B = \frac{40 \times 4}{6} = \frac{80}{3}\ kN$$

$M_A = 0$, $M_B = 0$ (Bending moment)

$$M_c = \frac{40 \times 4 \times 2}{6} = 53.34\ kN\ m$$

The $\dfrac{M}{EI}$ diagram is shown in Fig. 5.26b.

$$\theta_A = \frac{A(l - \bar{x})}{l}$$

$A \to$ Area of $\dfrac{M}{EI}$ diagram from A to B

$x \to$ Distance of centroid of area from A

$l \to$ span length

$$\theta_A = \frac{A(l - \bar{x})}{l} = \frac{1}{6}\left[\left(\frac{1}{2} \times 6 \times \frac{53.34}{EI}\right) \times \frac{8}{3}\right]$$

$$= \frac{71.1}{EI} \quad \text{(clockwise)} \quad \to \text{Slope at A}$$

$$\theta_B = \frac{-A\bar{x}}{l}$$

$$= \frac{1}{6}\left[\left(\frac{1}{2} \times 6 \times \frac{53.34}{EI}\right) \times \frac{10}{3}\right]$$

$$= \frac{-88.9}{EI} \quad \text{(anticlockwise)} \to \text{Slope at B}$$

Deflection under load at C:

$$y_c = \theta_A \, AC - A_{AC} \, \bar{x}_C$$

$$= \frac{71.1}{EI} \times 4 - \left(\frac{1}{2} \times 4 \times \frac{53.34}{EI}\right) \times \left(\frac{1}{3} \times 4\right)$$

$$= \frac{142.16}{EI} \quad \text{(Downward deflection)}$$

Maximum deflection:
(Refer Figs. 5.26c and 5.26d)

Maximum deflection occurs in the longer side

$$\theta_D = \theta_A - A_{AD}$$

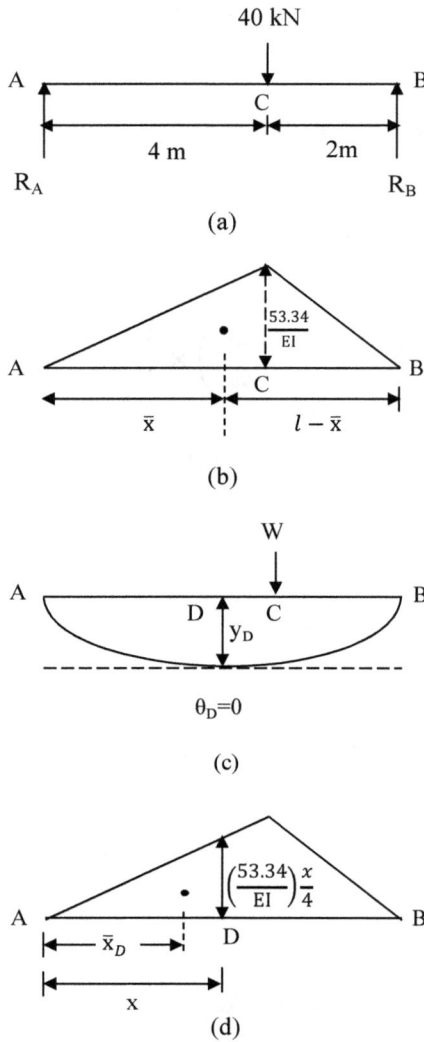

(a)

(b)

(c)

(d)

Fig. 5.26

$$= \frac{71.1}{EI} - \left(\frac{1}{2} \times x \times \frac{53.34}{EI} \times \frac{x}{4} \right)$$

At D maximum deflection occurs (say), the slope at this point must be zero.

$$\therefore \theta_D = 0$$

$$\rightarrow \quad \frac{71.1}{EI} - \left(\frac{1}{2} \times x \times \frac{53.34}{EI} \times \frac{x}{4} \right) = 0$$

$$\therefore x = 3.26 \text{ m}$$

$$\therefore y_{max} = y_B = \theta_A \, AD - A_{AD} \, \overline{x}_D$$

$$= \left(\frac{71.1}{EI} \right) x - \left(\frac{1}{2} \times x \times \frac{53.34}{EI} \times \frac{x}{4} \right) \left(\frac{1}{3} \times x \right)$$

$$= \frac{154.78}{EI} \quad \text{(downward deflection)}$$

(Positive deflection means Z is above tangent at A , i.e. deflection is downward)

Example 5.15 *A rectangular section steel beam 4m long, 75 mm wide and 100 mm deep is simply supported at its ends. If it is subjected to loads of 20 kN and 40 kN at distances of 2m and 3m respectively from the left end support, determine the deflection of beam at a distance 2.5m from left end. Also find maximum deflection. Take E_{steel} = 200 GPa.*

Solution:

$$I = \frac{bd^3}{12} = \frac{75 \times 100^3}{12} = 6.25 \times 10^6 \text{mm}^4$$

$$E = 200 \text{ GPa} = 2 \times 10^5 \text{ N/mm}^2$$

$$\therefore EI = 2 \times 10^5 \times 6.25 \times 10^6$$

$$= 1.25 \times 10^{12} \text{ N mm}^2 = 1250 \text{ kN m}^2 \text{Refer Fig. 5.27a.}$$

$$\sum M_B = 0$$

$$R_A \times 4 = 20 \times 2 + 40 \times 1,$$

$$\therefore R_A = 20 \text{ kN} \quad \text{and}$$

$$R_B = 60 - 20 = 40 \text{ kN}$$

Bending moment :

$M_A = 0, \qquad M_B = 0$

$M_C = 20 \times 2 = 40$ kN m

$M_D = 20 \times 3 - 20 \times 1 = 40$ kNm

The $\dfrac{M}{EI}$ diagram is shown in Fig. 5.27b.

Refer Fig. 5.27c

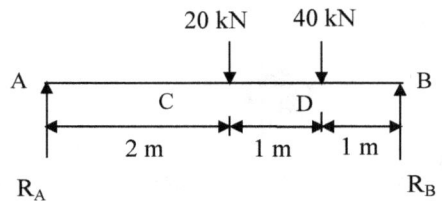

(a)

$$\theta_A = \frac{A(l - \bar{x})}{l} = \frac{1}{l}[A_1(l - \bar{x}_1) + A_2(l - \bar{x}_2) + A_3(l - \bar{x}_3)]$$

$$= \frac{1}{4}\left[\left(\frac{1}{2} \times 2 \times \frac{40}{EI}\right)\left(4 - \frac{4}{3}\right) + \left(1 \times \frac{40}{EI}\right)(1.5) + \left(\frac{1}{2} \times 1 \times \frac{40}{EI}\right) \times \frac{2}{3}\right] = \frac{45}{EI}$$

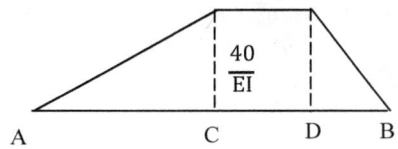

(b)

$$\bar{x}_1 = \frac{2}{3} \times 2 = \frac{4}{3} \text{ m}, \qquad \bar{x}_2 = 2.5 \text{ m}$$

$$\bar{x}_3 = 3 + \left(\frac{1}{3} \times 1\right) = \frac{10}{3} \text{ m}$$

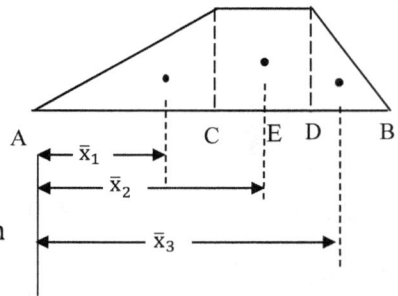

To find the deflection at 2.5 m from end A (say point E)

$$y_E = \theta_A AE - A_{AE} \bar{x}_E$$

(c)

Fig. 5.27

$$= \left(\frac{45}{EI} \times 2.5\right) - \left\{\left(\frac{1}{2} \times 2 \times \frac{40}{EI}\right)\left(2.5 - \frac{2}{3} \times 2\right) + \left(0.5 \times \frac{40}{EI}\right)\left(\frac{0.5}{2}\right)\right\}$$

$$= \frac{60.83}{EI}$$

$$\therefore y_E = \frac{60.83}{1250} = 0.0486 \text{ m}$$

$$\rightarrow \text{ Deflection at distance 2.5 m from left end}$$

To find maximum deflection (y_{max}):

Let maximum deflection occurs at a distance x from left end in CD part (say point F)

$$\theta_F = \theta_A - A_{AF} = 0 \qquad \text{(For maximum deflection)}$$

$$\therefore \frac{45}{EI} - \left\{\left(\frac{1}{2} \times 2 \times \frac{40}{EI}\right) + (x - 2)\frac{40}{EI}\right\} = 0 \qquad \rightarrow \qquad x = 2.125 \text{ m}$$

$$\therefore \text{ Maximum deflection } y_{max} = y_F = \theta_A . AF - A_{AF} . \bar{x}_F$$

$$\rightarrow \quad y_{max} = \left(\frac{45}{EI} \times 2.125\right)$$
$$- \left\{\left(\frac{1}{2} \times 2 \times \frac{40}{EI}\right)\left(0.125 + \frac{1}{3} \times 2\right)\right.$$
$$\left. + \left(0.125 \times \frac{40}{EI}\right)\left(\frac{0.125}{2}\right)\right\}$$

$$\therefore y_{max} = \frac{63.645}{1250} = 0.0509 \text{ m}$$

Example 5.16 *Obtain an expression for deflection and slope at free end B for the cantilever beam loaded as shown (Fig. 5.28a).*

Solution:

Bending moment between B and C is zero.

$$M_A = -\frac{Wl}{2}$$

$\frac{M}{EI}$ diagram is drawn (Fig. 5.28b)

Deflection at B:

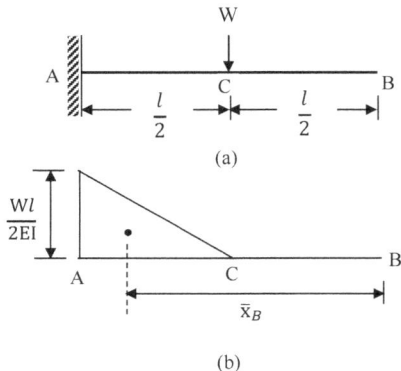

(a)

(b)

Fig. 5.28

$$y_B = -A_{AB}\, \bar{x}_B = -A_{AC}\, \bar{x}_B$$

$$= -\left(\frac{1}{2} \times \frac{l}{2} \times \frac{Wl}{2EI}\right)\left(\frac{l}{2} + \frac{2}{3} \times \frac{l}{2}\right)$$

$$= \frac{5Wl^3}{48\,EI} \qquad \rightarrow \qquad \text{Deflection at free end}$$

Slope at B:

$$\theta_B = -A_{AB} = -A_{AC} = -\left(\frac{1}{2} \times \frac{l}{2} \times \frac{Wl}{2EI}\right) = \frac{Wl^2}{8\,EI}$$
$$\rightarrow \quad \text{Slope at free end}$$

At Point C

$$y_C = -A_{AC}\, \bar{x}_C = -\left(\frac{1}{2} \times \frac{l}{2} \times \frac{Wl}{2EI}\right)\left(\frac{2}{3} \times \frac{l}{2}\right) = \frac{Wl^3}{24\,EI}$$

Note:

For cantilever subjected to point load at mid span

$$\text{Deflection at free end} = \frac{5Wl^3}{48EI}$$

$$\text{Slope at free end} = \frac{Wl^2}{48EI}$$

$$\text{Deflection under load} = \frac{Wl^3}{24EI}$$

$$\text{Slope under load} = \frac{Wl^2}{8EI}$$

When bending moment is zero for a portion, then slope remains unchanged in that portion.

Example 5.17   *Obtain an expression for deflection and slope at free end for cantilever beam loaded as shown (Fig. 5.29a).*

Solution:

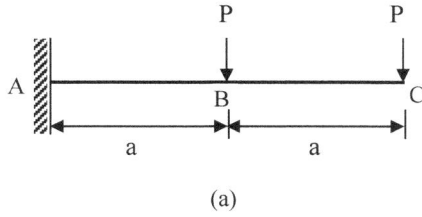

(a)

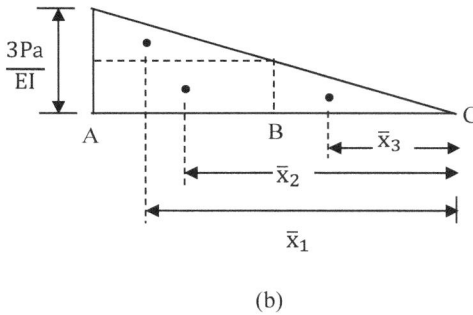

(b)

Fig. 5.29

Refer Fig. 5.29a

Bending moments

$M_c = 0$, $M_B = -Pa$

$M_A = -P \times a - P \times 2a = -3Pa$

$\dfrac{M}{EI}$ diagram is shown in Fig. 5.29b.

Deflection at C is

$y_C = \theta_A \, AC - A_{AC} \, \bar{x}_C$

$= 0 - A_{AC} \, \bar{x}_C = -(A_1 \bar{x}_1 + A_2 \bar{x}_2 + A_3 \bar{x}_3)$

$$= \left[\left(-\frac{1}{2} \times a \times \frac{2Pa}{EI} \right) \left(a + \frac{2}{3}a \right) + \left(-a \times \frac{Pa}{EI} \right) \left(a + \frac{a}{2} \right) \right.$$
$$\left. + \left(-\frac{1}{2} \times a \times \frac{Pa}{EI} \right) \left(\frac{2}{3}a \right) \right] = -\frac{7Pa^3}{2EI}$$

$$\therefore \ y_C = \frac{7Pa^3}{2EI} \quad \text{(Downward deflection)}$$

Slope at C is

$$\theta_C = \theta_A - A_{AC} = -A_{AC} = -(A_1 + A_2 + A_3)$$
$$= \left\{ \left(-\frac{1}{2} \times a \times \frac{2Pa}{EI} \right) + \left(-a \times \frac{Pa}{EI} \right) + \left(-\frac{1}{2} \times a \times \frac{Pa}{EI} \right) \right\} = \frac{-5Pa^2}{2EI}$$

Example 5.18 *A simply supported beam with overhang is loaded as shown (Fig. 5.30). If* $a = l/2$*, find the ratio P/W to make the deflection at the free end equal to zero.*

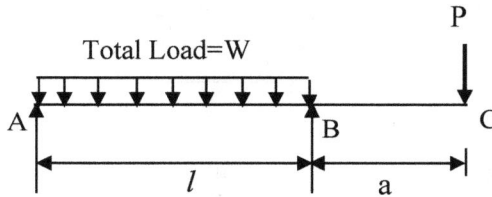

Fig. 5.30

Solution:

Bending moments are calculated for the two loads separately and two M/EI diagrams   are explained in Fig. 5.31.

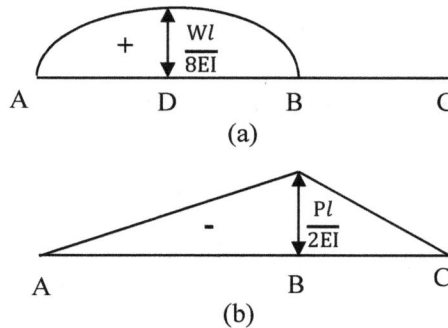

Fig. 5.31

$$M_D = \left(\frac{W}{2} \times \frac{l}{2} \right) - \left(\frac{W}{l} \times \frac{l}{2} \times \frac{l}{4} \right) = \frac{Wl}{8}$$

$$M_B = -\frac{Pl}{2} \quad (a = \frac{l}{2} \text{ given})$$

$$M_C = 0, \quad M_A = 0$$

Slope at B:

$$\theta_B = -\frac{A\bar{x}}{l} = -\frac{1}{l}(A_1\bar{x}_1 + A_2\bar{x}_2)$$

where A_1 and A_2 = Area between A and B for the diagrams respectively.

$$\therefore \theta_B = -\frac{1}{l}\left[\left(\frac{2}{3} \times l \times \frac{Wl}{8EI}\right)\frac{l}{2} + \left(\frac{-1}{2} \times l \times \frac{Pl}{2EI}\right)\frac{2}{3}l\right] = -\frac{Wl^2}{24EI} + \frac{Pl^2}{6EI}$$

Deflection at C

$$y_C = \theta_B \, BC - A_{BC} \, \bar{x}_C$$

$$= \left(-\frac{Wl^2}{24EI} + \frac{Pl^2}{6EI}\right)\frac{l}{2} - \frac{1}{2} \times \frac{l}{2} \times \frac{Pl}{2EI}\left(\frac{2}{3} \times \frac{l}{2}\right)$$

$$= -\frac{Wl^3}{48EI} + \frac{Pl^3}{8EI}$$

Given that $\quad y_c = 0$

$$\therefore \quad -\frac{Wl^3}{48EI} + \frac{Pl^3}{8EI} = 0$$

$$\rightarrow \quad \frac{Wl^3}{48EI} = \frac{Pl^3}{8EI}$$

$$\therefore \quad \frac{P}{W} = \frac{1}{6}$$

Example 5.19 *A beam ABC has a simply supported span AB=6 m and overhang BC=2 m. It carries a point load 20 kN at mid point of AB and uniformly distributed load of intensity 1.5 kN/m in the overhang. Find the deflection under point load and at the free end.*

Solution:

Refer Fig. 5.32a.

Due to point load of 20 kN alone

$M_D = 30$ kNm

Due to *udl* of 1.5 kN/m

$M_B = -1.5 \times 2 \times \dfrac{2}{2} = -3$ kNm

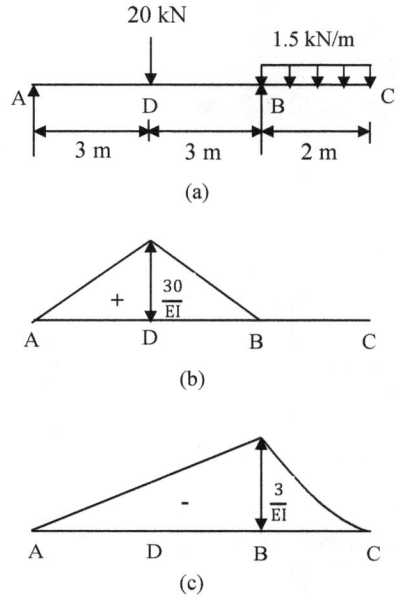

(a)

(b)

(c)

Fig. 5.32

$\dfrac{M}{EI}$ diagrams are shown for the point load and udl separately in Figs. 5.32 b and c, respectively.

$$\theta_A = \frac{A(l - \bar{x})}{l} = \frac{1}{l}[A_1(l - \bar{x}_1) + A_2(l - \bar{x}_2)]$$

$$= \frac{1}{6}\left[\left(\frac{1}{2} \times 6 \times \frac{30}{EI}\right) \times 3 + \left(\frac{-1}{2} \times 6 \times \frac{3}{EI}\right) \times \frac{1}{3} \times 6\right]$$

$$= \frac{42}{EI} \quad (\text{cw w. r. t. x axis})$$

$$\theta_B = \frac{-A\bar{x}}{l} = \frac{-1}{6}\left[\left(\frac{1}{2} \times 6 \times \frac{30}{EI}\right) \times 3 + \left(\frac{-1}{2} \times 6 \times \frac{3}{EI}\right) \times \frac{2}{3} \times 6\right]$$

$$= \frac{-39}{EI} \quad (\text{acw w. r. t. x axis})$$

To find deflection at D:

$$y_D = \theta_A\,AD - A_{AD}\,\bar{x}_D$$

$$= \left(\frac{42}{EI} \times 3\right) - \left[\left(\frac{1}{2} \times 3 \times \frac{30}{EI}\right)\left(\frac{1}{3} \times 3\right) + \left(\frac{-1}{2} \times 3 \times \frac{1.5}{EI}\right)\left(\frac{1}{3} \times 3\right)\right]$$

$$= \frac{83.25}{EI} \quad \text{(Downward deflection)}$$

To find deflection at C:

$$y_C = \theta_B\,BC - A_{BC}\,\bar{x}_C$$

$$= \left(\frac{-39}{EI} \times 2\right) - \left(\frac{-1}{3} \times 2 \times \frac{3}{EI}\right)\left(\frac{3}{4} \times 2\right)$$

$$= -\frac{75}{EI} \quad \text{(upward deflection)}$$

HIGHLIGHTS

Definitions

1. *Deflection Curve*: This is the longitudinal centroidal axis of beam after loading. This is also known as elastic curve.

2. *Flexural rigidity*: The product of Young's modulus (E) and moment of inertia (I) is termed as flexural rigidity (EI).

Concepts and Formulae

1. Curvature Equation (or Differential equation of elastic curve):

$$EI\,\frac{d^2y}{dx^2} = M$$

2. Maximum slope and maximum deflection in some standard cases:
(a) Cantilever:

(i) Subjected to point load at free end,

$$\theta_{max} = \frac{Wl^2}{2EI} \quad \text{at free end}$$

$$y_{max} = \frac{Wl^3}{3EI} \quad \text{at free end}$$

(ii) Subjected to uniformly distributed load over whole of its span

$$\theta_{max} = \frac{wl^3}{6EI} \quad \text{at free end}$$

$$y_{max} = \frac{wl^4}{8EI} \quad \text{at free end}$$

(iii) Subjected to couple at free end

$$\theta_{max} = \frac{Ml}{EI} \quad \text{at free end}$$

$$y_{max} = \frac{Ml^2}{2EI} \quad \text{at free end}$$

(iv) Subjected to uniformly varying load from zero at free end to w/unit length at fixed end.

$$\theta_{max} = \frac{wl^3}{24EI} \quad \text{at free end}$$

$$y_{max} = \frac{wl^4}{30EI} \quad \text{at free end}$$

(b) Simply supported beam:

(i) Subjected to point load at mid span,

$$\theta_{max} = \frac{Wl^2}{16EI} \quad \text{at support}$$

$$y_{max} = \frac{Wl^3}{48EI} \quad \text{at mid span}$$

(ii) Subjected to uniformly distributed load over entire span,

$$\theta_{max} = \frac{wl^3}{24EI} \text{ at support}$$

$$y_{max} = \frac{5}{384} \frac{wl^4}{EI} \text{ at mid span}$$

(iii) Subjected to uniformly varying load from zero at one end to w/unit length at other end

$$y_{max} = \frac{wl^4}{153 \ EI}$$

at x = (0.5193)l from the end where intensity of load is zero.

$$\theta_A = \frac{-7wl^3}{360 \ EI}, \qquad \theta_B = \frac{wl^3}{45 \ EI}$$

(iv) Subjected to couple at one end

$$y_{max} = \frac{Ml^2}{9\sqrt{3} \ EI} \qquad \text{at } x = (0.577)l \qquad \text{from the other end}$$

3. Macaulay's method is used to find slope and deflection in a beam subjected to asymmetrical point load, number of point loads and uniformly distributed load applied over part of the span.

In this method, a generalized bending moment equation is written. The bending moment equation is valid throughout the span by considering or neglecting certain terms. The same constants of integration are valid throughout the span.

4. For a simply supported beam of length l carrying asymmetrical point load W at distance 'a' from left end.

$$\text{Deflection under load, } y = \frac{Wa^2b^2}{3 \ EI \ l}$$

SHORT TYPE QUESTIONS

1. Deflection of a beam for a given load and span is inversely proportional to its ___

[Ans: Flexural rigidity]

2. The amount of deflection of a beam subjected to a type of loading
depends upon
(a) cross section   (b) bending moment
(c) either (a) or (b)   (d) both (a) and (b)

[Ans: (d)]

3. At the point of contraflexure
(a) the stress is zero   (b) the shear force is zero
(c) the bending moment is zero   (d) the slope is zero

[Ans: (c)]

4. The slope and deflection at a section in a loaded beam can be
found out by which of the following methods?
(a) Double integration method   (b) Moment method
(c) Macaulay's Method   (d) any of the above

[Ans. (d)]

5. Deflection of the free end of a cantilever beam subjected to concentrated load at the middle of the span is given by
(a) $\dfrac{Pl^3}{3EI}$ (b) $\dfrac{Pl^3}{24EI}$ (c) $\dfrac{Pl^3}{48EI}$ (d) $\dfrac{5Pl^3}{48EI}$

[Ans: (d)]

6. Moment area method is a method to determine _____ at a point.
(a) Bending moment   (b) shear force
(c) slope   (d) deflection

[Ans: (c) and (d)]

7. If a cantilever is loaded at the mid span, what will be the slope and
deflection at its free end. Given that slope under load is θ and deflection under load is y.

[Ans: slope = θ, deflection =(5/2)y]

8. Write the equation of Radius of curvature of elastic curve.

[Ans: $R = \dfrac{EI}{M}$ in terms of bending moment & flexural rigidity. Also $\dfrac{1}{R} = \dfrac{d^2y}{dx^2}$ in terms of coordinates (x,y)]

9. A simply supported beam of span *l* is carrying a point load W at the mid span. What is the deflection at the centre of the beam?

[Ans: $\dfrac{Wl^3}{48\,EI}$]

10. A cantilever beam carries load W uniformly distributed over its entire length. If the same load is placed at the free end of similar cantilever, then the ratio of maximum deflection in the first case to that in the second case will be

(a) $\dfrac{3}{8}$ (b) $\dfrac{8}{3}$ (c) $\dfrac{5}{8}$ (d) $\dfrac{8}{5}$

[Ans: (a)]

11. The maximum deflection of a simply supported beam of span *l* carrying *udl* of intensity w per unit length is

(a) $\dfrac{5Wl^4}{324EI}$

(b) $\dfrac{5Wl^4}{384EI}$

(c) $\dfrac{Wl^4}{384EI}$

(d) $\dfrac{Wl^4}{324EI}$

[Ans: (b)]

EXERCISE PROBLEMS

1. A beam of uniform section and constant depth is freely supported over a span of 3 m. If the point load at mid span is 30 KN and I_{xx} = 15.61 × 10⁻⁶ m⁴, calculate
(i) the central deflection
(ii) the slopes at the ends of the beam, Take E = 200 GPa.
[Ans. (i) 5.4 mm (ii) 0.309°]

2. A beam with a span of 4.5 m carries a point load of 30 kN at 3 m from the left support. If for the section I_{xx} = 54.97 × 10⁻⁶ m⁴ and E = 200 GPa, find
(i) the deflection under the load
(ii) the position and amount of maximum deflection

[Ans: (i) 4.09 mm downward (ii) 4.456 mm downward at 2.45 m from left end]

3. A simply supported beam of uniform rectangular cross section carries a point load at midspan. It is given that the ratio of depth to span of beam is 1:24 and the ratio of maximum deflection to the span is 1:500. Show that the maximum bending stress in the beam is $\frac{E}{2000}$ where E is the Young's modulus of the material of the beam.

4. A cantilever 150 mm wide and 200 mm deep projects 2 m out of a wall and is carrying a point load of 40 kN at the free end. Determine the slope and deflection of the cantilever at the free end. Take E = 2.1×10^5 MN/m^2.

[Ans: θ_{max} = 0.0038 radian, y_{max} = 5.079 mm]

5. A rectangular simply supported beam of length 2 m and cross section 100 mm × 200 mm is carrying uniformly distributed load of 10 kN/m throughout its span. Find the maximum slope and deflection. Take E = 2×10^4 N/mm^2.

[Ans: θ_{max} = 0.0025 radian at supports, y_{max} = 1.56 mm at mid span]

6. A cantilever 2.5 m long is carrying a load of 25 kN at free end and 35 kN at a distance of 1.3 m from the fixed end. Find the slope and deflection at the free end. Take E = 2×10^8 kN/m^2 and I = 15×10^{-4} m^4.

[Ans: θ_{max} = 0.00359 radian, y_{max} = 6.37 mm]

7. A beam AB of 8 m span is simply supported at the ends. It carries a point load of 10 kN at a distance of 1 m from the end A and a uniformly distributed load of 5 kN/m for a length of 2 m from the end B. If I = 10×10^{-6} m^4 determine
(i) deflection at mid span
(ii) maximum deflection
(iii) slope at the end A

[Ans: (i) 48.74 mm (downward), (ii) 8.75 m (downward), (iii) - 0.417°]

8. A cantilever of 2.5 m effective length carries a load of 25 kN at its free end. If the deflection at the free end is not to exceed 5 mm, what

must be the value of I for the section of the cantilever? Take E = 210 GN/m².

[Ans: 1.24×10 ⁻⁴ m⁴]

Chapter 6

Torsion

Learning Objectives

After going through this chapter, the reader will be able to
- differentiate between torsion and bending.
- explain the behaviour of structural members subjected to torsion.
- derive and explain torsion equation.
- design the shafts against torsion and bending and calculate the stresses.
- compute the stresses in composite shafts.
- analyse the statically indeterminate shafts

6.1 INTRODUCTION

In the last few chapters, we have studied axially loaded members and transversely loaded beams which are very common and important structural elements. For beams, we have discussed methods to calculate the bending moment and shear forces at any section, the internal stress distribution and the resulting deformations. The present Chapter deals with another important structural member – shaft transmitting power. Shafts are commonly used to transmit power from one point to many points. While shafts also act as beams as they are supported at points along their length and carry their own weight, they predominantly act as torsion elements. In this Chapter, torsion of statically determinate as well as indeterminate shafts will be taken up.

6.1.1 Definitions

Shaft: It is a structural member used to transmit mechanical power from one point to another. Shaft is subjected to torsion.

Torque (or *Twisting moment* or *Torsional moment*): This is the moment acting about the axis perpendicular to the plane of cross section. Torsional moment causes shear stress.

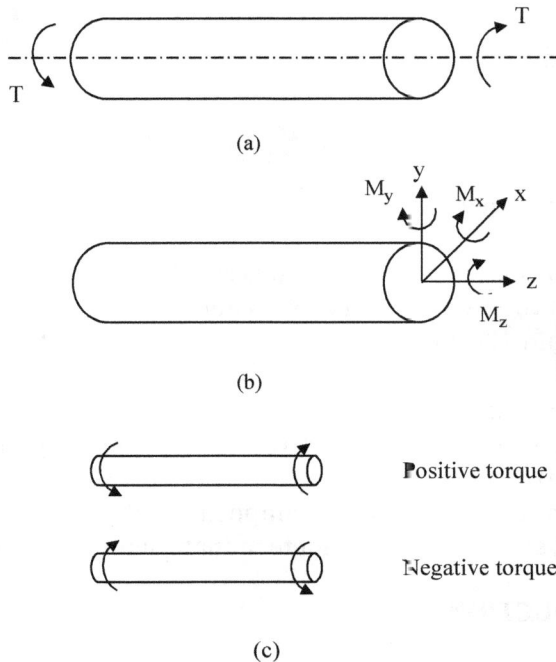

(a)

(b)

(c)

Fig. 6.1

Refer Fig. 6.1. It may be observed that the moment M_z about z axis (Fig. 6.1b) i.e. the axis perpendicular to the plane of cross section, is the torque. When a member is subjected to bending moment, bending occurs and normal stresses (bending stresses) are developed. But when a member is subjected to torque, twisting occurs and shear stresses are developed. As beam bends as an effect of bending moment, shaft twists as an effect of torsion.

Torque diagram: It represents the variation of torque along the length of the shaft. Following sign convention may be used while drawing the torque diagram:

Looking at the cross section of the member from the right, torque is positive if clockwise and negative if anticlockwise (Fig. 6.1c).

Pure Torsion: A member is said to be under pure torsion when it is subjected to pure torque without being associated with bending moment or axial force. Due to this pure torsion, the state of stress at any point in the cross section is one of pure shear.

6.2 DERIVATION OF TORSION FORMULA

When an external torque is applied to a shaft, it creates a corresponding internal torque within the shaft. In this section, we will develop an equation that relates this internal torque to the shear stress distribution on the cross section of circular shaft.

Assumptions:

1. Plane and circular cross sections of shaft remain plane and circular after torsion (no warping of cross section).
2. Torsion is uniform along the length of the shaft.
3. The cross sections of shaft rotate with radii remaining straight. It means shear strain is zero at the axis of shaft and increases linearly to a maximum value at the surface of shaft.
4. Material of the shaft follows Hooke's law.
5. Cross section of the shaft is uniform.
6. Material of the shaft is homogeneous and isotropic.
7. Torque is applied gradually.

Derivation:

Consider a shaft AB fixed at end A and free at end B, is subjected to Torque T as shown (Fig. 6.2).

Let
T = Applied torque
l = Length of shaft
θ = Angle of twist
ϕ = Shear strain
G = Modulus of rigidity
R = Radius of shaft

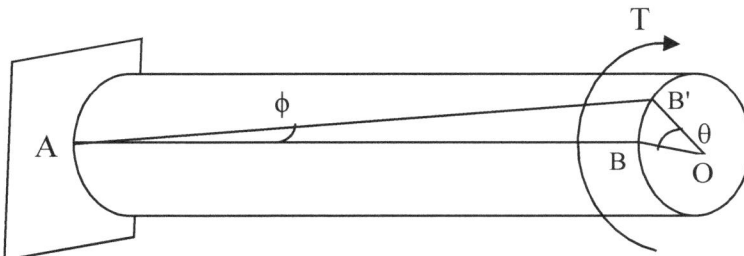

Fig. 6.2

Refer Fig. 6.2

AB = Before applying torque, AB' = After applying torque

OB = Radius of shaft before applying torque,

OB' = Radius of shaft after applying torque

The longitudinal and transverse sections of the shaft are represented in Fig. 6.3.

Refer Figs. 6.3a and 6.3b.

$$BB' = AB \tan \phi = l \tan \phi = l \times \phi$$

Again $BB' = R\theta$

$$\therefore l \phi = R\theta \quad \rightarrow \quad \phi = \frac{R\theta}{l}$$

From Hooke's law, shear stress is directly proportional to shear strain.

$$\tau \propto \phi \quad \rightarrow \quad \tau = G\phi = G \frac{R\theta}{l}$$

$$\frac{\tau}{R} = \frac{G\theta}{l}$$

Consider elemental area dA in the cross section of shaft (Fig. 6.3c).

τ_r = Shear stress at radial distance r, and
τ = Maximum shear stress at the surface of shaft (at radius R)

Shear stress on elemental area, $\tau_r = \frac{\tau}{R} r$

Force on elemental area, $dF = \tau_r dA = \frac{\tau}{R} r \, dA$

Elemental Torque, $dT = dF \times r = \frac{\tau}{R} r^2 \, dA$

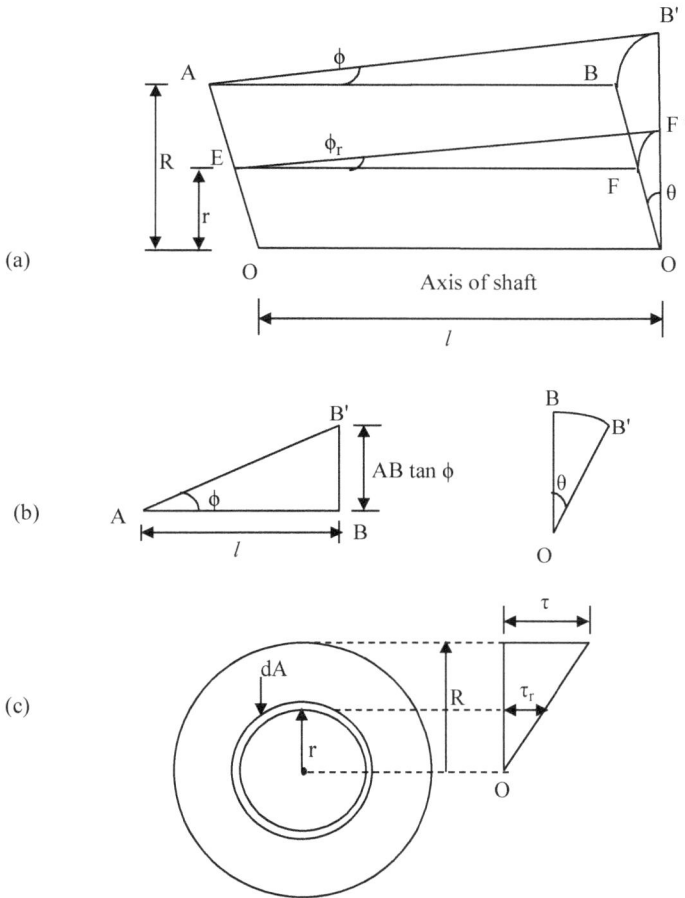

Fig. 6.3

Torque, $\quad T = \int dT = \dfrac{\tau}{R} \int r^2 \, dA = \dfrac{\tau}{R} \, J$

where $J = \int r^2 \, dA$ = polar moment of inertia

$\therefore \quad \dfrac{T}{J} = \dfrac{\tau}{R}$

$$\boxed{\dfrac{\tau}{R} = \dfrac{T}{J} = \dfrac{G\,\theta}{l}} \quad \longrightarrow \quad \textbf{Torsion Formula}$$

Note:

Polar moment of inertia:

For solid circular cross section (of diametr d), $J = I_x + I_y$

$$= 2\,\frac{\pi d^4}{64} = \frac{\pi d^4}{32}$$

($I_x, I_y \rightarrow$ Moment of inertia about x- and y- axes, respectively)

For hollow circular cross section (of outer diameter D and inner diameter d),

$$J = \frac{\pi(D^4 - d^4)}{32}$$

6.3 POWER TRANSMITTED BY SHAFT

Consider a shaft subjected to torque T and rotating at a speed of N rpm.

Angular velocity, $\omega = \dfrac{2\pi N}{60}$ rad/sec

Power = Torque $\times$ Angular velocity $= T\,\omega$

$$\boxed{\text{Power} = \frac{2\pi NT}{60}}$$

Mean torque shall be used for calculating power whereas maximum torque is used for calculating shear stress and angle of twist (Fig. 6.4).

$\therefore$ Power $= \dfrac{2\pi NT_{mean}}{60}$

{Power $\rightarrow$ Watt   where N $\rightarrow$ rpm and T $\rightarrow$ Nm}

and $\dfrac{\tau}{R} = \dfrac{T_{max}}{J} = \dfrac{G\theta}{l}$

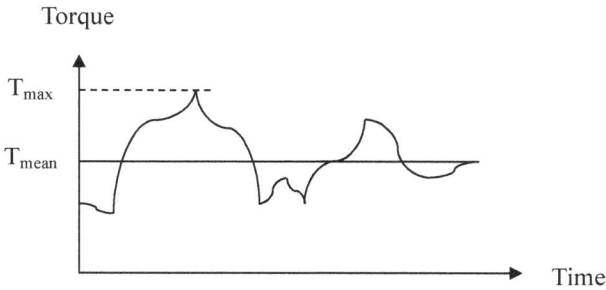

Fig. 6.4

6.4 POLAR SECTION MODULUS

In the case of bending stresses in beams, we have defined the term 'section modulus' (see chapter 4). On similar lines, polar section modulus is defined as the ratio of polar moment of inertia and the distance to the extreme fiber of section.

Polar section modulus, $\quad Z_p = \dfrac{J}{r_{max}}$

For solid circular section (of diametr d), $\quad Z_p = \dfrac{\pi d^4/32}{d/2} = \dfrac{\pi d^3}{16}$

For hollow circular section $\left(\begin{matrix} \text{of external diametr D and internal} \\ \text{diameter d} \end{matrix}\right)$,

$$Z_p = \dfrac{\pi(D^4 - d^4)/32}{D/2} = \dfrac{\pi}{16}\left(\dfrac{D^4 - d^4}{D}\right)$$

6.5 TORSIONAL MOMENT OF RESISTANCE

In the bending of beams, we have defined the term 'moment of resistance' (see chapter 4). On similar lines, torsional moment of resistance is defined as the maximum torque that can be carried by the section without exceeding the maximum permissible shear stress.

Torsional moment of resistance, $\quad T = \tau \times Z_p$

τ = Permissible shear stress, Z_p = Polar section modulus

The torsional moment of resistance is also called *torque carrying capacity* of the shaft.

Example 6.1 *A solid shaft in a rolling mill transmits 20 kW at 2Hz. Determine the diameter of the shaft if shear stress is not to exceed 40 MPa and angle of twist is limited to 6^0 in a length of 3 m. Use G=83 GPa.*

Solution:

$P = 20\,kW = 20 \times 10^3\,W$, $l = 3\,m$, $G = 83\,GPa$

$= 83 \times 10^9\,N/m^2$

Angular Velocity $= 2\,Hz = 2\,rev/sec =$
$(2 \times 60)\,rpm = 120\,rpm$

Power $= \dfrac{2\pi NT}{60}$

$20 \times 10^3 = \dfrac{2\pi \times 120 \times T}{60}$

$\rightarrow$ Torque, $T = 1591.5\ Nm$

Given $\tau \le 40\,MPa$, $\theta \le 6^0$

$\dfrac{\tau}{R} = \dfrac{T}{J} = \dfrac{G\theta}{l}$

Strength Consideration

$\dfrac{\tau}{R} = \dfrac{T}{J} = \dfrac{G\theta}{l}$

Stiffness Consideration

(i) Strength consideration: Shear stress shall not exceed 40 MPa

$\dfrac{\tau}{R} = \dfrac{T}{J}$ $\rightarrow$ $\dfrac{40 \times 10^6}{\left(\dfrac{D}{2}\right)} = \dfrac{1591.5}{\dfrac{\pi}{32} \times D^4}$ $\rightarrow$ $D = 0.0587\,m$

(ii) Stiffness consideration: Angle of twist shall not exceed 6^0

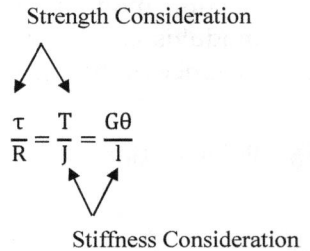

$$\frac{T}{J} = \frac{G\theta}{l} \quad \rightarrow \quad \frac{1591.5}{\frac{\pi}{32} \times D^4} = \frac{83 \times 10^9 \times 6 \times \frac{\pi}{180}}{3} \quad \rightarrow \quad D$$

$= 0.0486 \text{ m}$

∴ Safe diameter of shaft = 0.0587 m = 58.7 mm (bigger of the two values)

Example 6.2 *A hollow shaft with a diameter ratio 0.6 is required to transmit 400 kW at 120 rpm with a uniform twisting moment. The shearing stress in the shaft must not exceed 60 MPa and twist in a length of 2.5 m must not exceed 1°. Calculate the safe external diameter of shaft satisfying these conditions. Take modulus of rigidity for shaft material G = 80 GPa.*

Solution:

$$\frac{d}{D} = 0.6$$

$$\text{Power} = \frac{2\pi NT}{60}$$

$$\rightarrow \quad 400 \times 10^3 = \frac{2\pi \times 120 \times T}{60} \quad \rightarrow \quad T$$
$$= 31.83 \times 10^3 \quad Nm$$

Polar moment of inertia of hollow cylinder

$$J = \frac{\pi \times (D^4 - d^4)}{32} = \frac{\pi}{32} \times (D^4 - (0.6D)^4) = \frac{\pi}{32}[1 - (0.6)^4]D^4$$

$$= (0.0854) D^4$$

(i) Strength consideration:

$$\frac{\tau}{R} = \frac{T}{J} \quad \rightarrow \quad \frac{60 \times 10^6}{\frac{D}{2}} = \frac{31.83 \times 10^3}{(0.0854) D^4} \qquad \therefore \quad D = 0.145 \text{ m}$$

(ii) **Stiffness consideration:**

$$\frac{T}{J} = \frac{G\theta}{l} \quad \rightarrow \quad \frac{31.83 \times 10^3}{(0.0854) D^4} = \frac{80 \times 10^9 \times \left(1 \times \frac{\pi}{180}\right)}{2.5}$$

$\therefore D = 0.16$ m

Therefore, safe external diameter for the hollow shaft, D = 0.16 m

Example 6.3 *A hollow shaft having an inside diameter 60 % of its outside diameter is to replace a solid shaft transmitting same power at same speed. Calculate the percentage saving in material if the material to be used is also the same.*

Solution:

Both the shafts are made of same material, therefore allowable stress is same.

Both the shafts transmit same power at same speed means both are transmitting same torque.

D_1 = Diameter of solid shaft

d, D = Inner and outer diameters, respectively for hollow shaft

Consider solid shaft:

$$\frac{\tau}{R} = \frac{T}{J} \quad \rightarrow \quad \frac{\tau}{\frac{D_1}{2}} = \frac{T_s}{\frac{\pi D_1^4}{32}} \qquad \therefore T_s = \tau \frac{\pi D_1^3}{16}$$

Consider hollow shaft:

$$\frac{\tau}{R} = \frac{T}{J} \quad \rightarrow \quad \frac{\tau}{\left(\frac{D}{2}\right)} = \frac{T_h}{\frac{\pi[D^4 - (0.6D)^4]}{32}}$$

$$\therefore \quad T_h = \tau \frac{\pi[1 - (0.6)^4]}{16} D^3 = \frac{\pi}{16} \tau (0.8704) D^3$$

As $T_s = T_h$

$$\tau \frac{\pi D_1^3}{16} = \frac{\pi}{16} \tau (0.8704) D^3$$

$\therefore D_1 = (0.9547)\, D$

Now, $\dfrac{\text{Weight of solid shaft}}{\text{weight of hollow shaft}} = \dfrac{W_s}{W_h} = \dfrac{\gamma_s\, A_s\, l_s}{\gamma_h\, A_h\, l_h}$

$$= \frac{A_s}{A_h} = \frac{\frac{\pi}{4} D_1^2}{\frac{\pi[D^2 - (0.6D)^2]}{4}} = \frac{1}{1 - (0.6)^2} \left(\frac{D_1}{D}\right)^2$$

$$= 1.5625 \left(\frac{D_1}{D}\right)^2 = 1.5625 \left[\frac{(0.9547)\, D}{D}\right]^2 = 1.424$$

($l_s = l_h$ as same length, $\gamma_s = \gamma_h$ as same material)

Percentage saving in material $= \dfrac{W_s - W_h}{W_s} \times 100$

$$= \frac{(1.424)\, W_h - W_h}{(1.424)W_h} \times 100 = 29.77\,\%$$

Hollow shaft is lighter than solid shaft by 29.77 %

Example 6.4 *A solid cylindrical shaft is to transmit 450 kW power at 100 rpm*
    (i) Find its diameter if the maximum shear stress is not to exceed 60 N/mm²
    (ii) What percentage of saving in weight will be obtained if the above solid shaft is replaced by a hollow shaft?

Solution:

$$\text{Power} = \frac{2\pi N T}{60}$$

$$450 \times 10^3 = \frac{2\pi \times 100 \times T}{60} \qquad \rightarrow \qquad T = 42972 \quad \text{Nm}$$

$\rightarrow$ Torque transmitted

Case-I: (Solid shaft)

$$\frac{\tau}{R} = \frac{T}{J} \quad \rightarrow \quad \frac{60 \times 10^6}{\frac{D}{2}} = \frac{42972}{\frac{\pi D_1^4}{32}} \quad \rightarrow \quad D_1$$

$= 0.154$ m   (Diameter of solid shaft)

Case-II (hollow shaft)

$$\frac{\tau}{R} = \frac{T}{J} \quad \rightarrow \quad \frac{60 \times 10^6}{\frac{D}{2}} = \frac{42972}{\frac{\pi}{32}[D^4 - (kD)^4]}$$

where $k = \dfrac{d}{D} =$ ratio of inner to outer diameter of hollow shaft

$$\therefore D^3 = \frac{16 \times 42972}{60 \times 10^6} \times \left(\frac{1}{1 - k^4}\right)$$

$$\therefore D = 0.154 \left(\frac{1}{1 - k^4}\right)^{\frac{1}{3}} \qquad \text{(Outer diameter of hollow shaft)}$$

Now $\dfrac{\text{weight of solid shaft}}{\text{weight of hollow shaft}} = \dfrac{W_s}{W_h} = \dfrac{A_s}{A_h}$

$$= \frac{\frac{\pi}{4}(0.154)^2}{\frac{\pi[D^2 - (kD)^2]}{4}} = \frac{0.0237}{(1 - k^2)D^2}$$

$\therefore$ Percentage saving in weight $= \dfrac{W_s - W_h}{W_s} \times 100$

$$= \left\{1 - \frac{(1 - k^2)D^2}{0.0237}\right\} \times 100$$

(D = Outer diameter of hollow shaft)

Assuming $k = 0.6$, $D = 0.1612$ m

$\therefore$ Percentage saving in weight $= 1 - \dfrac{(1 - 0.6^2) \times (0.1612)^2}{0.0237}$

$$= 29.8 \% \quad \text{(Ans)}$$

Example 6.5 *Power of 300 kW is to be transmitted by a hollow shaft at 80 rpm. If the allowable shear stress is 60 N/mm² and the internal diameter is 0.6 of the external diameter, find the external and internal diameters assuming the maximum torque is 1.4 times the mean torque.*

Solution:

(Power is calculated on the basis of mean torque and shear stress is calculated on the basis of maximum torque)

$$\text{Power} = \frac{2\pi N T_{mean}}{60}$$

$$300 \times 10^3 = \frac{2\pi \times 80 \times T_{mean}}{60} \qquad \therefore T_{mean} = 3.58 \times 10^4 \text{ Nm}$$

$$T_{max} = 1.4 \times T_{mean} = 5.013 \times 10^4 \text{ Nm}$$

$$\frac{\tau}{R} = \frac{T_{max}}{J}$$

$$\frac{60 \times 10^6}{\left(\frac{D}{2}\right)} = \frac{5.013 \times 10^4}{\frac{\pi}{32}[D^4 - (0.6D)^4]}$$

$$\therefore D = 0.1697 \text{ m} \quad \rightarrow \quad \text{External diameter}$$

$$d = 0.1018 \quad \rightarrow \quad \text{Internal diameter}$$

Example 6.6 *Show that for a given shear stress, the minimum diameter required for a solid circular shaft to transmit P kW at N rpm can be expressed as*

$$d = constant \times \left(\frac{P}{N}\right)^{\frac{1}{3}}$$

What value of maximum shear stress has been used if the constant equals to 84.71 (d being in millimeters)?

Solution:

$$\text{Power} = \frac{2\pi NT}{60000} \text{ kW}$$

$$\therefore \ T = \frac{30000 \, P}{\pi N} \ Nm$$

From torsion formula, $\dfrac{\tau}{R} = \dfrac{T}{J}$

$$\rightarrow \quad \frac{\tau}{\left(\frac{d}{2}\right)} = \frac{\left(\frac{30000 \, P}{\pi N}\right)}{\frac{\pi}{32} d^4} \quad \text{or} \quad d^3 = \frac{16 \times 30000}{\pi^2 \tau} \left(\frac{P}{N}\right)$$

$$\therefore \ d = \text{constant} \times \left(\frac{P}{N}\right)^{\frac{1}{3}} \qquad \rightarrow \qquad \text{Diameter of shaft}$$

where constant $= \left(\dfrac{16 \times 30000}{\pi^2 \tau}\right)^{\frac{1}{3}}$ for d in meters

Given constant = 84.71 when d is in millimeters

$$\therefore \ \left(\frac{16 \times 30000}{\pi^2 \tau}\right)^{\frac{1}{3}} \times 1000 = 84.71$$

$$\rightarrow \quad \tau = 8 \times 10^7 \ N/m^2 \qquad \rightarrow \qquad \text{Maximum shear stress}$$

Example 6.7   *A hollow shaft of external diameter 120 mm transmits 300 kW power at 200 rpm. Determine the maximum internal diameter if the maximum stress in the shaft is not to exceed 60N/mm².*

Solution:

$P = 300 \ kW, \quad \tau = 60 \times 10^6 \ N/m^2, \quad N = 200 \ rpm$

$D = 120 \ mm = 0.12 \ m$

Let d = Internal diameter of hollow shaft

$$\text{Power} = \frac{2\pi NT}{60} \ watt$$

$$300 \times 10^3 = \frac{2 \times \pi \times 200 \times T}{60} \qquad\qquad \therefore \ T = 1.432 \times 10^4 \ N \, m$$

$$\frac{\tau}{R} = \frac{T_{max}}{J}, \quad \frac{60 \times 10^6}{\left(\frac{0.12}{2}\right)} = \frac{1.432 \times 10^4}{\frac{\pi}{32}[(0.12)^4 - (d)^4]}$$

$\therefore d = 0.088$ m $= 88$ mm $\rightarrow$ Internal diameter of hollow shaft

Example 6.8 *An aluminium wire of radius 5 mm is twisted by two revolutions over a length of 2 m. If G = 80 GPa, find the shear stress induced.*

Solution:

$\theta = 2$ revolution $= 2 \times 2\pi = 4\pi$ rad

$$\frac{\tau}{R} = \frac{G\theta}{l} \quad \rightarrow \quad \frac{\tau}{\left(\frac{10}{2}\right)} = \frac{80 \times 10^3 \times 4\pi}{2000}$$

$\therefore \tau = 2.5 \times 10^3 \ \dfrac{N}{mm^2} \quad \rightarrow \quad$ Shear stress induced

Example 6.9 *An axial hole of 100 mm radius is bored out from a 300 mm diameter solid circular shaft. What percentage of torsional strength is lost by this operation?*

Solution:

Solid Shaft:

$$\frac{\tau}{R} = \frac{T_{max}}{J} \quad \rightarrow \quad \frac{\tau}{\left(\frac{300}{2}\right)} = \frac{T_s}{\left(\frac{\pi \times 300^4}{32}\right)}$$

$\therefore T_s = (5.3 \times 10^6)\,\tau$

Hollow shaft:

$$\frac{\tau}{R} = \frac{T_{max}}{J} \quad \rightarrow \quad \frac{\tau}{\left(\frac{300}{2}\right)} = \frac{T_h}{\frac{\pi(300^4 - 200^4)}{32}}$$

$\therefore T_h = (4.25 \times 10^6)\,\tau$

As τ is same in both the cases,

Percentage loss in torsional strength $= \dfrac{T_s - T_h}{T_s} \times 100$

$$= \frac{5.3 \times 10^6\, \tau - 4.25 \times 10^6\, \tau}{5.3 \times 10^6\, \tau} \times 100 = 19.84\ \%$$

Example 6.10 *Two shafts, one of solid circular cross section and other of hollow circular cross section are made of same material and are of same cross sectional area. If the ratio of inner to outer diameter of hollow shaft is k, find the ratio of torsional strengths of the two shafts.*

Solution:

The solid shaft and hollow shaft are made of same material and are of same cross sectional area.

Allowable shear stress (τ) and specific weight (γ) are same for both the shafts.

Let D_1 = Diameter of solid shaft,
D = External diameter of hollow shaft
k = Ratio of internal to external diametrs of hollow shaft

$A_s = A_h$

$$\frac{\pi D_1^2}{4} = \frac{\pi [D^2 - (kD)^2]}{4}$$

$$D_1^2 = D^2(1 - k^2) \qquad \rightarrow \qquad D_1 = D\sqrt{(1 - k^2)}$$

Consider solid shaft:

$$\frac{\tau}{R} = \frac{T_{max}}{J} \quad \rightarrow \quad \frac{\tau}{\dfrac{D_1}{2}} = \frac{T_s}{\dfrac{\pi}{32} D_1^4} \quad \rightarrow \quad T_s = \frac{\pi}{16}\, \tau\, D_1^3$$

Consider hollow shaft:

$$\frac{\tau}{R} = \frac{T_{max}}{J} \quad \rightarrow \quad \frac{\tau}{\frac{D}{2}} = \frac{T_h}{\frac{\pi}{32}[D^4 - (kD)^4]} \quad \rightarrow \quad T_h = \frac{\pi}{16}\tau D^3(1 - k^4)$$

∴ Ratio of torsional strengths of solid and hollow shaft,

$$\frac{T_s}{T_h} = \frac{\frac{\pi}{16}\tau D_1^3}{\frac{\pi}{16}\tau D^3(1 - k^4)} = \left(\frac{1}{1 - k^4}\right)\frac{D_1^3}{D^3} = \left(\frac{1}{1 - k^4}\right) \times \frac{D^3(1 - k^2)^{\frac{3}{2}}}{D^3}$$

$$= \frac{\sqrt{1 - k^2}}{(1 + k^2)}$$

Note:

As $k < 1$, $\quad T_s < T_h$

Therefore, hollow shaft is stronger than solid shaft for same material and weight.

Example 6.11 *A hollow alloy steel shaft is to replace a solid mild steel shaft of 200 mm diameter. The external diameter of the hollow shaft is to be 200 mm. The power transmitted is to be increased by 20 % and the speed of rotation to be increased by 6 %. If the allowable shear stress for alloy steel is 22 % more than that for mild steel, determine the maximum internal diameter of the hollow shaft.*

Solution:

Material for solid shaft is mild steel and that for hollow shaft is alloy steel.

Given diameter of solid shaft, d = 200 mm

$\tau_h = 1.22 \times \tau_s$ (Shear stress) , $P_h = 1.2 \times P_s$ (Power) ,

$N_h = 1.06 \times N_s$ (speed)

Outer diameter of hollow shaft, $d_0 = 200$ mm

Let d_i = Maximum internal diameter of hollow shaft

$$\text{Power} = \frac{2\pi NT}{60}$$

$$\left[\frac{2\pi NT}{60}\right]_h = 1.2 \times \left[\frac{2\pi NT}{60}\right]_s$$

$$N_h \, T_h = 1.2 \times N_s \, T_s$$

$$(1.06 \times N_s) \, T_h = 1.2 \times N_s \times T_s$$

$$\therefore T_h = (1.132) \, T_s$$
where T_s and T_h
= torque transmitted by solid shaft and hollow shaft, respectively

$$T_s = \frac{\pi}{16} \tau_s d^3$$

$$T_h = \frac{\pi}{16} \tau_h \left[\frac{d_0^4 - d_i^4}{d_0}\right]$$

$$\therefore \frac{\pi}{16} \tau_h \left[\frac{d_0^4 - d_i^4}{d_0}\right] = 1.132 \times \frac{\pi}{16} \tau_s d^3$$

$$1.22 \times \tau_s \times \left[\frac{200^4 - d_i^4}{200}\right] = 1.132 \times \tau_s \times 200^3$$

$$\therefore d_i = 103.6 \text{ mm}$$

Example 6.12 *A 100 mm diameter solid alloy shaft is to be coupled with a hollow steel shaft of same external diameter. Find the maximum internal diameter of the steel shaft. Determine the speed at which the shaft is to be driven to transmit a power of 300 kW. The maximum allowable shear stresses for alloy steel are 50 N/mm² and 70 N/mm² respectively.*

Solution:

Consider alloy shaft:

$$\frac{\tau}{R} = \frac{T}{J} \qquad \rightarrow \qquad \frac{50}{\left(\frac{100}{2}\right)} = \frac{T}{\frac{\pi \times 100^4}{32}}$$

$\therefore T = 9.81 \times 10^6 \text{ Nmm}$

Consider steel shaft:

$$\frac{\tau}{R} = \frac{T}{J} \quad \rightarrow \quad \frac{70}{\left(\frac{100}{2}\right)} = \frac{9.81 \times 10^6}{\frac{\pi}{32}[100^4 - d^4]}$$

$\therefore d = 73.09 \text{ mm} \quad \rightarrow$ Inner diameter of steel shaft.

$T = 9.81 \times 10^6 \text{ Nmm} = 9.81 \times 10^3 \text{ Nm}$

$$\text{Power} = \frac{2\pi NT}{60} \quad \rightarrow \quad 300 \times 10^3 = \frac{2\pi \times N \times 9.81 \times 10^3}{60}$$

$\therefore N = 292 \text{ rpm} \quad \rightarrow \quad$ Speed of shaft

Example 6.13 *A propeller shaft of solid circular cross section and diameter d is joined by means of a collar of same materials that is securely bonded to both parts of shaft. What should be the diameter d_1 of the collar in order that the joint can transmit power as that of solid shaft ?*

Solution:

Consider solid shaft (of diameter d)

$$\frac{\tau}{R} = \frac{T}{J} \quad \rightarrow \quad \frac{\tau}{\frac{d}{2}} = \frac{T_s}{\frac{\pi}{32} \times d^4}$$

$\therefore T_s = \frac{\pi}{16} \tau d^3$

Consider collar (hollow shaft of outer diameter d_1 and inner diameter d):

$$\frac{\tau}{R} = \frac{T}{J} \quad \rightarrow \quad \frac{\tau}{\frac{d_1}{2}} = \frac{T_h}{\frac{\pi}{32}(d_1^4 - d^4)}$$

$\therefore T_h = \frac{\pi}{16} \tau \frac{(d_1^4 - d^4)}{d_1}$

Torque is the same for both shaft and collar.

$$\frac{\pi}{16} \tau d^3 = \frac{\pi}{16} \tau \frac{(d_1^4 - d^4)}{d_1}$$

$$\therefore d^3 = \frac{(d_1^4 - d^4)}{d_1}$$

Putting $d = kd_1$ (k = Ratio of inner and outer diameter of collar)

$$\therefore k^3 d_1^3 = \frac{d_1^4 - (kd_1)^4}{d_1} \qquad \rightarrow \qquad k^3 = 1 - k^4$$

$$\rightarrow \qquad k^4 + k^3 = 1$$

By trial, $k = 0.82$

$$\therefore \frac{d}{d_1} = 0.82 \qquad \rightarrow \qquad d_1 = (1.219) d$$

External diameter of collar $= 1.219 \times$ diameter of shaft

Example 6.14 *Evaluate the maximum torsional shear stress and angle of twist at points B and C for the hollow shaft loaded as shown (Fig. 6.5). Given $G = 10^{11} Pa$.*

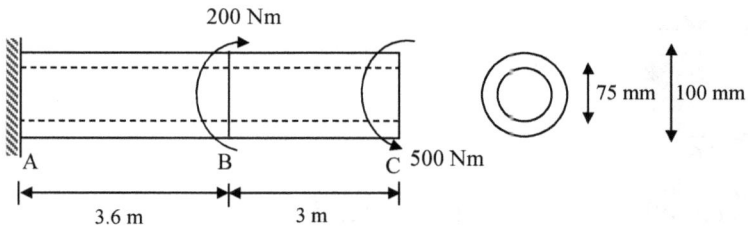

200 Nm

75 mm   100 mm

A   B   C   500 Nm

3.6 m   3 m

Fig. 6.5

Solution:

Active torque=500 (acw) – 200 (cw) =300 Nm (acw)

$\therefore$ Reactive torque at fixed support A is $T_A = 300$ Nm (cw)

$T_{AB} = 300$ Nm
(B rotates acw with respect to A)

$T_{BC} = 500$ Nm
(C rotates acw with respect to B)

The *torque diagram* is shown in Fig. 6.5a.

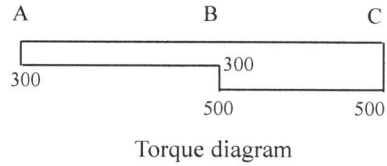

Fig. 6.5a

$$J = \frac{\pi}{32}(0.1^4 - 0.075^4) = 6.7 \times 10^{-6} \text{ m}^4$$

Maximum shear stress develops in BC portion

$$\frac{\tau}{R} = \frac{T}{J}$$

$$\therefore \tau = \frac{T_{BC}}{J} \times R = \frac{500}{6.7 \times 10^{-6}} \times \frac{0.1}{2} = 3.7 \times 10^6 \frac{N}{m^2}$$

Relative twist in AB portion

$$\theta_{AB} = \frac{T_{AB}\, l_{AB}}{G\,J} = \frac{300 \times 3.6}{10^{11} \times 6.7 \times 10^{-6}} = 1.61 \times 10^{-3} \text{ rad}$$

Similarly $\theta_{BC} = \dfrac{T_{BC}\, l_{BC}}{G\,J} = \dfrac{500 \times 3}{10^{11} \times 6.7 \times 10^{-6}} = 2.23 \times 10^{-3} \text{ rad}$

Angle of twist at A, $\quad \theta_A = 0 \quad$ (A is fixed)

Angle of twist at B, $\quad \theta_B = \theta_{AB} = 1.61 \times 10^{-3}$ rad

Angle of twist at C, $\quad \theta_C = \theta_B + \theta_{BC}$

$$= 1.61 \times 10^{-3} + 2.23 \times 10^{-3} = 3.84 \times 10^{-3} \text{ rad}$$

Note:

The quantity GJ is called *torsional rigidity* (in similar lines with flexural rigidity EI in beams).

Example 6.15 *A tapered shaft having radii r_1 and r_2 at the two ends and of length l is subjected to equal and opposite torque T at the ends (Fig. 6.6). Determine the angle of twist of one end relative to other.*

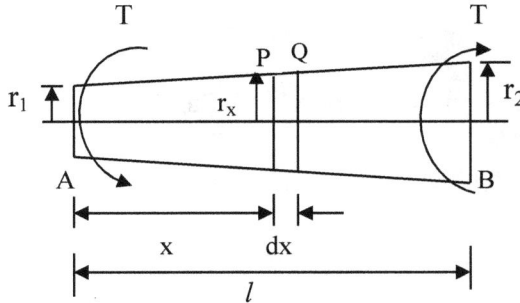

Fig. 6.6

Solution:

Refer Fig. 6.6

Radius of shaft at distance x from left end A is

$$r_x = r_1 + \frac{r_2 - r_1}{l} x = r_1 + c\,x$$
$$\text{where} \quad c = \frac{r_2 - r_1}{l}$$

Polar moment of inertia, $\qquad J_x = \dfrac{\pi\, d_x^4}{32} = \dfrac{\pi\, r_x^4}{2} = \dfrac{\pi (r_1 + c\,x)^4}{2}$

Let   dx = width of small segment PQ   (Fig. 6.6)

Relative twist between P and Q,

$$d\theta = \frac{T\,dx}{G\,J_x} = \frac{T\,dx}{G\,\dfrac{\pi(r_1 + c\,x)^4}{2}} = \frac{2T}{G\pi}\,\frac{dx}{(r_1 + c\,x)^4}$$

Relative twist between A and B ,

$$\theta = \int d\theta = \frac{2T}{G\pi} \int_0^l \frac{dx}{(r_1 + c\,x)^4} = \frac{2T}{G\pi}\left[\frac{(r_1 + c\,x)^{-3}}{-3} \times \frac{1}{c}\right]_0^l$$

$$= \frac{2T}{3G\pi c}\left[\frac{1}{(r_1 + c\,l)^3} - \frac{1}{(r_1)^3}\right] = -\frac{2T}{3G\pi\left(\frac{r_2 - r_1}{l}\right)}\left[\frac{1}{(r_2)^3} - \frac{1}{(r_1)^3}\right]$$

$$\therefore \theta = \frac{2Tl}{3G\pi(r_2 - r_1)}\left[\frac{(r_2)^3 - (r_1)^3}{r_1^3\, r_2^3}\right]$$

Example 6.16 *A composite shaft is made by rigidly and closely fitting a steel rod of 100 mm diameter inside a brass tube. If the torque applied to the composite shaft is to be equally shared by the two materials, find the outer diameter of tube. Given modulus of rigidity for steel and brass as 80 GPa and 40 GPa respectively. If the applied torque is 20000 Nm, calculate the maximum shear stress in each materials and the angle of twist in a length of 4 m.*

Solution:

Given $\quad T_s = T_b = \dfrac{20000}{2}$

$\qquad\qquad = 10000 \ \text{Nm} \qquad$ (Shafts share torque equally)

$\therefore G_s J_s = G_b J_b \qquad\qquad (T = GJ\,\theta/l)$

For steel: $\quad G_s = 80$ GPa, $\quad$ For brass: $\quad G_b = 40$ GPa

Let $\ d_o = $ Outer diameter of tube

$\therefore 80 \times 10^9 \times \dfrac{\pi}{32} \times (0.1)^4 = 40 \times 10^9 \times \dfrac{\pi}{32}[(d_o)^4 - (0.1)^4]$

$\therefore d_o = 0.1316 \text{ m} = 131.6 \text{ mm} \quad \rightarrow \quad$ Outer diameter of tube

$\dfrac{\tau}{R} = \dfrac{T}{J}$

For steel: $\quad \tau_s = \dfrac{10000 \times \left(\frac{0.1}{2}\right)}{\frac{\pi}{32} \times (0.1)^4} = 5.09 \times 10^7 \ \text{N/m}^2$

For brass: $\quad \tau_B = \dfrac{10000 \times \left(\frac{0.1316}{2}\right)}{\frac{\pi}{32} \times (0.1316^4 - 0.1^4)} = 3.35 \times 10^7 \ \text{N/m}^2$

Angle of Twist, $\theta = \dfrac{Tl}{GJ} = \dfrac{10000 \times 4}{80 \times \dfrac{\pi}{32} \times (0.1)^4} = 0.0509$ rad $= 2.918^0$

6.6 COMBINED BENDING AND TORSION

Consider a shaft of diameter D subjected to bending moment (M) and torsional moment (T) as shown (Fig. 6.7). The bending moment causes normal stress σ and the torsional moment causes shear stress τ.

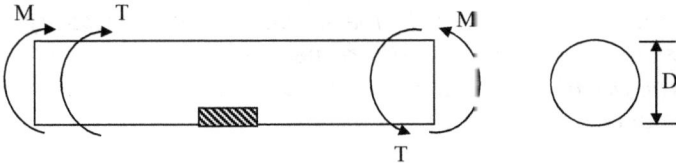

Fig. 6.7

Maximum bending stress, $\qquad \sigma = \dfrac{M}{I} y_{max} = \dfrac{M}{I} \times \dfrac{D}{2}$

For circular cross section $= J = I_x + I_y = I + I = 2\,I$

$\therefore \ \sigma = \dfrac{MD}{J}$

Maximum shear stress, $\qquad \tau = \dfrac{T}{J} R = \dfrac{T}{J} \times \dfrac{D}{2} = \dfrac{TD}{2J}$

Consider a small element at the bottom surface of the shaft (Fig. 6.7a).

$\sigma_x = \dfrac{MD}{J}, \qquad \sigma_y = 0, \qquad \tau = \dfrac{TD}{2J}$

$\therefore$ Principal stresses:

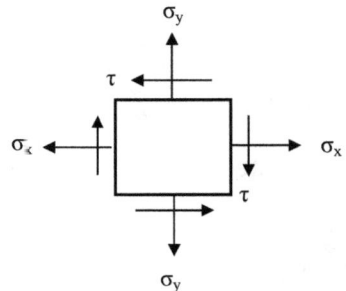

$\sigma_1, \sigma_2 = \dfrac{\sigma_y + \sigma_x}{2} \pm \sqrt{\left(\dfrac{\sigma_y - \sigma_x}{2}\right)^2 + \tau^2}$

Fig. 6.7a

$$= \frac{MD}{2J} \pm \sqrt{\left(\frac{MD}{2J}\right)^2 + \left(\frac{TD}{2J}\right)^2}$$

$$= \frac{D}{2J}\left(M \pm \sqrt{M^2 + T^2}\right)$$

$$\therefore \sigma_1 = \frac{D}{2J}\left[M + \sqrt{M^2 + T^2}\right] \quad \rightarrow \text{Major principal stress}$$

$$\sigma_2 = \frac{D}{2J}\left[M - \sqrt{M^2 + T^2}\right] \quad \rightarrow \text{Minor principal stress}$$

Maximum shear stress

$$\tau_{max} = \frac{\sigma_1 - \sigma_2}{2} = \frac{D}{2J}\sqrt{M^2 + T^2}$$

6.6.1 Equivalent Bending Moment (M_e)

Equivalent Bending Moment is the bending moment which causes the same maximum normal stress as that caused by the combined bending moment and torsional moment.

Maximum normal stress caused by equivalent bending moment (M_e) is

$$\sigma_{max} = \frac{M_e}{I} y_{max} = \frac{M_e}{I} \frac{D}{2} = \frac{M_e D}{J}$$

Maximum normal stress caused by combined bending moment and torsional moment is

$$\sigma_{max} = \frac{D}{2J}\left[M + \sqrt{M^2 + T^2}\right]$$

$$\therefore \frac{M_e D}{J} = \frac{D}{2J}\left[M + \sqrt{M^2 + T^2}\right]$$

$$\therefore \quad M_e = \frac{1}{2}\left[M + \sqrt{M^2 + T^2}\right] \quad \rightarrow \quad \text{Equivalent Bending Moment}$$

6.6.2 Equivalent Torsional Moment (T_e)

Equivalent Torsional Moment is the torsional moment which causes the same maximum shear stress as that caused by the combined bending moment and torsional moment.

Maximum shear stress caused by equivalent torsional moment (T_e) is

$$\tau_{max} = \frac{T_e}{J} R = \frac{T_e}{J} \frac{D}{2} = \frac{T_e D}{2J}$$

Maximum shear stress caused by combined bending moment and torsional moment is

$$\tau_{max} = \frac{D}{2J} \sqrt{M^2 + T^2}$$

$$\therefore \frac{T_e D}{2J} = \frac{D}{2J} \sqrt{M^2 + T^2}$$

$$\therefore T_e = \sqrt{M^2 + T^2} \quad \rightarrow \quad \text{Equivalent Torque}$$

Example 6.17 *A circular shaft 4 m long and 80 mm diameter carries a pulley of weight 2000 N at the centre of the shaft and is subjected to a twisting moment of 2000 Nm. Determine the values of principal stresses and maximum shear stress of the shaft which is freely supported at its ends.*

Solution:

Polar moment of inertia, $\quad J = \frac{\pi}{32} (0.08)^4 = 4.02 \times 10^{-6} \text{ m}^4$

Maximum bending moment $\quad M = \frac{Wl}{4} = \frac{2000 \times 4}{4} = 2000 \text{ Nm}$

Given T = 2000 Nm

Principal stresses:

$$\sigma_1, \sigma_2 = \frac{D}{2J} \left[M \pm \sqrt{M^2 + T^2} \right]$$

$$= \frac{0.08}{2 \times 4.02 \times 10^{-6}} \left[2000 \pm \sqrt{2000^2 + 2000^2} \right]$$

$\sigma_1 = 4.8 \times 10^7 \ \text{N/m}^2 \quad \rightarrow \quad$ Major principal stress

$\sigma_2 = -8.24 \times 10^6 \ \text{N/m}^2 \rightarrow$ Minor principal stress

Maximum shear stress

$$\tau_{max} = \frac{\sigma_1 - \sigma_2}{2} = \frac{D}{2J} \sqrt{M^2 + T^2} = \frac{0.08}{2 \times 4.02 \times 10^{-6}} \sqrt{2000^2 + 2000^2}$$

$$= 2.813 \times 10^7 \ \text{N/m}^2$$

Example 6.18 *A hollow shaft is subjected to a torque of 40×10⁶ Nmm and bending moment of 30×10⁶ Nmm. The internal diameter of shaft is half of external diameter. If the maximum shear stress is not to exceed 82 N/mm², find the diameter of the shaft.*

Solution:

Given, $\quad T = 40 \times 10^6 \ \text{Nmm}, \qquad M = 30 \times 10^6 \ \text{Nmm}$

$$J = \frac{\pi}{32} \left[D^4 - \left(\frac{D}{2} \right)^4 \right] = 0.092 \ D^4$$

$$\tau_{max} = \frac{D}{2J} \sqrt{M^2 + T^2}$$

$$82 = \frac{D}{2 \times J} \sqrt{(40 \times 10^6)^2 + (30 \times 10^6)^2}$$

$\therefore D = 149 \quad \text{mm} \rightarrow$ Outer diameter of shaft

$\therefore$ Inner diameter $= \dfrac{149}{2} = 74.5 \ \text{mm}$

Example 6.19 *The maximum normal stress and maximum shear stress analyzed for a shaft of 150 mm diameter under a combined bending and torsion were found to be 120 MN/m² and 80 MN/m² respectively. Find the bending moment and the torque to which the shaft*

is subjected to. If the maximum shear stress be limited to 100 MN/m², find by how much the torque can be increased by keeping bending moment constant.

Solution:

$$\sigma_1 = 120 \ \frac{N}{mm^2}, \qquad \tau_{max} = 80 \ \frac{N}{mm^2},$$

$$J = \frac{\pi}{32} \times 150^4 = 4.97 \times 10^7 mm^4$$

Case I

Maximum normal stress, $\sigma_1 = \dfrac{D}{2J}\left[M + \sqrt{M^2 + T^2}\right]$

$$120 = \frac{150}{2 \times 4.97 \times 10^7}\left[M + \sqrt{M^2 + T^2}\right]$$

$$\therefore \left[M + \sqrt{M^2 + T^2}\right] = 7.952 \times 10^7 \qquad\qquad ---(1)$$

Maximum shear stress $\tau_{max} = \dfrac{D}{2J}\sqrt{M^2 + T^2}$

$$80 = \frac{150}{2 \times 4.97 \times 10^7}\sqrt{M^2 + T^2}$$

$$\therefore \sqrt{M^2 + T^2} = 5.301 \times 10^7 \qquad\qquad\qquad ---(2)$$

Solving (1) and (2), $M = 2.65 \times 10^7$ Nmm, $T = 4.59 \times 10^7$ Nmm

Case II

M is kept constant, $\tau_{max} = 100$ N/mm², $M = 2.65 \times 10^7$ Nmm

$$\tau_{max} = \frac{D}{2J}\sqrt{M^2 + T^2}$$

$$100 = \frac{150}{2 \times 4.97 \times 10^7}\sqrt{(2.65 \times 10^7)^2 + T^2}$$

$T = 6.07 \times 10^7$ Nmm

∴ Increase in torque $= 6.07 \times 10^7 - 4.59 \times 10^7 = 1.48 \times 10^7$ Nmm $= 14.8$ kNm

6.7 STATICALLY INDETERMINATE SHAFTS

For statically indeterminate shafts, internal torque at a given cross section cannot be determined from the equilibrium equations alone. To solve such problems of statically indeterminate shafts, compatibility equations are also required in addition to equilibrium equations.

Let us consider a shaft as shown in Fig. 6.8a for which the free body diagram is explained in Fig. 6.8b.

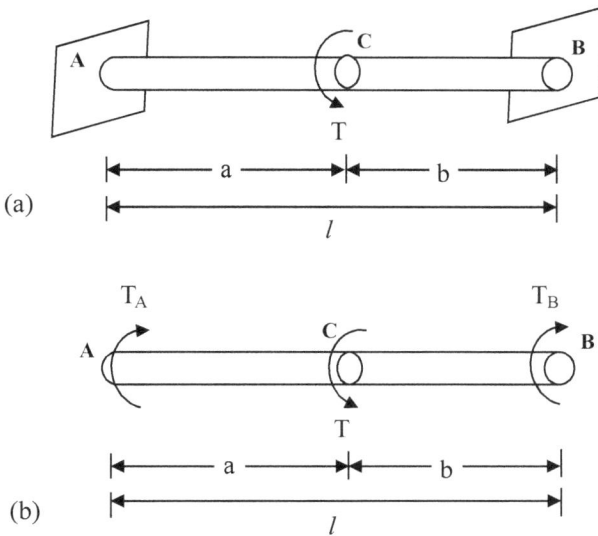

Fig. 6.8

$T =$ Applied torque at C

T_A, T_B
$=$ Reactive torques at supports A and B, respectively (see Fig. 6.8b)

Now,   (i) Equilibrium equation: $T_A + T_B = T$
        (ii) Compatibility equation: $\theta_{AC} = \theta_{BC}$

$(\theta_{AC}, \theta_{BC} = $ Angle of twist in portions AC and BC, respectively)

$$\therefore \frac{T_A\, a}{G\,J} = \frac{T_B\, b}{G\,J} \quad \rightarrow \quad T_A = T_B\left(\frac{b}{a}\right)$$

$$\therefore T_B\left(\frac{b}{a}\right) + T_B = T \qquad\qquad \therefore T_B\left(\frac{b+a}{a}\right) = T$$

$$\therefore \quad T_B = \frac{T\,a}{l} \qquad \text{and} \qquad T_A = \frac{T\,b}{l} \qquad (l = a + b)$$

Note:

Refer Fig. 6.9

(i) when $a = b = \dfrac{l}{2}$

$T_A = T_B = \dfrac{T}{2}$

(ii) when $a < b$

$T_A > T_B$

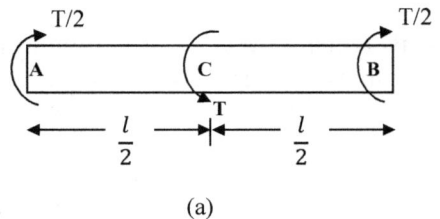

(a)

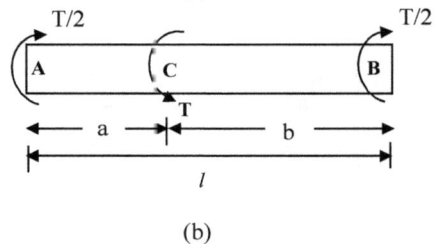

(b)

Fig. 6.9

Example 6.20 *A hollow shaft ACB of outside diameter 50 mm and inside diameter 38 mm is held against rotation at ends A and B as shown (Fig. 6.10). Horizontal forces P are applied at the ends of the vertical arms. Determine the allowable value of force P if maximum permissible shear stress in the shaft is 40 MPa.*

Solution:

$$J = \frac{\pi}{32}[(0.05)^4 - (0.038)^4] = 4.08 \times 10^{-7}\,\text{m}^4$$

Compatibility equation:

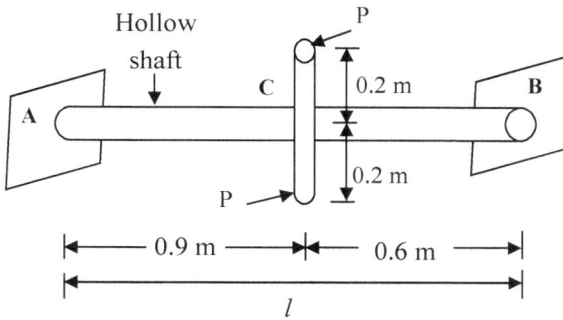

Fig. 6.10

$$\theta_{AC} = \theta_{CB} \qquad \rightarrow \qquad \left(\frac{Tl}{GJ}\right)_{AC} = \left(\frac{Tl}{GJ}\right)_{CB}$$

$$\frac{T_A \times 0.9}{GJ} = \frac{T_B \times 0.6}{GJ} \qquad \rightarrow \qquad T_A = \frac{2}{3}T_B$$

Equilibrium equation:

$$T_A + T_B = T \qquad \rightarrow \qquad \frac{2}{3}T_B + T_B = 0.4P$$

$$\therefore T_B = 0.24\,P \qquad \text{and} \quad T_A = 0.16\,P$$

Maximum shear stress develops in CB parts

Consider CB part:

$$\frac{T_{CB}}{J} = \frac{\tau}{R}$$

$$\rightarrow \qquad \frac{0.24P}{4.08 \times 10^{-7}} = \frac{40 \times 10^6}{\left(\frac{0.05}{2}\right)}$$

$$\therefore P = 2720\ N$$

Example 6.21 *A shaft fixed at both ends is subjected to torque as shown (Fig.6.11). Find the resisting torque at supports and the maximum shear stress induced.*

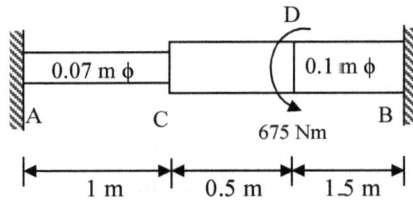

Fig. 6.11

Solution:

Compatibility equation:

$$\theta_{AD} = \theta_{DB} \qquad \text{(Relative twist in AD and BD portions are same)}$$

$$\theta_{AC} + \theta_{CD} = \theta_{DB}$$

$$\left(\frac{Tl}{GJ}\right)_{AC} + \left(\frac{Tl}{GJ}\right)_{CD} = \left(\frac{Tl}{GJ}\right)_{DB}$$

$$\frac{T_A \times 1}{G \times \dfrac{\pi \times (0.07)^4}{32}} + \frac{T_A \times 0.5}{G \times \dfrac{\pi \times (0.1)^4}{32}} = \frac{T_B \times 1.5}{G \times \dfrac{\pi \times (0.1)^4}{32}}$$

$$\therefore T_A = (0.3215)T_B$$

Equilibrium Equation:

$$T_A + T_B = T$$

$$(0.3215)T_B + T_B = 675$$

$$T_B = 510.76 \text{ Nm} \quad \text{and} \quad T_A = 164.24 \text{ Nm}$$

Maximum shear stress occurs in either AC part or BD part

AC part:

$$\frac{\tau}{R} = \frac{T_{AC}}{J} \quad \rightarrow \quad \frac{\tau}{\left(\dfrac{0.07}{2}\right)} = \frac{164.24}{\dfrac{\pi}{32} \times (0.07)^4}$$

$$\therefore \tau_{AC} = 2.43 \times 10^6 \text{ N/m}^2$$

DB Part:

$$\frac{\tau}{R} = \frac{T_{DB}}{J} \quad \text{or,} \quad \frac{\tau}{\left(\frac{0.1}{2}\right)} = \frac{510.76}{\frac{\pi}{32} \times (0.1)^4}$$

$$\therefore \tau_{DB} = 2.6 \times 10^6 \text{ N/m}^2$$

In CD part, maximum shear stress will not be developed because torque is less and diameter is more.

$$\therefore \tau_{max} = 2.6 \times 10^6 \text{ N/m}^2 \text{ occurs in DB part}$$

Example 6.22 *Two shafts made up of same materials are to be connected by a flange coupling as shown in Fig. 6.12. There is an initial mismatch in the hole location of 6°. Determine the maximum shear stress in each shaft. G=84 GPa.*

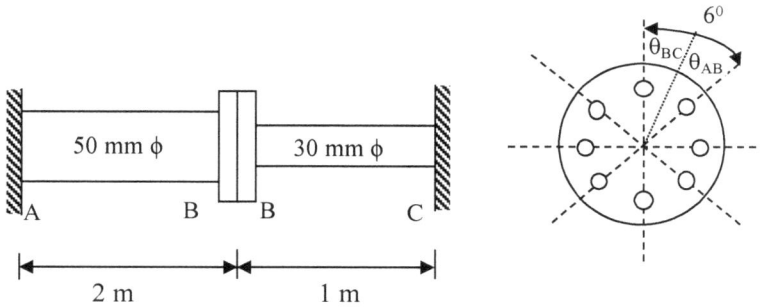

Fig. 6.12

Solution:

Refer Fig. 6.12

Polar moment of inertia:

$$J_{AB} = \frac{\pi}{32}(0.05)^4 = 6.136 \times 10^{-7} \text{ m}^4$$

$$J_{BC} = \frac{\pi}{32}(0.03)^4 = 7.95 \times 10^{-8} \text{ m}^4$$

Internal torques are equal and opposite to each other.

$T_{AB} = T_{BC} = T$

Given $\theta_{AB} + \theta_{BC} = 6 \times \dfrac{\pi}{180}$

$\left(\dfrac{Tl}{GJ}\right)_{AB} + \left(\dfrac{Tl}{GJ}\right)_{BC} = 6 \times \dfrac{\pi}{180}$

$\dfrac{T \times 2}{80 \times 10^9 \times 6.136 \times 10^{-7}} + \dfrac{T \times 1}{80 \times 10^9 \times 7.95 \times 10^{-8}} = 6 \times \dfrac{\pi}{180}$

$\therefore T = 555.5 \ \text{Nm}$

Shear stress in AB portion:

$$\tau_{AB} = \dfrac{T}{J_{AB}} \times R_{AB} = \dfrac{555.5}{6.136 \times 10^{-7}} \times \dfrac{0.05}{2} = 22.6 \times 10^6 \ \text{N/m}^2$$
$$= 22.6 \ \text{MPa}$$

Shear stress in BC portion:

$$\tau_{BC} = \dfrac{T}{J_{BC}} \times R_{BC} = \dfrac{555.5}{7.95 \times 10^{-8}} \times \dfrac{0.03}{2} = 104.8 \times 10^6 \ \text{N/m}^2$$
$$= 104.8 \ \text{MPa}$$

Example 6.23 *A hollow circular tube fits over the end of a solid circular bar as shown in Fig. 6.13. The far ends of both are fixed. A hole through the solid bar makes an angle 2° with a line through two holes in hollow tube. The solid bar is twisted until the holes are aligned and then a pin is inserted through the holes. When the system returns to static equilibrium, determine the stresses induced in the two bars. Given G=80 GPa.*

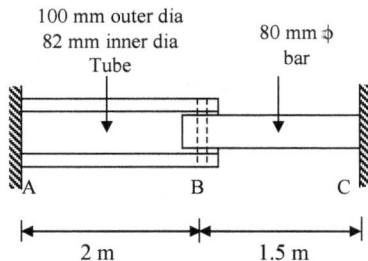

100 mm outer dia
82 mm inner dia
Tube

80 mm ⌀
bar

A   B   C

2 m   1.5 m

Fig. 6.13

Solution:

$$J_{AB} = \frac{\pi}{32}[0.1^4 - 0.082^4] = 5.37 \times 10^{-6} \text{ m}^4$$

$$J_{BC} = \frac{\pi}{32}[0.08^4] = 4 \times 10^{-6} \text{m}^4$$

Internal torques developed in portions AB and BC are equal to each other.

Compatibility equation:

Relative twist in AB part + Relative twist in BC part $= \left(2 \times \dfrac{\pi}{180}\right)$

$$\left(\frac{Tl}{GJ}\right)_{AB} + \left(\frac{Tl}{GJ}\right)_{BC} = 2 \times \frac{\pi}{180}$$

$$\frac{T \times 2}{80 \times 10^9 \times 5.37 \times 10^{-6}} + \frac{T \times 1.5}{80 \times 10^9 \times 4 \times 10^{-6}} = 2 \times \frac{\pi}{180}$$

$$\therefore T = 3745 \text{ Nm}$$

Consider AB part:

$$\frac{\tau}{R} = \frac{T}{J} \quad \rightarrow \quad \frac{\tau_{AB}}{\left(\frac{0.1}{2}\right)} = \frac{3745}{5.37 \times 10^{-6}} \qquad \therefore \tau_{AB}$$
$$= 34.8 \times 10^6 \text{ N/mm}^2$$

Consider BC part:

$$\frac{\tau}{R} = \frac{T}{J} \quad \rightarrow \quad \frac{\tau_{BC}}{\left(\frac{0.08}{2}\right)} = \frac{3745}{4 \times 10^{-6}} \qquad \therefore \tau_{BC}$$
$$= 37.5 \times 10^6 \text{ N/mm}^2$$

Angle of twist:

$$\theta_{AB} = \left(\frac{T\,l}{GJ}\right)_{AB} = \frac{3745 \times 2}{80 \times 10^9 \times 5.37 \times 10^{-6}} = 0.017 \text{ rad} = 0.99°$$

$$\theta_{BC} = \left(\frac{T \, l}{G \, J}\right)_{BC} = \frac{3745 \times 1.5}{80 \times 10^9 \times 4 \times 10^{-6}} = 0.0175 \text{ rad} = 1.01°$$

$$\theta_{AB} + \theta_{BC} = 2°$$

HIGHLIGHTS

Definitions

1. *Pure Torsion*: In this case shaft is subjected to torque only without being associated with any bending moment or axial force.

2. *Polar moment of Inertia*: It is the moment of inertia of cross section about the axis which is perpendicular to the cross section.

3. *Polar modulus of section*: It is defined as the ratio of polar moment of inertia to extreme radial distance of the fibre from the centre.

4. *Torsional rigidity*: It is defined as the product of modulus of rigidity (G) and the polar moment of inertia (J). It may be looked as the torsion required to produce unit angle of twist in unit length.

Concepts and Formulae

1. Torsion formula: $\dfrac{\tau}{R} = \dfrac{T}{J} = \dfrac{G\theta}{l}$

2. Polar moment of inertia:

$$J = \frac{\pi}{32} d^4 \qquad \text{for solid shaft (of diameter d)}$$

$$J = \frac{\pi}{32} [D^4 - d^4]$$
for hollow shaft (of outer diameter D and inner diameter d)

3. Torque transmitted by shaft:

$$T = \frac{\pi}{16} \tau \, d^3 \quad \text{for solid shaft}$$

$$T = \frac{\pi}{16} \tau \left(\frac{D^4 - d^4}{D}\right) \quad \text{for hollow shaft}$$

4. Power transmitted by shaft

$$P = \frac{2\pi NT}{60} \quad \text{watt} \quad \text{where } N \to \text{rpm}, \quad T \to \text{Nm}$$

5. For the same material, length and torque to be transmitted, weight of hollow shaft is less than that of a solid shaft or in other words solid shaft is heavier than hollow shaft for same material, length and torque.

6. Combined bending and torsion:

Equivalent bending moment, $\quad M_e = \dfrac{1}{2}\left[M + \sqrt{M^2 + T^2}\right]$

Equivalent torque, $\quad T_e = \sqrt{M^2 + T^2}$

7. Torsion of tapered shaft: $\quad \theta = \dfrac{2Tl}{3G\pi(r_2 - r_1)}\left[\dfrac{r_2^3 - r_1^3}{r_2^3 r_1^3}\right]$

SHORT TYPE QUESTIONS

1. What is the polar moment of inertia of the cross section of a circular shaft of diameter d ?

$$\left[\text{Ans: } \frac{\pi}{32}d^4\right]$$

2. A hollow shaft of same cross sectional area as that of a solid shaft can resist
(a) less torque   (b) more torque   (c) equal torque

[Ans: (b)]

3. Torsional rigidity of a bar is the product of its _____ and modulus of rigidity.

[Ans: polar moment of inertia]

4. The magnitude of shear stress induced in a shaft due to applied torque varies
(a) from zero at center to maximum at circumference.
(b) from maximum at center to zero at circumference.
(c) from maximum at centre to minimum at circumference but not zero.
(d) from minimum at centre and not zero to maximum at circumference.

[Ans. (a)]

5. Pure torsion of a circular shaft produces
(a) longitudinal normal stress in the shaft.
(b) only direct shear stress in the transverse section of shaft.
(c) circumferential shear stress on a surface element of shaft.
(d) a longitudinal shear stress and a circumferential shear stress on a surface element of the shaft.

[Ans: (b)]

6. A hollow prismatic bar of circular section is subjected to a torsional moment. The maximum shear stress occurs at
(a) the inner wall of cross section   (b) the middle of the thickness
(c) the outer surface of shaft

[Ans: (c)]

7. A circular shaft is subjected to a torque T. The maximum shear stress developed is τ. What will be the value of maximum tensile stress developed?

[Ans: τ]

8. There are two shafts of same material, one is solid having diameter d and other is hollow having outer diameter d and inner diameter d/2. What will be the ratio of strength of solid shaft to that of hollow shaft.

[Ans: 16/15]

9. The equivalent bending moment under combined action of bending moment M and torque T is
(a) $\left[\sqrt{M^2 + T^2}\right]$
(b) $\frac{1}{2}\left[\sqrt{M^2 + T^2}\right]$

(c) $\left[M + \sqrt{M^2 + T^2}\right]$

(d) $\frac{1}{2}\left[M + \sqrt{M^2 + T^2}\right]$

[Ans: (d)]

10. A round shaft of diameter d and length l fixed at both ends A and B is subjected to a twisting moment T at C at a distance $l/4$ from A. The torsional stresses in the parts AC and CB will be
(a) equal
(b) in the ratio of 1:3
(c) in the ratio of 3:1
(d) indeterminate

[Ans: (c)]

EXERCISE PROBLEMS

1. The working condition to be satisfied by a shaft transmitting power are (i) the shaft must not twist more than 1^0 in a length of 15 times diameter (ii) the shear stress must not exceed 80 MN/m². What is the actual working stress and diameter of the shaft to transmit 736 kW at 200 rpm? Take shear modulus as 80 GPa.
[Ans: Dia. 156.66 mm, shear stress = 46.5 N/mm²]

2. A solid shaft transmits 250 kW at 100 rpm. If the shear stress is not to exceed 75 N/mm², what should be the diameter of the shaft? If this shaft is to be replaced by a hollow one whose internal diameter is 0.6 times its outer diameter, determine the size and the percentage saving in weight the maximum shear stress being the same.
[Ans: d_0 = 123.03 mm, d_i = 73.82 mm, % saving = 29.795 %]

3. A solid steel shaft is to transmit 75 kW at 200 rpm. Taking allowable shear stress as 70 MPa, find the suitable diameter for the shaft if the maximum torque transmitted is 30% more than the mean torque. Also find the outside diameter of the hollow shaft whose inside diameter is 0.7 of the outside which can replace the solid shaft.
[Ans: d = 0.0697 m, d_0 = 0.0764 m]

4. Find the strength of a hollow circular shaft of external diameter 50 mm, internal diameter 40 mm and length 1 m. Take allowable shear stress 100 N/mm² and angle of twist 2^0. Given G = 1 × 10⁵ N/mm².

[Ans: 1.09 kNm]

5. A hollow shaft of diameter ratio 3/5 transmits 600 kW at 110 rpm. Maximum torque is 12% greater than the mean torque. Shear stress should not exceed 60 MN/m² and the twist in 3 m length should not exceed 1⁰. Calculate the maximum size of the shaft. G = 80 GN/m².

[Ans: d_0 = 180 mm, d_i = 108 mm]

6. Two shaft of same material and same length are subjected to same torque. One shaft is of a solid circular cross section and the other is of hollow circular cross section having inside diameter equal to 2/3 of the outside diameter and in each shaft the maximum shear is same. Compare the weight of the two shafts.

[Ans: $\dfrac{W_H}{W_S}$ = 0.643]

7. A hollow shaft is of 120 mm external diameter and diameter ratio 0.6. If the maximum shear stress in the shaft is limited to 100 MPa and allowable twist is 1⁰ per metre length, find the maximum power that can be transmitted to the shaft, if it is to rotate at 100 rpm. Take G = 8 × 10⁴ MPa.

[Ans: 259 kW]

Chapter 7

Thin Cylinders and Spheres

Learning Objectives

After going through this chapter, the reader will be able to
- analyze thin walled cylindrical and spherical pressure vessels.
- calculate the stresses, strains and change in capacity of thin shells due to internal fluid pressure.
- calculate the stresses and strains in cylindrical shells with hemispherical ends.
- analyze stresses in conical water tanks.
- design wire wound cylinders.

7.1 INTRODUCTION

Pressure vessels (or shells) are used to carry fluids. The pressure vessels are essential structural elements in offshore structures, submarines and airspace crafts. The shape of the pressure vessels may be cylindrical, spherical, conical and so on. Pressure vessels are made of cast iron, sheet steel and nonferrous alloys. The cylindrical shells are generally made by rolling the sheet metal and joining the ends by riveting or welding. It can also be made joint less (solid drawn). Examples of pressure vessels include boiler shell, air compressors, water tank, pipeline carrying fluid under pressure etc. The wall thickness of the shell is small as compared to its diameter. A shell is known to be a thin shell when the ratio of its internal diameter to wall thickness is more than 20 otherwise it is considered as a thick shell. In this Chapter, we will deal with the analysis of thin pressure vessels.

7.2 STRESSES IN THIN CYLINDRICAL SHELLS

The following stresses are developed in the pressure vessels (or shells).

(i) Hoop stress or circumferential stress (σ_c) : This is the tensile stress acting along the circumference of the cylinder.

(ii) Longitudinal stress (σ_l): This is the tensile stress acting along the length of the cylinder. It develops only when the cylinder has closed ends.

(iii) Radial stress (σ_r): This is the compressive stress acting along the radius of the cylinder. Radial stress is small in thin shell and is generally neglected.

For an element in the wall of the cylindrical shell, the various stresses are explained in Fig. 7.1.

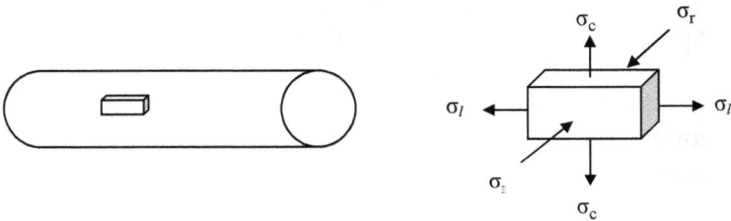

Fig. 7.1

7.2.1 Assumptions

The following assumptions are made in the analysis of stresses in thin shells:

(i) Material of the shell is homogeneous and isotropic.
(ii) Circumferential stress is uniform across the wall thickness.
(iii) Bending of walls is neglected.
(iv) Radial stress is small and is neglected.

7.3 CIRCUMFERENTIAL STRESS IN THIN CYLINDRICAL SHELL

Consider a thin cylindrical shell subjected to internal fluid pressure.

Refer Fig. 7.2

Let P = internal fluid pressure,
   d = internal diameter of cylinder
   t = wall thickness,
   l = length of cylinder

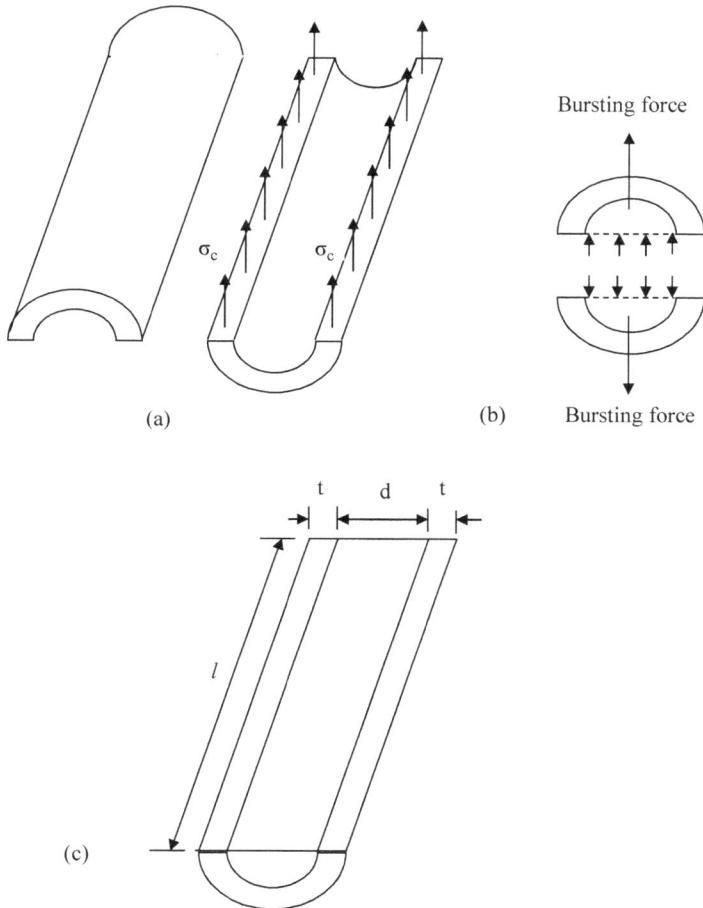

Fig. 7.2

Bursting force = internal fluid pressure × area = P × l × d

Resisting force = circumferential stress × area

= σ_c × (2 × l × t)

Considering the equilibrium of one half of the cylinder (Fig. 7.2).

Bursting force = Resisting force

P × l × d = σ_c × (2 × l × t)

$$\boxed{\sigma_c = \frac{Pd}{2t}}$$ → Circumferential stress or
Hoop Stress

7.4 LONGITUDINAL STRESS IN THIN CYLINDRICAL SHELL

Refer Fig. 7.3

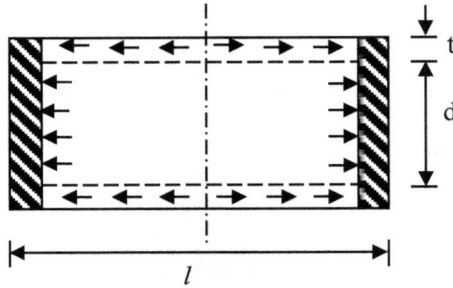

Fig. 7.3

The cylinder under consideration has its two ends covered with two end plates connected to them as shown.

Bursting force = internal fluid pressure × area = $P \times \frac{\pi}{4} d^2$

Resisting force = longitudinal stress × area = $\sigma_l \times (\pi dt)$

For equilibrium,

Bursting force = Resisting force

$$P \times \frac{\pi}{4} d^2 = \sigma_l \times (\pi dt)$$

$$\boxed{\sigma_l = \frac{Pd}{4t}}$$ → Longitudinal stress

Note:

1. Both the circumferential stress and longitudinal stress are independent of the length of cylinder.

2. In case of thin cylindrical shell subjected to internal fluid pressure, the tensile circumferential stress developed is twice that of the longitudinal stress. Therefore, in no case, the hoop stress (or circumferential stress) be greater than the permissible stress for the material of the cylinder.

7.5 CIRCUMFERENTIAL STRAIN IN THIN CYLINDRICAL SHELL

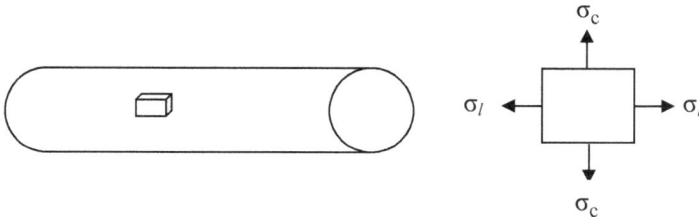

Fig. 7.4

Consider a small element in the wall of cylindrical shell subjected to internal fluid pressure. Figure 7.4 explains the circumferential and longitudinal stresses acting on the element when the cylinder is subjected to internal fluid pressure.

Let
σ_c = circumferential stress
σ_l = longitudinal stress
v = Poisson's ratio
E = Young's modulus
t = thickness of cylinder

Now, Circumferential strain (or Hoop strain)

$$\epsilon_c = \frac{\sigma_c}{E} - v\frac{\sigma_l}{E} = \frac{Pd}{2tE} - v\frac{Pd}{4tE}$$

$$\boxed{\epsilon_c = \frac{Pd}{2tE}(1 - 0.5\,v)} \longrightarrow \quad \text{Hoop Strain}$$

Circumferential strain (or Hoop strain) is the ratio of increase in circumference to the original circumference. In other words

$\epsilon_c = \dfrac{\delta d}{d}$ (circumference depends upon diameter)

$\therefore$ Increase in diameter, $\delta d = \epsilon_c \times d = \dfrac{Pd}{2tE}(1 - 0.5\,v) \times d$

$$\boxed{\delta d = \dfrac{Pd^2}{2tE}(1 - 0.5\,v)}$$ →   Increase in diameter of cylinder

7.6 LONGITUDINAL STRAIN IN THIN CYLINDRICAL SHELL

Refer Fig. 7.4

Longitudinal Strain, $\epsilon_l = \dfrac{\sigma_l}{E} - v\dfrac{\sigma_c}{E} = \dfrac{Pd}{4tE} - v\dfrac{Pd}{2tE}$

$$\boxed{\epsilon_l = \dfrac{Pd}{2tE}(0.5 - v)}$$ →   Longitudinal Strain

Longitudinal strain is the ratio of increase in length of cylinder to its original length.

$\epsilon_l = \dfrac{\delta l}{l}$

Increase in length, $\delta l = \epsilon_l \times l = \dfrac{Pd}{2tE}(0.5 - v) \times l$

$$\boxed{\delta l = \dfrac{Pdl}{2tE}(0.5 - v)}$$ →   Increase in length of cylinder

7.7 VOLUMETRIC STRAIN IN THIN CYLINDRICAL SHELL

To find the increase in internal volume of cylinder, it is required to find out the volumetric strain (ϵ_v).

Let

d, l = inner diameter and length of the cylinder, respectively before application of fluid pressure
d', l' = final diameter and length of cylinder, respectively
δd, δl = increase in diameter and length, respectively

Final inner diameter = initial diameter + increase in diameter

$$d' = d + \delta d = d\left(1 + \frac{\delta d}{d}\right) = d(1 + \epsilon_c)$$

Final length = Initial length + increase in length

$$l' = l + \delta l = l\left(1 + \frac{\delta l}{l}\right) = l(1 + \epsilon_l)$$

Now volumetric strain, $\quad \epsilon_v = \dfrac{\delta V}{V} = \dfrac{\text{final volume} - \text{initial volume}}{\text{initial volume}}$

Initial volume, $\quad V = \dfrac{\pi}{4}d^2 l$

Final volume, $\quad V' = \dfrac{\pi}{4}(d')^2 l'$

$$\therefore \ \epsilon_v = \frac{V' - V}{V} = \frac{\frac{\pi}{4}(d')^2 l' - \frac{\pi}{4}d^2 l}{\frac{\pi}{4}d^2 l} = \frac{(d')^2 l' - d^2 l}{d^2 l}$$

$$= \frac{d^2(1 + \epsilon_c)^2\, l(1 + \epsilon_l) - d^2 l}{d^2 l}$$

$$= (1 + \epsilon_c)^2(1 + \epsilon_l) - 1 = (1 + \epsilon_c^2 + 2\epsilon_c)(1 + \epsilon_l) - 1$$

Neglecting higher powers and product of ϵ_c and ϵ_l

$$\boxed{\epsilon_v = 2\epsilon_c + \epsilon_l} \quad\longrightarrow\quad \text{Volumetric Strain}$$

Now, $\quad \epsilon_v = 2\left[\dfrac{Pd}{2tE}(1 - 0.5\,v)\right] + \dfrac{Pd}{2tE}(0.5 - v) = \dfrac{Pd}{2tE}(2.5 - 2\,v)$

$\therefore$ Increase in volume $\quad \delta V = \dfrac{Pd}{2tE}(2.5 - 2\,v) \times \dfrac{\pi}{4}d^2 l$

$$\boxed{\delta V = \dfrac{P\pi d^3 l}{8tE}(2.5 - 2\,v)} \quad\longrightarrow\quad \text{Increase in volume of cylinder}$$

This is the amount of fluid that can additionally be pumped into the cylindrical shell due to increase of its volume.

Again due to compressibility effect of fluid, some amount of fluid can be additionally pumped into the cylinder (due to decrease in volume of fluid).

Decrease in volume of fluid $\; = \dfrac{PV}{K} \qquad (K \rightarrow \text{Bulk modulus})$

Therefore, total volume of fluid that can additionally be pumped into the cylindrical shell is the sum of increase in volume of cylinder and the decrease in volume of fluid.

Note:

Principal stresses in cylindrical shell:

Major principal stress, $\qquad \sigma_1 = \sigma_c = \dfrac{Pd}{2t}$

Minor principal stress, $\qquad \sigma_2 = \sigma_l = \dfrac{Pd}{4t}$

Maximum shear stress in cylindrical shell, $\qquad \tau_{max} = \dfrac{\sigma_1 - \sigma_2}{2} = \dfrac{Pd}{8t}$

7.8 RIVETED JOINTS IN CYLINDRICAL SHELL

As discussed in the beginning of this chapter, cylindrical shells are made by rolling the sheet and joining its ends. Figure 7.5 explains the longitudinal and circumferential joints in a cylindrical shell. It may be observed that circumferential joint is made to get required length of cylinder while longitudinal joint is made to get required diameter of the shell. Circumferential stress is influenced by efficiency of longitudinal joint (η_l) while longitudinal stress is influenced by efficiency of circumferential joint (η_c).

Considering the joint efficiencies

$$\sigma_c = \frac{Pd}{2t\,\eta_l} \quad \text{and} \quad \sigma_l = \frac{Pd}{4t\,\eta_c}$$

For no joints, $\quad \sigma_c = \dfrac{Pd}{2t} \quad \text{and} \quad \sigma_l = \dfrac{Pd}{4t}$

Circumferential joint

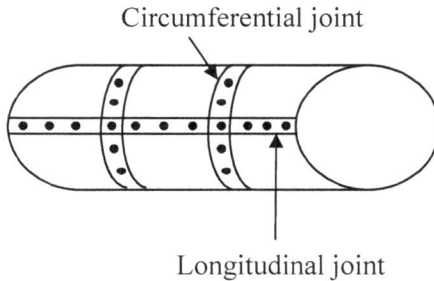

Longitudinal joint

Fig. 7.5

Example 7.1 *A thin walled cylinder of I m inside diameter and 3 m long is subjected to an internal pressure of 2 MN/m². The thickness of the cylinder is 30 mm. calculate the hoop and longitudinal stresses and the change in internal volume. Take E = 210 GN/m² and Poisson's ratio = 0.3.*

Solution:

d = 1 m, l = 3 m, t = 0.03 m, P = 2×10⁶ N/m², E = 210×10⁹ N/m² , ν = 0.3

Hoop stress, $\quad \sigma_c = \dfrac{Pd}{2t} = \dfrac{2 \times 10^6 \times 1}{2 \times 0.03} = 33.33 \times 10^6 \ \text{N/m}^2$

Longitudinal stress,
$$\sigma_l = \frac{Pd}{4t} = \frac{2 \times 10^6 \times 1}{4 \times 0.03} = 16.67 \times 10^6 \ \text{N/m}^2$$

Increase in volume, $\delta V = \dfrac{P\pi d^3 l}{8tE}(2.5 - 2\,v)$

$$= \frac{2 \times 10^6 \times \pi \times 1^3 \times 3}{8 \times 0.03 \times 210 \times 10^9}(2.5 - 2 \times 0.3) = 7.1 \times 10^{-4}\,m^3$$

Example 7.2 *A thin cylindrical shell 1.5 m internal diameter, 2.4 m long, plates 25 mm thick is under internal fluid pressure of 1 N/mm². Assuming the end plates are rigid, find the changes in length, diameter and volume. E = 206000 N/mm²,v = 0.267.*

Solution:

Change in diameter

$$\delta d = \frac{Pd^2}{2tE}(1 - 0.5\,v) = \frac{1 \times 1500^2}{2 \times 25 \times 206000}(1 - 0.5 \times 0.267)$$
$$= 0.189\ mm$$

Change in length

$$\delta l = \frac{Pdl}{2tE}(0.5 - v) = \frac{1 \times 1500 \times 2400}{2 \times 25 \times 206000}(0.5 - 0.267) = 0.0814\ mm$$

Change in volume, $\delta V = \dfrac{P\pi d^3 l}{8tE}(2.5 - 2\,v)$

$$= \frac{1 \times \pi \times 1500^3 \times 2400}{8 \times 25 \times 206000}(2.5 - 2 \times 0.267) = 1.214 \times 10^6\ mm^3$$

$$= 0.001214\ m^3$$

Example 7.3 *A closed thin cylindrical pressure vessel having a radius of 1200 mm and wall thickness of 15 mm is subjected to an internal pressure of 1 MPa. E= 200 GPa, Poisson's ratio = 0.25. Calculate the*
(i) hoop and longitudinal stresses
(ii) maximum shear stress
(iii) change in diameter

Solution:

Given   radius =1200 mm $\therefore$ d = 2400 mm
t = 15 mm,   P = 1 MPa = 1 N/mm², E = 200 GPa = 200×10³ N/mm²

(i) Hoop stress, $\sigma_c = \dfrac{Pd}{2t} = \dfrac{1 \times 2400}{2 \times 15} = 80 \text{ N/mm}^2$

Longitudinal stress, $\sigma_l = \dfrac{Pd}{4t} = \dfrac{1 \times 2400}{4 \times 15} = 40 \text{ N/mm}^2$

(ii) Principal stresses are $\sigma_1 = 80 \text{ N/mm}^2$, $\sigma_2 = 40 \text{ N/mm}^2$

$\therefore$ Maximum shear stress ,
$$\tau_{max} = \frac{\sigma_1 - \sigma_2}{2} = \frac{80 - 40}{2} = 20 \text{ N/mm}^2$$

(iii) Change in diameter, $\delta d = \dfrac{Pd^2}{2tE}(1 - 0.5\, v)$

$$= \frac{1 \times 2400^2}{2 \times 15 \times 200 \times 10^3}(1 - 0.5 \times 0.25) = 0.84 \text{ mm}$$

Example 7.4 *A water main of 800 mm diameter contains water at a pressure head of 100 m. Find the thickness of the metal required for the water main. Given that permissible stress is 20 N/mm².*

Solution:

Pressure head = P/w   where P is pressure and w = specific weight of water

w = 9810 N/m³

Pressure, P = 9810×100 = 981×10³ N/m²

Circumferential stress, $\sigma_c = \dfrac{Pd}{2t} = \dfrac{981 \times 10^3 \times 0.8}{2 \times t}$

Given allowable stress = 20×10⁶ N/m²

$\therefore$ $\dfrac{981 \times 10^3 \times 0.8}{2 \times t} = 20 \times 10^6$

$t = 0.019$ m $= 19$ mm $\rightarrow$ Thickness of metal required

Example 7.5 *A cylindrical tank is of 2 m diameter, 3 m long and 15 mm thick. Find the increase in capacity when the pressure inside is 1.5 N/mm² with*
(i) no axial compressive load on the tank.
(ii) an axial compressive load of 470 kN acting on it.
Find also the maximum shear stress to which the material of the cylinder is subjected to. Take E = 2×10⁵ N/mm², Poisson's ratio = 0.3.

Solution:

$d = 2$ m, $l = 3$ m , $t = 15$ mm

(i) No axial load

Change in volume, $\delta V = \dfrac{P\pi d^3 l}{8tE}(2.5 - 2v)$

$$= \frac{1.5 \times 10^6 \times \pi \times 2^3 \times 3}{8 \times 0.015 \times 2 \times 10^{11}}(2.5 - 2 \times 0.3) = 8.95 \times 10^{-3} \text{ m}^3$$

Maximum shear stress ,
$$\tau_{max} = \frac{Pd}{8t} = \frac{1.5 \times 10^6 \times 2}{8 \times 0.015} = 25 \times 10^6 \text{ N/m}^2$$

(ii) Axial compressive load of 470 kN acting

Hoop stress, $\sigma_c = \dfrac{Pd}{2t} = \dfrac{1.5 \times 10^6 \times 2}{2 \times 0.015} = 100 \times 10^6 \text{ N/m}^2$

Longitudinal stress, $\sigma_l = \dfrac{Pd}{4t} - \dfrac{F}{\pi dt}$

$$= \frac{1.5 \times 10^6 \times 2}{4 \times 0.015} - \frac{470000}{\pi \times 2 \times 0.015} = 45.01 \times 10^6 \text{ N/m}^2$$

Hoop strain,
$$\epsilon_c = \frac{\sigma_c}{E} - v\frac{\sigma_l}{E} = \frac{100 \times 10^6}{2 \times 10^{11}} - 0.3 \times \frac{45.01 \times 10^6}{2 \times 10^{11}} = 43.2 \times 10^{-5}$$

Longitudinal strain,

$$\epsilon_c = \frac{\sigma_l}{E} - \nu\frac{\sigma_c}{E} = \frac{45.01 \times 10^6}{2 \times 10^{11}} - 0.3 \times \frac{100 \times 10^6}{2 \times 10^{11}} = 7.5 \times 10^{-5}$$

Volumetric strain,

$$\epsilon_v = 2\epsilon_c + \epsilon_l = 2 \times 43.2 \times 10^{-5} + 7.5 \times 10^{-5} = 93.9 \times 10^{-5}$$

∴ Change in volume, $\delta V = \epsilon_v \times V = 93.9 \times 10^{-5} \times \left(\frac{\pi}{4} \times 2^2 \times 3\right)$

$$= 8.85 \times 10^{-3} \text{ m}^3$$

Major principal stress, $\sigma_1 = 100 \times 10^6 \text{ N/m}^2$

Minor principal stress, $\sigma_2 = 45.01 \times 10^6 \text{ N/m}^2$

Maximum shear stress ,

$$\tau_{max} = \frac{\sigma_1 - \sigma_2}{2} = \frac{100 \times 10^6 - 45.01 \times 10^6}{2} = 27.4 \times 10^6 \text{ N/m}^2$$
$$= 27.4 \text{ N/mm}^2$$

Example 7.6 *A cylindrical water tank 6 m diameter with its axis vertical is made from steel plate that is 3 mm thick. Find the maximum height to which the tank may be filled so that the circumferential stress is limited to 60 MPa.*

Solution:

d = 6 m , t = 3 mm = 0.003 m

Let h = height to which tank is filled.

Maximum water pressure,

P = w × h = 9810 h (w = specific weight of water)

Maximum circumferential stress = 60 MPa

$$\sigma_c = \frac{Pd}{2t}$$

$$60 \times 10^6 = \frac{9810\,h \times 6}{2 \times 0.003}$$

$\therefore$ h = 6.11 m

Example 7.7 *A boiler shell is to withstand an internal pressure of 1.5 N/mm² gauge. The efficiencies of longitudinal and circumferential joints are 75% and 40% respectively. If the tensile stress of the plate is not to exceed 100 N/mm², find the safe diameter of the shell. Thickness of the plate = 15 mm.*

Solution:

t = 15 mm, η_l = 0.75, η_c = 0.4

Allowable stress = 100 N/mm²

$$\sigma_c = \frac{Pd}{2t\,\eta_l}$$

$$100 = \frac{1.5 \times d}{2 \times 15 \times\ 0.75} \qquad \rightarrow \qquad d = 1500\ \text{mm}$$

$$\sigma_l = \frac{Pd}{4t\,\eta_c}$$

$$100 = \frac{1.5 \times d}{4 \times 15 \times\ 0.4} \qquad \rightarrow \qquad d = 1600\ \text{mm}$$

For safety, the stress should not be more than 100 N/mm².

Therefore, safe diameter is the lesser of the two values.
d = 1500 mm

Example 7.8 *A cylindrical pressure vessel has an internal diameter of 1.6 m. It is subjected to internal fluid pressure of 0.03 N/m². Plate thickness is 15 mm. Material of the vessel has ultimate tensile stress of 6 N/m². The efficiencies of circumferential and longitudinal joints are 50% and 80% respectively. Find the factor of safety.*

Solution:

$$\text{Allowable stress} = \frac{\text{Ultimate tensile stress}}{\text{Factor of safety}}$$

$$\sigma_{\text{all}} = \frac{6}{F}$$

Considering hoop stress to be allowable stress

$$\sigma_c = \frac{Pd}{2t\,\eta_l}$$

$$\frac{6}{F} = \frac{0.03 \times 1.6}{2 \times 0.015 \times 0.8} \qquad \rightarrow \qquad F = 3$$

Considering longitudinal stress to be allowable stress

$$\sigma_l = \frac{Pd}{4t\,\eta_c}$$

$$\frac{6}{F} = \frac{0.03 \times 1.6}{4 \times 0.015 \times 0.5} \qquad \rightarrow \qquad F = 3.75$$

True factor of safety is the lesser of the two values.

∴ Factor of safety = 3

Example 7.9 *A thin cylindrical compressed air drum is 1.9 m diameter with plates 12.7 mm thick. The efficiencies of longitudinal and circumferential joints are 85% and 45% respectively. If the tensile stress in the plate is limited to 100 N/mm², find the maximum safe air pressure.*

Solution:

$$\text{Hoop Stress,} \quad \sigma_c = \frac{Pd}{2t\,\eta_l} = \frac{P \times 1900}{2 \times 12.7 \times 0.85} = 88\,P\ \text{N/mm}^2$$

$$\text{Longitudinal Stress,} \quad \sigma_l = \frac{Pd}{4t\,\eta_c} = \frac{P \times 1900}{4 \times 12.7 \times 0.45}$$

$$= 83.11\,P\ \text{N/mm}^2$$

Considering hoop stress to be allowable stress

$88 P = 100 \quad \rightarrow \quad P = 1.136 \ \text{N/mm}^2$

Considering longitudinal stress to be allowable stress

$83.11 P = 100 \quad \rightarrow \quad P = 1.203 \ \text{N/mm}^2$

$\therefore$ Safe air pressure P = 1.136 N/mm²

7.9 STRESSES IN THIN SPHERICAL SHELLS

Consider a thin spherical shell of inner diameter d and thickness t is subjected to internal fluid pressure P (Fig. 7.6).

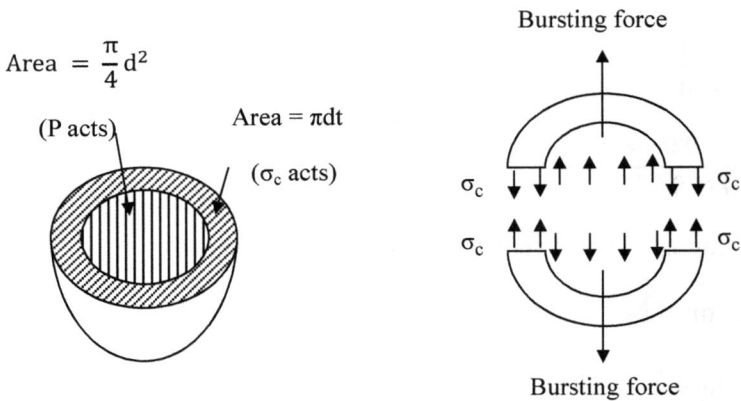

Fig. 7.6

Now bursting force $= P \times \dfrac{\pi}{4} d^2$

Resisting force $= \sigma_c \times \pi dt$

For equilibrium,   Resisting force = Bursting force

$\sigma_c \times \pi dt = P \times \dfrac{\pi}{4} d^2$

$$\boxed{\sigma_c = \frac{Pd}{4t}}$$ Circumferential stress or Hoop Stress
                                          for spherical shell

7.10 CIRCUMFERENTIAL STRAIN IN THIN SPHERICAL SHELL

Consider a small element in the wall of spherical shell under internal fluid pressure. The stresses acting on this element are represented in Fig. 7.7.

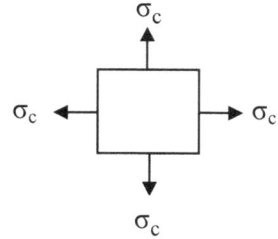

Circumferential strain (or Hoop strain)

$$\epsilon_c = \frac{\sigma_c}{E} - v\frac{\sigma_c}{E} = \frac{\sigma_c}{E}(1-v)$$

Fig. 7.7

$$\boxed{\epsilon_c = \frac{Pd}{4tE}(1-v)} \longrightarrow \text{Circumferential strain or Hoop Strain}$$

Hoop strain, $\epsilon_c = \dfrac{\delta d}{d}$

∴ Increase in diameter, $\delta d = \epsilon_c \times d$

$$\boxed{\delta d = \frac{Pd^2}{4tE}(1-v)} \longrightarrow \text{Increase in diameter of spherical shell}$$

7.11 VOLUMETRIC STRAIN IN THIN SPHERICAL SHELL

To find the increase in internal volume of spherical shell, it is required to find out the volumetric strain (ϵ_v).

Let

d = initial inner diameter , d'= final inner diameter
δd = change in inner diameter

$$d' = d + \delta d = d\left(1 + \frac{\delta d}{d}\right) = d(1 + \epsilon_c)$$

Initial volume, $\quad V = \dfrac{\pi d^3}{6}$

Final volume, $V' = \dfrac{\pi(d')^3}{6} = \dfrac{\pi d^3 (1 + \epsilon_c)^3}{6}$

$\therefore$ Volumetric strain, $\epsilon_v = \dfrac{V' - V}{V}$

$$= \dfrac{\dfrac{\pi d^3 (1 + \epsilon_c)^3}{6} - \dfrac{\pi d^3}{6}}{\dfrac{\pi d^3}{6}} = (1 + \epsilon_c)^3 - 1$$

$$= (1 + \epsilon_c{}^3 + 3\epsilon_c{}^2 + 3\epsilon_c) - 1$$

Neglecting higher powers of ϵ_c

$$\boxed{\epsilon_v = 3\epsilon_c} \quad\longrightarrow\quad \text{Volumetric Strain}$$

$$\epsilon_v = \dfrac{3Pd}{4tE}(1 - v)$$

Increase in volume $\delta V = \dfrac{3Pd}{4tE}(1 - v) \times \dfrac{\pi d^3}{6}$

$$\boxed{\delta V = \dfrac{P\pi d^4}{8tE}(1 - v)} \quad\longrightarrow\quad \begin{array}{l}\text{Increase in Volume} \\ \text{of spherical shell}\end{array}$$

Now volume of fluid that can additionally be pumped into the spherical shell

= increase in volume of sphere + the decrease in volume of fluid

$$= \dfrac{P\pi d^4}{8tE}(1 - v) + \dfrac{PV}{K}$$

$(V = \dfrac{\pi d^3}{6} \rightarrow$ Volume of sphere , $K \rightarrow$ Bulk Modulus$)$

Example 7.10 *A spherical shell of 0.4 m diameter and 10 mm wall thickness is completely filled with fluid at atmospheric pressure. Additional fluid is pumped in till the pressure increases by 0.05 N/m². Find the volume of additional fluid. Take v = 0.25, E = 1000 N/m².*

Solution:

Bulk modulus K is not available and hence fluid can be considered incompressible.

$$\delta V = \frac{P\pi d^4}{8tE}(1-v) = \frac{0.05 \times \pi \times 0.4^4}{8 \times 0.01 \times 1000}(1-0.25) = 3.76 \times 10^{-5} \text{ m}^3$$

Example 7.11 *A thin spherical copper shell of diameter 0.3 m and thickness 1.6 mm is just full of water at atmospheric pressure. Find how much the internal pressure be increased by pumping in 25000 mm³water. Take v = 0.286, E = 100000 N/mm², K = 2200 N/mm²*

Solution:

Volume of spherical shell,
$$V = \frac{\pi d^3}{6} = \frac{\pi 300^3}{6} = 1.413 \times 10^7 \text{ mm}^3$$

Volume of water that can additionally be pumped in

= increase in volume of sphere + decrease in volume of water

$$= \frac{P\pi d^4}{8tE}(1-v) + \frac{PV}{K}$$

or $25000 = \dfrac{P \times \pi \times 300^4}{8 \times 1.6 \times 100000}(1-0.286) + \dfrac{P \times 1.413 \times 10^7}{2200}$

$P = 1.2179 \text{ N/mm}^2$

7.12 CYLINDRICAL SHELL WITH HEMISPHERICAL ENDS

Fig. 7.8 shows a thin cylindrical shell with hemispherical ends.

Let

t_1 = thickness of cylindrical shell
t_2 = thickness of spherical shell
d = inner diameter of cylindrical shell and spherical shell
E = Modulus of elasticity
v = Poisson's ratio
P = internal fluid pressure

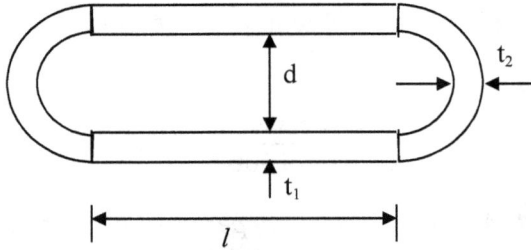

Fig. 7.8

Increase in volume, $\delta V = (\delta V)_{cylinder} + (\delta V)_{sphere}$

$$\delta V = \frac{P \pi d^3 l}{8 t_1 E}(2.5 - 2v) + \frac{P \pi d^4}{8 t_2 E}(1 - v)$$

To avoid bending of walls, increase in diameter of cylinder and sphere should be same i.e. hoop strain shall be equal for both the cylinder and sphere.

$$(\epsilon_c)_{cylinder} = (\epsilon_c)_{sphere}$$

$$\frac{Pd}{2t_1 E}(1 - 0.5v) = \frac{Pd}{4t_2 E}(1 - v) \quad \rightarrow \quad \frac{1}{t_1}(1 - 0.5v) = \frac{1}{2t_2}(1 - v)$$

$$\boxed{\frac{t_2}{t_1} = \frac{1 - v}{2 - v}} \longrightarrow \text{Condition to avoid bending of walls}$$

Note:

The thickness of cylinder should be more than that of sphere to avoid bending of walls.

Example 7.12 *A boiler drum consists of a cylindrical portion 2.4 m long, 1.2 m diameter and 0.025 m thick closed by hemispherical ends. In a hydraulic test to 8 MPa, how much additional water will be pumped in after initial filling at atmospheric pressure? Assume that circumferential strain at the junction of cylinder and sphere is same for both. Take E = 210 GPa, v = 0.3, K = 2.24 GPa.*

Solution:

Cylinder : d = 1.2 m, l = 2.4 m, t_1 = 0.025 m
Sphere : d = 1.2 m, t_2 = unknown

As circumferential strain (ε_c) is same for both

$$\frac{t_2}{t_1} = \frac{1 - v}{2 - v}$$

$$\rightarrow \quad t_2 = \frac{1 - 0.3}{2 - 0.3} \times 0.025 = 0.0102 \text{ m}$$

Now increase in volume of vessel,

$$\delta V = \frac{P\pi d^3 l}{8 t_1 E}(2.5 - 2v) + \frac{P\pi d^4}{8 t_2 E}(1 - v)$$

$$= \frac{8 \times 10^6 \times \pi \times 1.2^3 \times 2.4}{8 \times 0.025 \times 210 \times 10^9}(2.5 - 2 \times 0.3)$$
$$+ \frac{8 \times 10^6 \times \pi \times 1.2^4}{8 \times 0.0102 \times 210 \times 10^9}(1 - 0.3)$$

$$= 6.843 \times 10^{-3} \text{ m}^3$$

Volume of vessel,

$$V = \frac{\pi}{4}d^2 l + \frac{\pi d^3}{6} = \left(\frac{\pi}{4} \times 1.2^2 \times 2.4\right) + \left(\frac{\pi \times 1.2^3}{6}\right)$$
$$= 3.62 \text{ m}^3$$

Decrease in volume of water

$$= \frac{PV}{K} = \frac{8 \times 10^6 \times 3.62}{2.24 \times 10^9} = 0.0116 \text{ m}^3$$

∴ Additional fluid that can be pumped in $= 6.843 \times 10^{-3} + 0.0116 = 0.0184$ m³

7.13 MEMBRANE STRESSES

Consider a closed thin walled vessel whose form is a surface of revolution (Fig. 7.9a).

It is subjected to internal pressure P.

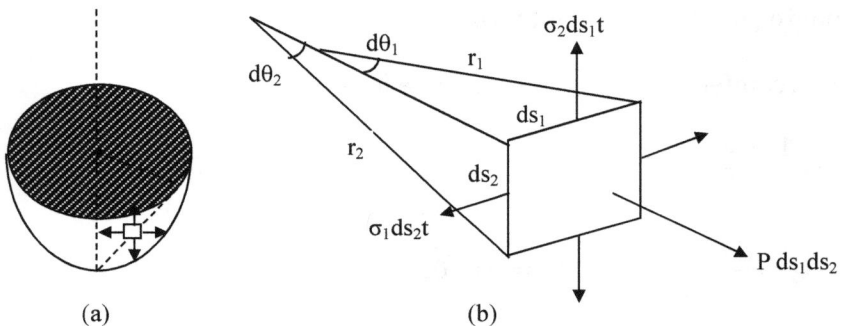

(a)   (b)

Fig. 7.9

The wall thickness t of this vessel is very small compared to its radii of curvature and therefore, the wall does not offer any resistance to bending. Hence the wall acts like a thin membrane. The stresses in the membrane are taken to be uniform and are acting tangentially to the middle surface of wall. These stresses are called membrane stresses.

A small membrane element is shown (Fig. 7.9b).

Let
σ_1 = Stress intensity in circumferential direction (hoop stress)
σ_2 = Stress intensity along the meridian (meridional stress)
r_1 = Radius of curvature of circumferential arc
r_2 = Radius of curvature of meridional arc
$d\theta_1$ = Angle subtended by circumferential elemental arc
$d\theta_2$ = Angle subtended by meridional elemental arc
ds_1 = Length of circumferential elemental arc
ds_2 = Length of meridional elemental arc

Resolving the forces (Fig. 7.9b)

$$\sigma_1 \, ds_2 \, t \, d\theta_1 + \sigma_2 \, ds_1 \, t \, d\theta_2 = P \, ds_1 ds_2$$

$$\sigma_1 \frac{ds_1 ds_2 \, t}{r_1} + \sigma_2 \frac{ds_1 ds_2 \, t}{r_2} = P \, ds_1 ds_2 \quad \left(d\theta_1 = \frac{ds_1}{r_1} \, , d\theta_2 = \frac{ds_2}{r_2} \right)$$

$$\boxed{\frac{\sigma_1}{r_1} + \frac{\sigma_2}{r_2} = \frac{P}{t}} \longrightarrow \text{Relation between hoop stress and meridional stress}$$

For spherical vessel, $r_1 = r_2 = r$, $\sigma_1 = \sigma_2 = \sigma$

$$\therefore \quad \sigma = \frac{Pr}{2t}$$

For cylindrical vessel , $r_2 = \infty$

$$\frac{\sigma_1}{r_1} + \frac{\sigma_2}{\infty} = \frac{P}{t}$$

$$\therefore \sigma_1 = \frac{Pr_1}{t} \qquad \rightarrow \quad \text{Hoop stress}$$

7.14 STRESSES IN A CONICAL WATER TANK

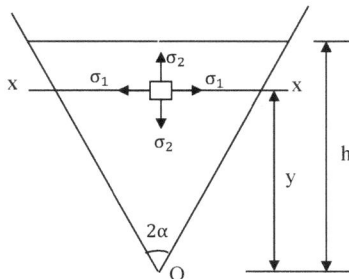

Fig. 7.10

Figure 7.10 shows a conical tank filled with water upto a depth h.

Consider any level x-x at a distance y from O as shown (Fig. 7.10).

At this level, radius of meridional arc $r_2 = \infty$

$$\frac{\sigma_1}{r_1} + \frac{\sigma_2}{\infty} = \frac{P}{t} \quad \rightarrow \quad \sigma_1 = \frac{Pr_1}{t} \quad (r_1 = \text{radius of circumferential arc})$$

Here $P = w(h - y) \rightarrow$ water pressure at level x-x   (w=specific weight)

$$r_1 = \frac{y \tan \alpha}{\cos \alpha} \qquad \text{(refer Fig. 7.10a)}$$

$$\boxed{\sigma_1 = \frac{w(h - y)y \tan \alpha}{t \cos \alpha}} \quad \longrightarrow \quad \text{Hoop Stress at level x-x}$$

Maximum Hoop stress:

For maximum Hoop stress, $\dfrac{d\sigma_1}{dy} = 0$

$$\frac{d}{dy}\left[\frac{w \tan \alpha \ (hy - y^2)}{t \cos \alpha}\right] = 0 \qquad \Rightarrow \qquad y = \frac{h}{2}$$

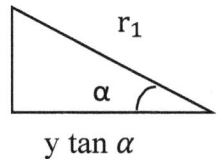

Fig. 7.10a

$$(\sigma_1)_{max} = \frac{w \tan \alpha}{t \cos \alpha}\left(h \times \frac{h}{2} - \frac{h^2}{4}\right)$$

$$\boxed{(\sigma_1)_{max} = \frac{w\,h^2 \tan \alpha}{4\,t \cos \alpha}} \quad \longrightarrow \quad \text{Maximum Hoop Stress}$$

$$\sigma_2\, 2\,\pi\, y \tan \alpha \ t \cos \alpha = w[\pi(y \tan \alpha)^2](h - y) + w\left[\pi(y \tan \alpha)^2 \frac{y}{3}\right]$$

(weight of water is resisted by the vertical component of induced meridional tension on the circumference x-x)

$$\boxed{\sigma_2 = \frac{w \tan \alpha}{2\,t \cos \alpha}\left(hy - \frac{2}{3}y^2\right)} \quad \longrightarrow \quad \begin{array}{l}\text{Meridional Stress}\\ \text{at level x-x}\end{array}$$

Maximum Meridional stress:

For maximum meridional stress $\quad \dfrac{d\,\sigma_2}{dy} = 0$

$$\frac{d}{dy}\left[\frac{w\tan\alpha}{2\,t\cos\alpha}\left(hy - \frac{2}{3}y^2\right)\right] = 0 \quad \Rightarrow \quad y = \frac{3h}{4}$$

$$(\sigma_2)_{max} = \frac{w\,\tan\alpha}{2t\,\cos\alpha}\left\{h\times\frac{3h}{4} - \frac{2}{3}\left(\frac{3h}{4}\right)^2\right\}$$

$$\boxed{(\sigma_2)_{max} = \frac{3wh^2\tan\alpha}{16\,t\,\cos\alpha}} \longrightarrow \begin{array}{l}\text{Maximum Meridional} \\ \text{Stress}\end{array}$$

7.15 WIRE WINDING OF THIN CYLINDERS

Wire winding of thin cylinders is made to increase their pressure carrying capacity so that the chances of bursting of the cylinder are reduced. The cylinder is wound with layers of wire kept under tension. A wire tightly wound around the cylinder being itself under tension gives rise to compressive stresses in the cylinder which neutralizes the tensile stresses in the cylindrical shell to a great extent. Fluid pressure inside the shell increases the tensile stresses in the wire around the cylinder.

The resultant circumferential stress in the cylinder is the sum of the initial compressive stress due to wire winding and further tensile stress due to internal pressure. The resultant circumferential stress in the wire is the sum of two tensile stresses developed due to (i) wire winding under tension and (ii) internal pressure in the cylinder. Therefore, the pressure carrying capacity of the cylindrical shell is increased.

Figure 7.11 shows a cylindrical shell closely wound with wire under tensile stress.

Let
d = cylinder diameter
t = cylinder thickness

d_w = wire diameter
l = length of cylinder
n = number of turns of wire in length l of cylinder (n = l/d_w)
σ_w = initial tensile stress in wire
σ_c = compressive circumferential stress developed in cylinder

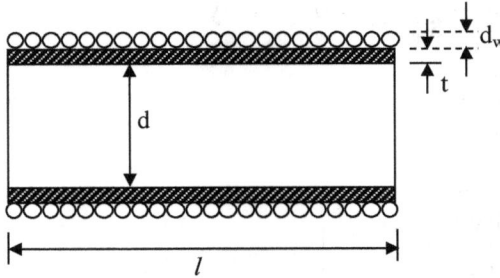

Fig. 7.11

Without internal fluid pressure: (Refer Fig. 7.12a)

Tensile force exerted by wire $= \sigma_w \times 2 \times \dfrac{\pi}{4} \times d_w^2 \times n$

Compressive force developed in cylinder $= \sigma_c \times 2tl$

For equilibrium –

$$\sigma_c \times 2tl = \sigma_w \times 2 \times \frac{\pi}{4} \times d_w^2 \times n$$

$$\boxed{\sigma_c = \frac{\pi\, d_w}{4\,t}\, \sigma_w}$$

With internal fluid pressure: (Refer Fig. 7.12b)

Let P = internal fluid pressure
σ_c' = Circumferential stress developed in cylinder
σ_w' = Stress developed in wire
σ_l' = Longitudinal stress developed in cylinder

Bursting force = P d l

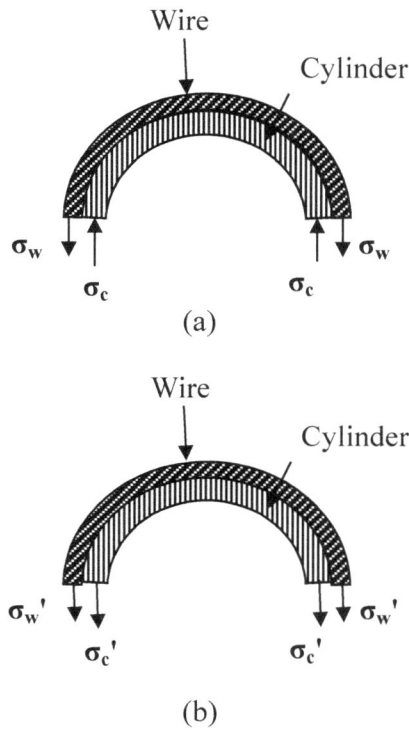

(a)

(b)

Fig. 7.12

Resisting force $= \left(\sigma_w' \times 2 \times \dfrac{\pi}{4} \times d_w^2 \times n\right) + (\sigma_c' \times 2tl)$

For equilibrium $\left(\sigma_w' \times 2 \times \dfrac{\pi}{4} \times d_w^2 \times n\right) + (\sigma_c' \times 2tl) = P\,d\,l$

$\left(\sigma_w' \times \dfrac{\pi}{2} \times d_w\right) + (\sigma_c' \times 2t) = P\,d \qquad --- (1)$

Compatibility equation:

Circumferential strain in cylinder = Circumferential strain in wire

$\dfrac{\sigma_c'}{E} - v\dfrac{\sigma_l'}{E} = \dfrac{\sigma_w'}{E_w} \qquad (E, E_w$

$= $ Young's modulus for cylinder and wire respectively)

$$\rightarrow \quad \frac{\sigma'_c}{E} - v\frac{Pd}{4tE} = \frac{\sigma'_w}{E_w} \qquad\qquad ---(2)$$

Solving equations (1) and (2) the stresses σ'_c and σ'_w can be computed.

Resultant stress in wire = $\sigma_w + \sigma'_w$

Resultant stress in cylinder = $\sigma_c - \sigma'_c$

Example 7.13 *A cast iron cylinder of 200 mm inner diameter and 12.5 mm thick is closely wound with a layer of 4 mm diameter steel wire under a tensile stress of 55 MN/m². Determine the stresses set up in the cylinder and steel wire if water under a pressure of 3 MN/m² is admitted in the cylinder. Take E_c = 100 GN/m², E_s = 200 GN/m², v = 0.25.*

Solution:

d = 200 mm = 0.2 m,   t = 12.5 mm = 0.0125 m ,   P = 3 MN/m²
d_w = 4 mm = 0.004 m, σ_w = 55 MN/m²

Before admitting water into cylinder:

Initial circumferential stress in the cylinder

$$\sigma_c = \frac{\pi\, d_w}{4\, t}\, \sigma_w = \frac{\pi \times 0.004}{4 \times 0.0125} \times 55 = 13.82 \text{ MN/m}^2$$

After admitting water into cylinder:

For equilibrium, bursting force and resisting force are equal

$$\sigma'_w \times \frac{\pi}{2} \times d_w + \sigma'_c \times 2t = P\, d$$

$$\sigma'_w \times \frac{\pi}{2} \times 0.004 + \sigma'_c \times 2 \times 0.0125$$
$$= 3 \times 10^6 \times 0.2 \qquad\qquad ----(1)$$

Circumferential strain in cylinder = Circumferential strain in wire

$$\frac{\sigma'_c}{E} - \nu \frac{Pd}{4tE} = \frac{\sigma'_w}{E_w}$$

$$\rightarrow \quad \frac{\sigma'_c}{100 \times 10^9} - 0.25 \times \frac{3 \times 10^6 \times 0.2}{4 \times 0.0125 \times 100 \times 10^9} = \frac{\sigma'_w}{200 \times 10^9}$$

$$\rightarrow \quad \sigma'_c - 3 \times 10^6 = \frac{\sigma'_w}{2} \qquad\qquad --- (2)$$

Solving (1) and (2)

$$\sigma'_c = 17 \times 10^6 \text{ N/m}^2 = 17 \text{ MN/m}^2$$

$$\sigma'_w = 28 \times 10^6 \text{ N/m}^2 = 28 \text{ MN/m}^2$$

Resultant stresses:

Resultant stress in cylinder = $\sigma'_c - \sigma_c = 17 - 13.82 = 3.18$ MN/m^2 (tensile)

Resultant stress in wire = $\sigma'_w + \sigma_w = 28 + 55 = 83$ MN/m^2 (tensile)

HIGHLIGHTS

Definitions

1. Hoop stress (or circumferential stress): This is the tensile stress acting along the circumference of the cylinder.

2. Longitudinal stress: This is the tensile stress acting along the length of the cylinder. It develops only if the cylinder has closed ends.

3. Radial stress: This is the compressive stress acting along the radius of the cylinder.

Concepts and Formulae

1. A pressure vessel is a thin one if the ratio of its internal diameter to thickness ≥ 20

2. Thin cylinders are frequently required to operate under pressure upto 30 MPa. For high pressure such as 250 Mpa or more, thick cylinders are used.

3. Hoop stress (or circumferential stress) $\sigma_c = \dfrac{Pd}{2t}$

Longitudinal stress $\sigma_l = \dfrac{Pd}{4t}$ P → Internal fluid pressure

These formulas are valid for seamless shell i.e. solid drawn shell.

4. For built up shell (having longitudinal and circumferential joints)

$$\sigma_c = \frac{Pd}{2t\,\eta_l} \quad \text{and} \quad \sigma_l = \frac{Pd}{4t\,\eta_c}$$

5. Hoop strain, $\epsilon_c = \dfrac{Pd}{2tE}(1 - 0.5\,v)$

Increase in diameter, $\delta d = \dfrac{Pd^2}{2tE}(1 - 0.5\,v)$

Longitudinal strain, $\epsilon_l = \dfrac{Pd}{2tE}(0.5 - v)$

Increase in length, $\delta l = \dfrac{Pdl}{2tE}(0.5 - v)$

6. For thin cylindrical shell:

Volumetric Strain, $\epsilon_v = 2\epsilon_c + \epsilon_l$

Increase in Volume, $\delta V = \dfrac{P\pi d^3 l}{8tE}(2.5 - 2\,v)$

7. Total volume of fluid that can additionally be pumped into the cylindrical shell

= increase in volume of cylinder + decrease in volume of fluid

$$= \frac{P\pi d^3 l}{8tE}(2.5 - 2v) + \frac{PV}{K}$$

Neglecting compressibility of fluid, $\frac{PV}{K} = 0$

8. For safe condition, plate thickness $t \geq \frac{Pd}{2\,\sigma_{all}}$

σ_{all} = allowable tensile stress for shell material

9. Maximum shear stress in cylindrical shell,
$$\tau_{max} = \frac{\sigma_1 - \sigma_2}{2} = \frac{Pd}{8t}$$

10. For thin spherical shell:

Hoop stress (or circumferential stress), $\sigma_c = \frac{Pd}{4t}$

There is no longitudinal stress

Increase in Diameter, $\delta d = \frac{Pd^3}{4tE}(1 - v)$

Volumetric Strain, $\epsilon_v = 3\epsilon_c$ ($\epsilon_c \to$ Hoop strain)

Increase in Volume, $\delta V = \frac{P\pi d^4}{8tE}(1 - v)$

Volume of fluid that can additionally be pumped into the shell

$$= \frac{P\pi d^4}{8tE}(1 - v) + \frac{PV}{K}$$

11. For cylindrical shell with hemispherical ends, condition to avoid bending of walls

$$\frac{t_2}{t_1} = \frac{1 - \nu}{2 - \nu}$$

(t_1 and t_2 = thicknesses of cylinder and sphere, respectively)

12. For wire wound thin cylinder:

Without internal fluid pressure, $\sigma_c = \dfrac{\pi\, d_w}{4\, t}\, \sigma_w$

σ_c = compressive circumferential stress in cylinder
σ_w = initial tensile stress in wire
d_w = wire diameter
t = cylinder wall thickness

With internal fluid pressure —

$$\sigma'_w \times \frac{\pi}{2} \times d_w + \sigma'_c \times 2t \;=\; P\,d \qquad -----(1)$$

$$\frac{\sigma'_c}{E} - \nu\frac{Pd}{4tE} \;=\; \frac{\sigma'_w}{E_w} \qquad\qquad -----(2)$$

Solving equations (1) and (2) the stresses σ'_c and σ'_w can be computed. Then

Resultant stress in wire = $\sigma_w + \sigma'_w$

Resultant stress in cylinder = $\sigma_c - \sigma'_c$

σ_c' = Circumferential stress developed in cylinder ,

σ_w' = Stress developed in wire

E and E_w
= Young's modulus for the cylinder and wire, respectively

SHORT TYPE QUESTIONS

1. The maximum stress in a cylinder subjected to internal fluid pressure occurs at ____ surface.

[Ans. inner]

2. The stress in the wall of a cylinder normal to its longitudinal axis due to forces acting along the circumference is known as ____ stress.

[Ans. hoop]

3. In a thin walled pressure vessel subjected to internal fluid pressure, the hoop stress is ____ the longitudinal stress.

[Ans. twice]

4. When a thin cylindrical shell is subjected to internal fluid pressure which of the following stress is developed in its wall.
(a) circumferential stress   (b) longitudinal stress
(c) both 'a' and 'b'   (d) none of these

[Ans. c]

5. A thin cylindrical shell of diameter 'd' and wall thickness 't' is subjected to internal fluid pressure P. If Poisson's ratio is v what is the value of circumferential strain?

$$[Ans. \ \frac{Pd}{2tE} (1 - 0.5 v)]$$

6. The maximum stress produced in a thin cylindrical shell is ____ times that in a thin spherical shell having same diameter, thickness and internal pressure.

[Ans. two]

7. From design point of view, spherical pressure vessels are preferred over cylindrical pressure vessels because they
(a) are cost effective in fabrication
(b) have uniform higher circumferential stress
(c) have uniform lower circumferential stress
(d) have a large volume for the same quantity of material used.

[Ans. (d)]

8. When a thin cylinder of diameter d and thickness t is pressurized with an internal pressure of P then (v = Poisson's ratio and E = modulus of elasticity)

(a) the circumferential strain is equal to $\dfrac{Pd}{2tE}(0.5 - v)$

(b) the longitudinal strain is equal to $\dfrac{Pd}{2tE}(1 - 0.5\,v)$

(c) the longitudinal stress is equal to $\dfrac{Pd}{2t}$

(d) the ratio of circumferential strain to longitudinal strain is equal to $\dfrac{1 - 0.5\,v}{0.5 - v}$

$$[\text{Ans. (d)}]$$

9. _____ stress can be neglected compared to other stresses in a thin shell.

$$[\text{Ans. Radial}]$$

10. What is the ratio of volumetric strain to hoop strain in a thin spherical shell?

$$[\text{Ans. three}]$$

11. Circumferential and longitudinal strains in a cylindrical boiler under steam pressure are ε_1 and ε_2 respectively. Change in the volume of boiler cylinder per unit volume will be

(a) $\varepsilon_1 + 2\,\varepsilon_2$ (b) $\varepsilon_1\,\varepsilon_2^{\,2}$ (c) $2\varepsilon_1 + \varepsilon_2$ (d) $\varepsilon_1^{\,2}\,\varepsilon_2$

$$[\text{Ans. (c)}]$$

12.   A cylinder is said to be thin if the ratio of its diameter and thickness is more than _____ .

$$[\text{Ans. 20}]$$

EXERCISE PROBLEMS

1. A gas cylinder is of 1.5 m internal diameter and 30 mm thickness. Find the allowable pressure of the gas inside the cylinder if the tensile stress in the material is not to exceed 100 MN/m².

$$[\text{Ans. } P = 4 \text{ MN/m}^2]$$

2. Calculate the thickness of metal necessary for a cast iron main 800 mm in diameter for water at a pressure head of 100 m if the maximum permissible tensile stress is 20 MN/m² and weight of water is 10 kN/m³.

[Ans. t = 20 mm]

3. A boiler is subjected to an internal pressure of 2 N/mm². The thickness of the plate is 20 mm and allowable tensile stress is 120 N/mm². Efficiency of the longitudinal joint is 90% and that of circumferential joint is 40%. Find out suitable diameter of the shell.

[Ans. 192 cm]

4. A cylindrical vessel whose ends are closed by means of rigid flange plates 3 mm thick. The internal length and diameter of the vessel are 500 mm and 250 mm respectively. Determine the longitudinal and circumferential stresses in the cylindrical shell due to an internal fluid pressure of 3 MN/m². Also calculate the increase in the length, diameter and volume of the vessel. Take E = 200 GPa and v = 0.3.

[Ans. σ_c = 125 MPa, σ_l = 62.5 MPa, δd = 0.1328 mm, δl = 0.0625 mm, δV = 2.91×10⁴ mm³]

5. A thin cylindrical shell 2 m long has 200 mm diameter and thickness of metal 10 mm. It is filled completely with a fluid at atmospheric pressure. If an additional 25000 mm³ fluid is pumped in, find the pressure developed and hoop stress developed. Find also the changes in the diameter and length.

[Ans. P = 4.188 N/mm², σ_c = 41.88 N/mm², δd = 0.0356 mm, δl = 0.0837 mm]

6. A spherical shell is of diameter 800 mm and thickness 12 mm. determine the pressure which causes an increase in volume of 40000 mm³. Take E = 200 GPa and v = 0.3.

[Ans. P = 0.852 N/mm²]

7. A spherical vessel 1 m diameter and 8 mm thick filled with water under a pressure of 10 bar. Find the change in volume of the sphere. Also calculate the volume of additional amount of water to be pumped when pressure is changed from 10 bar to 15 bar. Take v = 0.3, E = 200 GPa and K for water 2.4 GPa.

[Ans. δV = 184 cm³, additional volume of water = 169.62 cm³]

8. A 200 mm diameter cast iron pipe has thickness of 12 mm and is closely wound with a layer of 5 mm diameter steel wire under a tensile stress of 60 MPa. If now water under a pressure of 3.5 MPa is admitted into the pipe, find the stresses induced in the pipe and steel wire. Take E_{CI} = 100 GPa, E_s = 200 GPa, Poisson's ratio = 0.3.

[Ans. 0.27 MPa (compressive) in pipe, 89.97 MPa (tensile) in wire]

Index

A

axial load, 17, 31, 53, 61-62, 66-67, 76, 78, 162, 412

axially loaded member, 361

B

bar of uniform strength, 62, 78

beams
  composite, 60-61, 65-66, 69, 72, 77, 81, 219, 270-272, 274, 283-284, 361, 383
  composite beams, 219, 270
  of uniform strength, 62, 78,

219, 268, 270, 283, 285, 287
  types of, 2, 161-162, 164-166

bending moment
  sign convention, 89, 133, 168-169, 294, 362

bulk modulus, 47-48, 51, 53-54, 57, 59, 76, 80, 82, 408, 418-419

C

cantilever beam, 163, 169-170, 214, 227, 244, 295, 297, 299-300, 302, 323, 341, 347-348, 356-357

complementary shear stress, 87, 153, 248

Y

www.ingramcontent.com/pod-product-compliance
Lightning Source LLC
Chambersburg PA
CBHW071315210326
41597CB00015B/1231